LECTINS AND PATHOLOGY

LECTINS AND PATHOLOGY

Edited by

Michel Caron

*Biochimie Cellulaire des Hémopathies Lymphoïdes
Université Paris Nord, France*

and

Annie–Pierre Sève

Hôpital Saint–Louis, Paris, France

**ho
ap**

harwood academic publishers
Australia • Canada • France • Germany • India • Japan
Luxembourg • Malaysia • The Netherlands • Russia • Singapore
Switzerland

Amsteldijk 166
1st Floor
1079 LH Amsterdam
The Netherlands

British Library Cataloguing in Publication Data

A catalogue record for this book is available from the British Library.

ISBN: 90-5702-491-8

CONTENTS

CONTRIBUTORS

Shiro Akahani
Department of Otolaryngology
Osaka Teishin Hospital
2-6-40 Karasugatsuji, Tennouji-ku
Osaka City, Osaka 543
Japan

Sonia Beeckmans
Laboratorium voor Scheikunde
  der Proteïnen
Instituut voor Moleculaire Biologie
  en Biotechnologie
Vrije Universiteit Brussel
Paardenstraat 65
B-1640 St-Genesius-Rode
Belgium

Najma Bhat
Division of Geographic Medicine
  and Infectious Diseases
Tufts University School of Medicine
750 Washington Street
Boston, MA 02111
USA

Nicolai V. Bovin
Shemyakin Institute of Bioorganic
  Chemistry
Russian Academy of Sciences
ul. Miklukho-Maklaya 16/10
Moscow
Russia

Robert Bresalier
Cancer Research Laboratory
Henry Ford Health Sciences Center
Detroit, MI 48202
USA

Ulrich Brinck
Zentrum Pathologie
Georg-August-Universität Göttingen
Robert-Koch-Straße 40
D-37075 Göttingen
Germany

Michel Caron
Biochimie Cellulaire des Hémopathies
  Lymphoïdes
Université Paris Nord
74, rue Marcel Cachin
93017 Bobigny cedex
France

Vincent Castronovo
Metastasis Research Laboratory
Pathology Building B23
Sart Tilman
B-4000 Liège
Belgium

Françoise De Cupere
Laboratorium voor Scheikunde
  der Proteïnen
Instituut voor Moleculaire Biologie
  en Biotechnologie
Vrije Universiteit Brussel
Paardenstraat 65
B-1640 St-Genesius-Rode
Belgium

John M.S. Forrest
Scottish Crop Research Institute
Invergowrie
Dundee DD2 5DA
UK

Hans-J Gabius
Institut für Physiologische Chemie
Tierärztliche Fakultät
Ludwig-Maximilians-Universität
München
Veterinärstraße 13
D-80539 München
Germany

Sigrun Gabius
Hämatologisch-Onkologische
  Schwerpunktpraxis
Sternstraße 12
D-83022 Rosenheim
Germany

Klaus Kayser
Abteilung Pathologie
Thoraxklinik Amalienstraße 5
D-69126 Heidelberg
Germany

Fu-Tong Liu
La Jolla Institute for Allergy
    and Immunology
10355 Science Center Drive
San Diego, CA 92121
USA

Michèle Mouricout
Institut de Biotechnologie
Faculté des Sciences
Université de Limoges
123, Avenue Albert Thomas
87060 Limoges
France

Werner E.G. Müller
Institut für Physiologische Chemie
Abteilung für Angewandte
    Molekularbiologie
Johanns Gutenberg-Universität Mainz
Duesbergweg 6
D-55099 Mainz
Germany

Pratima Nangia-Makker
Karmanos Cancer Institute
Wayne State University
School of Medicine
110 East Warren Avenue
Detroit, MI 48201
USA

Avraham Raz
Karmanos Cancer Institute
Wayne State University
School of Medicine
110 East Warren Avenue
Detroit, MI 48201
USA

Heinz C. Schröder
Institut für Physiologische Chemie
Abteilung für Angewandte
    Molekularbiologie
Johanns Gutenberg-Universität Mainz
Duesbergweg 6
D-55099 Mainz
Germany

Annie-Pierre Sève
INSERM U496
Maturation et Differenciation
    en Pathologie Humaine
Centre G. Hayem
Hôpital St-Louis
1, avenue Claude Vellefaux
75010 Paris
France

Carlo Unverzagt
Institut für Organische Chemie
Ludwig-Maximilians-Universität
    München
Karlstraße 23
D-80333 München
Germany

Frédéric A. Van Den Brûle
Metastasis Research Laboratory
Pathology Building B23
Sart Tilman
B-4000 Liège
Belgium

Edilbert Van Driessche
Laboratorium voor Scheikunde
    der Proteïnen
Instituut voor Moleculaire Biologie
    en Biotechnologie
Vrije Universiteit Brussel
Paardenstraat 65
B-1640 St-Genesius-Rode
Belgium

Bruno Védrine
Institut de Biotechnologie
Faculté des Sciences
Université de Limoges
123, Avenue Albert Thomas
87060 Limoges
France

Honorine D. Ward
Division of Geographic Medicine
  and Infectious Diseases
Tufts University School of Medicine
750 Washington Street
Boston, MA 02111
USA

Pamela Zambenedetti
Dipartimento di Biologia
Università di Padova
via Trieste 75
35131 Padova
Italy

Paolo Zatta
Centro CNR per lo Studio della
Biochimica e della Fisiologia delle
  Metalloproteine
Università di Padova
via Trieste 75
35131 Padova
Italy

1. LECTINS AND PATHOLOGY: AN OVERVIEW

MICHEL CARON[1] and ANNIE-PIERRE SÈVE[2]

[1]*Biochimie Cellulaire des Hémopathies Lymphoïdes, Université Paris Nord, 74 rue Marcel Cachin, 93017 Bobigny Cedex, France*
[2]*INSERM U496, Maturation et Differenciation Cellulaire en Pathologie Humaine, Centre G. HAYEM, 1 avenue Claude Vellefaux, Hôpital Saint-Louis, 75010 Paris, France*

Although known for several decades, it is only during the last five or so years that lectins have aroused the interest of a very broad section of investigators in the biological and medical sciences. There are many reasons for the current interest in lectins. Prominent among these is the fact that the study of lectins now appears to be in a period of transition from the earlier descriptive phase, which was concerned with purification and structure, to a period when we are, perhaps, beginning to have some understanding of their fundamental role in physiological and pathological processes.

Lectins are widely distributed in nature, including in microorganisms, in plants where they were firstly described, and all species of animal kingdom. The list of molecules that fall into this category grows daily: extracellular matrix components, many cell surface receptors, proteins involved in bacterial and viral pathogenesis are but a few examples. Lectins are interesting proteins not only because they are ubiquitous, but they also could interact with carbohydrates and the interactions wich occur could easily be disrupted. These interactions allow them to be active partners in the multimeric organisation of complexes which take place in the cell physiology. Thus, investigators with interest as diverse as cell-cell interactions, cancer invasion and metastasis, inflammation, and immunology have become concerned with lectins.

The voluminous literature generated by this interest in lectins has made it extremely difficult for the nonspecialist to keep abreast of the latest developments. The objective of this book is to summarize the current state of knowledge on the importance of lectins in pathology. In an attempt to accomplish this goal, this book includes chapters devoted to selected topics written by active investigators, experts in their respective fields. Coverage is by no means encyclopedic; rather the thrust is to emphasize the recent advances. During the past decades the great interest of these proteins was there capacity to be used as interesting markers to study various pathologies, and vegetable lectins such as Concanavalin A or Wheat Germ Agglutinin were used by the pathologists. Therefore, the first chapters deal primarily with lectins and lectin-binding molecules as tools.

Since several years now there is an increasing amount of data concerning the involvement of endogenous (animal or human) lectins in various physiological activities. In fact, some of them have, so far, a well known biological role. Then, most of the chapters emphasizes the utility of the elucidation of the function of endogenous lectins in the understanding of many diseases and pathogenesis.

Knowledge about carbohydrate recognition domains of these lectins enables to distinguish at least five families, more or less homogeneous, of animal lectins (Table 1). Among them, two major families can be distinguished: the soluble and ß-galactoside-specific one (S-type) and the Ca++ dependant one (C-type) (Drickamer and Taylor, 1993). A number of C-type mammalian lectins are involved in receptor-mediated endocytosis of glycoproteins. It is the case of hepatic lectin which plays a role in the clearance of glycoproteins form the circulatory system (Hudgin *et al.*, 1974; Kawasaki and Ashwell, 1976). Other C-type lectins have been implicated in cellular recognition processes including adhesion, metastasis, ... This concept is well illustrated by the regulation of leukocyte trafficking, which relies on the specific interaction between C-type lectins (selectins) and the oligosaccharidic structure sialyl Lewis[x] (Varki, 1993). The role of this interaction is now well understood in the case of leucocytes adhesion to vascular endothelium and the involvement of selectins in the process of the slop (Mcever and Cummings, 1997).

However, precise understanding of the biological function of many mammalian lectins, and especially those of the S-type, remains unclear. These lectins, which do not require divalent cations for their activity, are widely distributed among animal tissues. They consist of several isolectins with conserved amino acid sequences. They are currently classified as members of the "galectin super family" (Barondes *et al.*, 1994). The existence of a galectin family of proteins in not only vertebrates but also in invertebrates such as nematodes (Hirabayashi and Kasai, 1992) and sponge (Pfeifer *et al.*, 1993) was reported. It is generally accepted that galectins are multifunctional proteins. An intriguing point is that the role of the galectins appears to be different, depending on the cells where they are studied, of the physiological or pathological state of these cells, and of the galectin molecules localisation (membrane, cytoplasm or nucleus) (Chadli *et al.*, 1997; Hubert and Sève 1994; Mehul and Hughes, 1997).

One of the interest of endogenous lectins is there ability to interact not only through glycoprotein-lectin interactions but also through protein-protein interaction (Gabius, 1994). In the nucleus, it was described that two lectins, namely, galectin 3 and CBP70 interact by a protein-protein interaction. This interaction is disrupted when a competitive sugar recognizes the CRD of galectin 3 (Sève *et al.*, 1993; Sève *et al.*, 1994). This observation emphasizes the possible role of endogenous lectins such as galectins in the multimeric assembly of polypeptides. The possible role of galectins in the prion disease is discussed in this book. In addition, galectins 1 and 3, found in the nucleus, play a role in splicing (Dagher *et al.*, 1995; Vyakarnam *et al.*, 1997). In other chapters of this book, functional role of galectins in pathology, namely, in the interactions at the membrane level with laminin and in tumor metastasis, are discussed.

Another crucial point in the recent years was the correlation which was demonstrated between galectins and apoptosis. Exogenous (plant, invertebrate) lectins have already been used as tools able to induce apoptosis in different cell types (Kim *et al.*, 1993; Kong *et al.*, 1996). More recently it was shown that some endogenous galectins, which display functional significance in vivo, can trigger or inhibit programmed cell death (Akahani *et al.*, 1997; Lutomski *et al.*, 1997; Perillo *et al.*, 1995). Recently intra-cytoplasmic galectin 3 was described as an anti-apototic molecule which contains the NWGR amino acid sequence highly conserved in the BH1 domain of Bcl2 gene family. Moreover, galectine 3 was described to react with

Table 1   Classification of animal lectins

Family	Different groups	Members	Carbohydrate specificity
C-type	Group 1: proteoglycans	Aggrecan, versican,...	Gal, Fuc,...
	Group 2: type II receptors	CD23, CD72, hepatic lectin,...	Mainly galactosides
	Group 3: collectins	MBP, conglutinin,...	Man, Fuc,...
	Group 4: selectins	L, E, P selectins	Fucosylated epitopes
	Group 5: lymphocyte antigens	CD69, NK cell lectins	Mostly unknown
	Group 6: macrophage Man receptor		Man
	Miscellaneous proteins	Tetranectin, PSP,...	Various
S-type	Proto	Galectins 1, 2, 5, 7	ß-galactosides
(galectins)	Chimera	Galectin 3	
	Tandem-repeat	Galectins 4, 6, 8, 9	
	Miscellaneous	Gal10, CBP67,...	
I-type		CD22, CD33, MAG,...	sialic/hyaluronic acid,...
P-type		Mannose 6-Ph receptors	Mannose 6-P
Pentraxins		CRP, SAP,...	Sulfated/phosphorylated saccharides
Miscellanous		Heparin-binding lectins	Glycosaminoglycans
("N-type")		Hyaluronectin	Hyaluronate
		Tumor necrosis factor	Diacetylchitobiose
		Interleukin 1	Uromodulin
		Interleukin 2	High mannose
		CBP70	GlucNac
		CSL	Mannose

Bcl2, the protein-protein interaction which occurs between these proteins being disrupted by the adjunction of a sugar interacting with galectin 3 (Tang, Hsu and Liu, 1996). At the membrane level recent data described the role of galectin 1 in apoptosis. In that case, once again the glycoprotein interaction between galectin and a putative ligand, namely CD45, could allow the multimeric assembly of membrane protein which permit the transductional signal to act (Lutomski *et al.*, 1997; Perillo *et al.*, 1995).

Last but not least, in the final chapters of this book, comprehensive description of the current state of knowledge of the importance of lectins in bacteriology and parasitology is presented. Together, the chapters of this volume document major progress in lectin research, in an authoritative and unique fashion accessible to specialists in both protein biochemistry and human and animal pathology, as well as to a wide circle of graduate students and other scholars in diverse fields of biomedicine.

REFERENCES

Akahani, S., Nanghia-Makker, P., Inohara, H., Kim, H.C. and Raz A. (1997). Galectin 3: A novel antiapoptotic molecule with a functional BH1 (NWGR) domain of Bcl-2 family. *Cancer Research*, **57**, 5272–5276.

Barondes, S. H., Cooper, D.N.W., Gitt, M.A. and Leffler, H. (1994). Galectins : structure and function of a large family of animal lectins. *J. Biol. Chem.*, **269**, 20807–20810.

Chadli, A., LeCaer J-P., Bladier, D., Joubert-Caron, R. and Caron, M. (1997). Purification and characterization of a human brain galectin-1 ligand. *J. Neurochem.*, **68**, 1640–1647.

Dagher, S.F., Wang, J.L. and Patterson, R.J. (1995). Identification of galectin-3 as a factor in pre-mRNA splicing. *Proc. Natl. Acad. Sci. USA*, **92**, 1213–1217.

Drickamer, K. and Taylor, M.E. (1993). Biology of animal lectins. *Ann. Rev. Cell Biol.*, **9**, 237–264.

Gabius, H.-J. (1994). Non-carbohydrate binding partners/domains of animal lectins. *Int. J. Biochem.*, **26**, 469–477.

Hirabayashi, J., Satoh, M. and Kasai, K. (1992). Evidence that caenorhabditid elegans 32 kDa B-galactoside-binding protein is homologous to vertebrate β-galactoside-binding lectins. *J. Biol. Chem.*, **267**, 15485–15490.

Hubert, J. and Sève, A-P. (1994). Nuclear lectins. *Lectins: Biology, Biochemistry, Clinical Biochemistry.*, **10**, 220–226.

Hudgin, R.L., Pricer, W.E. and Ashwell, G (1974). The isolation and properties of a rabbit liver binding protein specific for asialoglycoproteins. *J. Biol. Chem.*, **249**, 5536–5543.

Kawasaki, T. and Ashwell, G (1976). Chemical and physical properties of an hepatic membrane protein that specifically binds asialoglycoproteins. *J. Biol. Chem.*, **251**, 1296–1302.

Kim, M., Rao, M.V., Tweardy, D.J., Prakash, M., Galili, U. and Gorelik, E. (1993). Lectin induced apoptosis of tumour cells. *Glycobiology*, **3**, 447–453.

Kong, S. K., Suen, Y.K., Chan, Y.M., Chan, C.W., Choy, Y.M., Fung, K.P. and Lee, C.Y. (1996). Concanavaline A-induced apoptosis in murine macrophages through a Ca2+ independent pathway. *Death and differenciation*, **3**, 307–314.

Lutomski, D., Fouillit, M., Bourin, P., Mellottée, D., Denize, N., Pontet, M., Bladier, D., Caron, M. and Joubert-Caron, R. (1997). Externalization and binding of galectin-1 on cell surface of K562 cells upon erythroid differentiation. *Glycobiology*, **7**, 1193–1199.

Mcever, R.P. and Cummings, R.D. (1997). Perspectives series: cell adhesion in vascular biology. Role of PSGL-1 binding to selectins in leucocytes recruitment. *J. Clin. invest.*, **100**, 485–491.

Mehul, B. and Hughes, RC (1997). Plasma membrane targetting, vesicular budding and release of galectin 3 from the cytoplasm of mammalian cells during secretion. *J. Cell Sci.*, **110**, 1169–1178.

Perillo, N., Pace, K.E., Seilhame, J.J. and Baum, L.G. (1995). Apoptosis of T-cells mediated by galectin-1. *Nature*, **378**, 736–738.

Pfeifer, K., Haasemann, M., Gamulin, V., Bretting, H., Farhenlolz, F. and Müller, W.E.G. (1993). S-type lectins occur also in invertebrates: High conservation of the carbohydrate recogntion domain in the lectin genes from the marine sponge Geodia cydodium. *Glycobiology*, **3**, 179–184.

Sève, A.-P., Felin, M., Doyennette-Moyne, M-A., Sahraoui, T., Aubery, M. and Hubert, J. (1993). Evidence for a lactose mediated association between two nuclear-binding proteins. *Glycobiology*, **3**, 23–30.

Sève, A.-P., Hadj-Sahraoui, Y., Felin, M., Doyennette-Moyne, M-A., Aubery, M. and Hubert, J. (1994). Evidence for a lactose mediated association between the carbohydrate-binding proteins CBP35 and CBP70 in membrane depleted nuclei. *Exp. Cell Res.*, **213**, 191–197.

Varki, A. (1993). Biological roles of oligosaccharides : all the theories are correct. *Glycobiology*, **3**, 97–130.

Vyakarnam, A., Dagher, S.U., Wang, J.L. and Patterson, R.J. (1997). Evidence for a role for galectin-1 in pre-mRNA splicing. *Mol. cell. Biol.*, **17**, 4730–4737.

Yang, R.-Y., Hsu, D.K. and Liu, F-T. (1996). Expression of galectin-3 modulates T-cell growth and apoposis. *Proc. Natl. Acad. Sci. USA*, **93**, 6737–6742.

2. CARRIER-IMMOBILIZED CARBOHYDRATE LIGANDS: DESIGN OF THE LECTIN-DETECTING TOOLS AND THEIR CLINICAL APPLICATIONS WITH FOCUS ON HISTOPATHOLOGY

HANS-J. GABIUS[1], NICOLAI V. BOVIN[2], ULRICH BRINCK[3], SIGRUN GABIUS[4], CARLO UNVERZAGT[5] and KLAUS KAYSER[6]

[1]*Institut für Physiologische Chemie, Tierärztliche Fakultät, Ludwig-Maximilians-Universität München, Veterinärstraße 13, D-80539 München, Germany*
[2]*Shemyakin Institute of Bioorganic Chemistry, Russian Academy of Sciences, ul. Miklukho-Maklaya 16/10, Moscow, Russia*
[3]*Zentrum Pathologie, Georg-August-Universität Göttingen, Robert-Koch-Straße 40, D-37075 Göttingen, Germany*
[4]*Hämatologisch-Onkologische Schwerpunktpraxis, Sternstraße 12, D-83022 Rosenheim, Germany*
[5]*Institut für Organische Chemie, Ludwig-Maximilians-Universität München, Karlstraße 23, D-80333 München, Germany*
[6]*Abteilung Pathologie, Thoraxklinik, Amalienstraße 5, D-69126 Heidelberg, Germany*

INTRODUCTION

Biological information transfer governs the social behavior of cells and thus the organization of organs and tissues. Defects and disorders in the exquisitely adjusted regulatory processes will cause disease states of varying degrees of severity. Any reliable delineation of a molecular aberration will be the basis to devise, if possible, efficient therapeutic strategies to correct the pinpointed dysregulation. Since histochemical analysis can provide detailed information on the presence of the defined epitopes under scrutiny and their spatial distribution, its application is an important step to unravel the inherent mechanisms which establish the complex structures at the different levels of cellular organization. Generally, antibodies afford the opportunity to describe the localization of a distinct determinant. The monitoring of expression of a gene for a certain protein can also be shifted to the level of ribonucleic acid, referring to the popular technique of *in situ* hybridization. When considering these two approaches, information storage is implicitly performed by two well-investigated code systems, i.e. the ensembles of four nucleotides and twenty-one proteinogenic amino acids. Quite frequently, textbooks nourish the notion that vital information transfer is exclusively confined to these two biological alphabets. However, recent work in the fields of glycosciences has convincingly proven the inadequacy of this preconception (for an overview, see Gabius and Gabius, 1993, 1997). We are now not only witnessing the emergence of the concept of a third code system. Moreover, it is becoming obvious that its coding capacity surpasses by orders of magnitude that of the familiar oligo- and polymeric structures consisting of nucleotides and amino acids. This newly recognized code system is established by a panel of monosaccharides.

Several special properties of their structure endow carbohydrates with the ability to form oligomeric isomers beyond the limit given by the peptide bond or the 3',5'-phosphodiester connection of nucleotides. Explicitly, the linkage position for a glycosidic bond can be variable, potentially involving hydroxyl groups in positions 1, 2, 3, 4 or 6 for a hexapyranose unit, and any involvement of the anomeric position offers the alternative between the α- or β-anomeric configurations. To graphically illustrate the nominal coding capacity of the sugar code, the number of isomers has been calculated for a hexamer from an ensemble of six letters, yielding 6.4×10^7 hexapeptides and 1.44×10^{15} hexasaccharides (Laine, 1997). Since the repertoire of intermolecular forces within receptor-ligand interactions, predominantly hydrogen bonding and van der Waals interactions, will fully apply to any compound of the three code systems (Cambillau, 1995; Rini, 1995; Bundle, 1997; Siebert *et al.* 1997), information transfer can readily be facilitated — when using a protein as receptor - with ligands from any mentioned category. Thus, the function to act as information-bearing code units has already been ascribed at an early stage of structural analysis to the sugar part of cellular glycoconjugates besides other roles in influencing physicochemical properties and structural aspects including conformational stabilization and increased resistance against proteolytic degradation (Cook 1986; Montreuil 1995). It is consequently pertinent to briefly summarize the structural principles of glycosylation and then to introduce major recognition molecules for the carrier-bound carbohydrate signals.

GLYCOSYLATION — MORE THAN A MERELY STRUCTURAL EXTENSION OF PROTEINS AND LIPIDS?

The same reason which is pivotal for the attractivity of carbohydrates as code units impedes the structural characterization of oligomers, i.e. the theoretically enormous variability with respect to isomer formation (Laine, 1997). To address this problem, appropriate biophysical and chemical protocols have been developed with admirable pertinacity (Hounsell, 1997). The necessity to not only memorize a sequence, but also the linkage position, the anomeric configuration and the nature of the actual enantiomers quite often deters colleagues to be intrigued by carbohydrate structures. Historically, work in this area has thus been primarily centered around (homo)polymers such as starch. Starting with structural work on the sugar part of ovine submaxillary mucin in 1959, of orosomucoid in 1971 and of human serotransferrin in 1974 and fueled by the introduction of high-field ^{1}H-NMR spectroscopy to the determination of glycan primary structure in 1977, our knowledge in this field has impressively been broadened (Montreuil, 1995; Sharon and Lis, 1997).

Carbohydrates are attached to the protein backbone via various linkage types, the structures of the main linkage regions being shown in Figure 1. Whereas the modification of a hydroxyl group of serine or threonine by a N-acetylglucosamine moiety will not sustain any enzymatic elongation beyond this monosaccharide, the other two shown glycosylation types are the starting points for glycan chain synthesis (Varki, 1993; Hounsell *et al.*, 1996; Hart, 1997; Sharon and Lis, 1997). Their generation is performed in a stepwise manner by a battery of glycosyltransferases primarily in the Golgi cisternae (Roth, 1987; Brockhausen and Schachter, 1997; Colley, 1997;

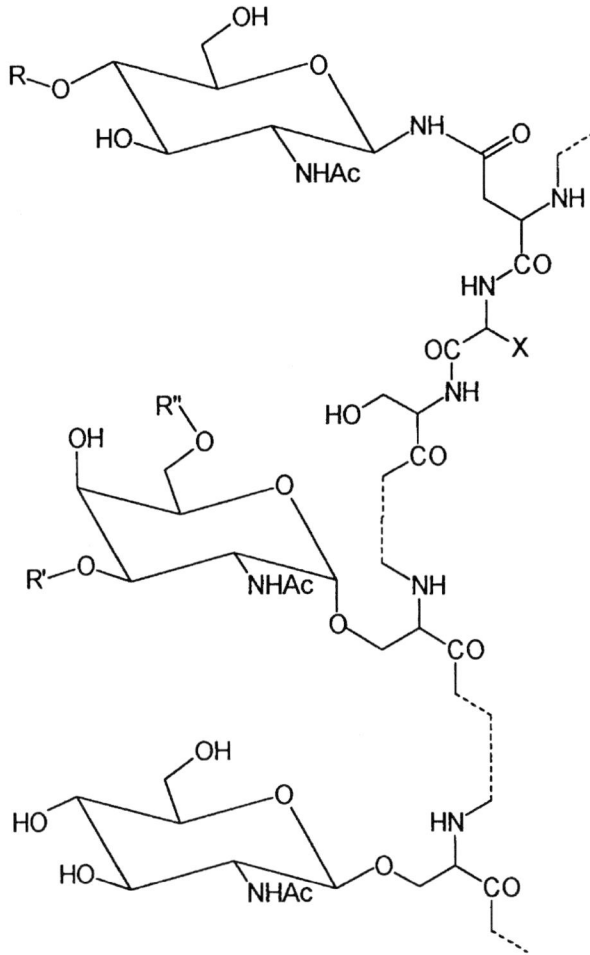

Figure 1 Illustration of common types of covalent linkages in glycoproteins between suitable side chain groups of the peptide backbone and a sugar unit, namely the covalent conjugate of β-N-acetylglucosamine (GlcNAc) with the amide nitrogen atom of an asparagine residue located in the sequon Asn-X-Ser which is the glycosylation signal for the attachment of the large family of N-linked sugar chains, of α-N-acetylgalactosamine with an oxygen atom of the hydroxyl groups of either serine or threonine which is typical for most of the O-linked sugar chains and of α-N-acetylglucosamine also with the hydroxyl groups of serine or threonine yielding O-GlcNAc-modified proteins without further elongation of the sugar part in this case.

Pavelka, 1997). N-Linked glycosylation is initiated in the endoplasmic reticulum by a transfer of a preformed lipid-anchored precursor (glc$_3$man$_9$glcNAc$_2$) whose processing by glycosidases leads to the common core structure (Figure 2). Interestingly, the transient presence of glucose molecules on the glycoprotein is a crucial part of a quality-control system in the endoplasmic reticulum which retains the nascent product in this region, until molecular chaperones with affinity to this

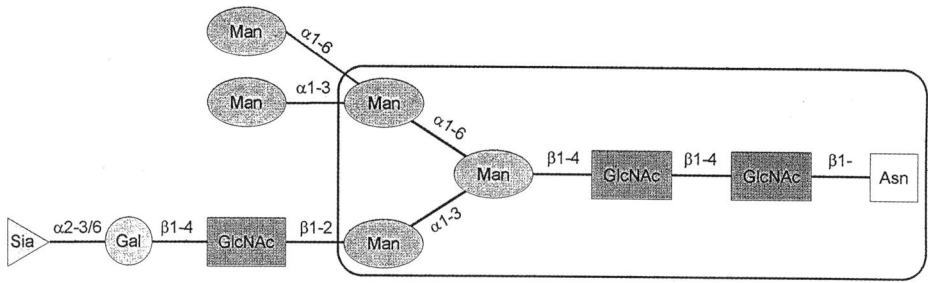

Figure 2 Illustration of the core unit of N-linked sugar chains (box) and of its extension by antennae which are formed either exclusively with mannose units (high-mannose type) or chains containing N-acetyllactosamine groups as backbone (complex type). The depicted hybrid type harbors both kinds of chain structure.

glycointermediate (i.e. $glc_1man_9glcNAc_2$) have completed their assistance in attaining correct folding of the nascent glycoprotein (Gahmberg and Tolvanen, 1996; Helenius *et al.*, 1997). This protein-carbohydrate interaction is not exceptional. Further routing of diverse glycoproteins to different cellular compartments or to various cell types is likewise guided by a distinct sugar signal, often involving unique modifications such as the 6'-phosphorylation of a mannose residue or the 4'-sulfation of a N-acetylgalactosamine residue (Varki, 1993, 1996; Roche and Monsigny, 1996; Gabius, 1997a; Hooper *et al.* 1997; Sharon and Lis, 1997). Well-documented disorders of glycan synthesis in e.g. I-cell disease or leucocyte adhesion deficiency type II and the deliberate genetic manipulation of glycosylation are also revealing how a deviation from normal processing can entail clinically apparent consequences (Varki, 1993; Stanley and Ioffe, 1995; Hathaway and Shur, 1997; McDowell and Gahl, 1997).

Similar to certain parts of the carbohydrate antennae which are attached to proteins the sugar part of glycolipids and proteoglycans can participate in recognitive interactions (Hakomori and Igarashi, 1995; Sandhoff and Kolter, 1995; Kopitz, 1997; Kresse, 1997). Three classes of proteins are capable of this interaction, accounting for deciphering the carbohydrate code-based message. The first group includes all enzymes dealing with sugars as substrates. The binding to enzymes, namely glycosyltransferases and glycosidases, is required to shape and remodel the glycochains. Since cell surface glycosylation is considered as a characteristic signature of organisms with profound differences for example between yeast, plant, insect or mammalian glycosylation, foreign glycoproteins are often potent elicitors of an immune response towards such disparities, leading to the production of sugar-specific immunoglobulins. They establish the second category of carbohydrate-binding proteins. Whereas these two classes are relatively homogeneous, harboring either a sugar target-directed enzymatic activity or the characteristic immunoglobulin structure, the third class is rather heterogeneous. It comprises all carbohydrate-binding proteins which are neither enzymes with sugar specificity nor immunoglobulins (Barondes, 1988; Gabius, 1994). Their occurrence has been verified in organisms of all branches of the evolutionary tree, the tools for their

Table 1 Methods used in the Search for Lectins

Tools	Parameter
multivalent glycans and (neo)glyco-conjugates or defined cell populations	Carbohydrate-dependent inhibition of lectin-mediated glycan precipitation or cell agglutination
labelled (neo)glycoconjugates and	
— matrix-immobilized extract fractions or purified proteins	signal intensity
— cell populations	labeling intensity
— tissue sections	staining intensity
— animal	biodistribution of signal intensity
(neo)glycoconjugate-drug chimera and cell populations	cellular responses (cell viability etc.)
matrix-immobilized (neo)glycoconjugates and	
— cell populations	carbohydrate-inhibitable cell adhesion
— cell extracts	carbohydrate-elutable proteins
homology searches with	
— computer programs and knowledge of structural aspects of carbohydrate recognition domains	homology score in sequence alignment or knowledge-based modeling
— lectin motif-reactive antibody	extent of cross-reactivity

from Gabius 1997a

detection having become more refined over the decades compared to the initial use of erythrocytes in haemagglutination assays (Sharon and Lis, 1987; Gabius, 1997a).

HOW TO DETECT LECTINS

General Considerations

The definition of a lectin pinpoints the property of sugar binding. This activity is the clue to detect lectins. As summarized in Table 1, carbohydrate structures are often employed in search of complementary binding sites. To avoid the problem of microheterogeneity of natural glycoproteins, it is preferable to enlist the service of a chemically prepared product which harbors a convenient label besides the

H.-J. GABIUS *et al.*

Table 2 Current Categories for Classification of various Animal Lectins

Family	Structural Motif	Carbohydrate Ligand	Modular Arrangement
C-type	conserved CRD	variable (mannose, galactose, fucose, heparin tetrasaccharide)	yes
I-type	immunoglobulin-like CRD	variable ($man_6glcNAc_2$, HNK-1 epitope, hyaluronic acid, a2,3/a2,6-sialyllactose)	yes
galectins (S-type)	conserved CRD	β-galactosides	variable
pentraxins	pentameric subunit arrangement	4,6-cyclic acetal of β-galactose, galactose, sulfated and phosphorylated monosaccharides	yes
P-type	homologous, not yet strictly defined CRD	mannose-6-phosphate-containing glycoproteins	yes

CRD: carbohydrate recognition domain, from Gabius, 1997a

sugar part. The properties of the carrier, the ligand and the density of its presentation can be tailored with a remarkable level of sophistication to the actual purpose, leading to neoglycoproteins and neoglycoconjugates (Lee and Lee, 1991, 1994a, 1997; Bovin and Gabius, 1995). A hallmark of these substances is the cluster effect, the tremendous (non-linear) affinity enhancement with a concentration increase of the ligand in a multivalent neoglycoconjugate. Binding of the probe invariably also requires multivalency at the level of the receptor to actually profit from the spatial arrangement of the sugar moieties and to bring about tight binding. The measured gains in the free energy can thus be rationalized by the assumption that the geometries of the receptor-ligand placement will ensure a snugly fit of clustered sites without a dramatic entropic penalty (Lee and Lee, 1991, 1994a, 1997; Gabius, 1998). The adequate density of carbohydrate recognition domains on cell surfaces has been assessed for several C-type animal lectins, e.g. the asialoglycoprotein receptors of hepatocytes and macrophages, the mannose receptor of macrophages and the collectin mannan-binding lectin (Lee and Lee, 1994b). In addition to this special subclass of animal lectins binding assays with carrier-immobilized ligands readily facilitate to detect a receptor, if present, for any carbohydrate structure. As usual, technical advances are a key factor to broaden the insight into expression of biomolecules. Currently, it is well-accepted to depict five categories of animal lectins (Table 2). The structural motif of the carbohydrate recognition domain is the decisive feature for the placement of any newly purified lectin (Drickamer, 1993; Powell and Varki, 1995; Gabius, 1997a; Kishore *et al.* 1997). Since even at the current state of analysis various activities will not fit into the accepted scheme, the necessity of its further extension and elaboration is indisputable.

In the area of histochemistry dealing with glycoconjugates and their receptors, referred to as glycohistochemistry, plant and invertebrate lectins are widely employed as tools for the analysis of the occurrence and localization of various glycoepitopes (Damjanov, 1987; Danguy et al., 1988, 1997; Walker, 1989; Spicer and Schulte, 1992; Kannan and Nair, 1997). Although interpretations for an assumed physiological relevance of the detected expression patterns should be given very cautiously due to the origin of the tools, the descriptive value of this approach is obvious. Having coined this technique "lectin histochemistry", the next step in glycohistochemistry deals with the search for endogenous binding sites for a custom-made saccharide structure. It has been referred to as "reverse lectin histochemistry" (Gabius et al., 1993; Danguy et al., 1995). In principle, this technique allows to monitor the presence of accessible receptor sites for a defined sugar structure, if they maintain their reactivity after the processing of the specimen under investigation. Adequate specificity controls to prove sugar-dependent binding, its sensitivity to minor structural alterations in the ligand and the lack of reactivity by the carrier or the label (e.g. by presence of receptors for albumin or biotin) and to reveal similarity of glyco- and immunohistochemical staining patterns in suitable cases with abundant expression of only one type of lectin for a certain ligand have underscored the feasibility of the approach (Bardosi et al., 1989; Kuchler et al., 1990, 1992; Gabius and Bardosi, 1991; Gabius et al., 1994a; Kayser et al., 1994a,b; Danguy, 1995; Danguy et al., 1995, 1997; Kannan and Nair, 1997). Its evaluation attests the wide distribution of carbohydrate receptors in tissues and lends wings to refinements of lectin histochemistry.

To address the fully justified concern that fine-specificities of endogenous lectins and of the laboratory tools from plants and invertebrates (exogenous lectins) will not be identical, the final step in glycohistochemistry is the application of the tissue lectins themselves as histochemical tool (Gabius and Bardosi, 1991; Gabius et al., 1993). The combined use of a lectin-specific antibody or of a neoglycoprotein to visualize the presence of the lectin and of the lectin itself for monitoring of accessible binding sites is expected to be instrumental for enhancing our knowledge on functional aspects of protein-carbohydrate interactions in situ (for descriptions of the experimental protocols of these techniques, see Gabius and Gabius, 1993). To attain this aim in reverse lectin histochemistry, the different parameters of probe design are to be fully exploited. They comprise the preparation of the ligand, its conjugation to the carrier and the nature of the polymeric backbone.

Chemical Preparation of the Carbohydrate Ligand and its Conjugation to a Protein Carrier

Ligand design should take structural and geometric factors into consideration. It starts with the use of rather simple structures, taking advantage of the primary recognition of a glycoligand by a lectin via a mono- or disaccharide (Cambillau, 1995; Rini, 1995; Bundle, 1997; Gabius, 1997a, 1998; Siebert et al., 1997). Unlike three documented cases of glycosyltransferases which impart modifications in the sugar antenna only after substrate selection via sequential recognition of the appropriate acceptor site on the oligosaccharide and a peptide motif of the protein part (Baenziger 1994), a similarly stringent requirement for a two-site check involving a part of the protein is only rarely documented for carbohydrate-binding proteins such as the basic fibroblast growth factor (Gallagher and Turnbull, 1992). It can be assumed that even naturally occurring mono- or disaccharides as part of a neoglycoprotein and their spatial arrangement with appropriate density may suffice to delineate the presence of lectins with a notable degree of selectivity. Again, the emphasized and biologically advantageous variability of oligosaccharide structures poses a formidable problem to the organic chemist. It calls for elegant strategies of synthesis to solve it. In this sense, developments in carbohydrate chemistry are of pivotal interest to lectinologists and histochemists eager to move to reverse lectin histochemistry (Gabius, 1988).

This intention can e.g. be prompted by the initial step of lectin histochemistry. When a certain epitope of interest is localized in a tissue section, for example the T-antigen by the peanut agglutinin, and the ensuing question concerning the presence of endogenous sites with affinity to this disaccharide is to be answered, access to this substance is expected from chemical synthesis (Gabius *et al.*, 1990a). As exemplarily outlined in Figure 3, a cleverly devised pathway is indispensable to facilitate the desired configuration of the linkage point for the glycosidic bond and its anomericity. Extending the general strategy to generate a glycosyl donor such as O-glycosyl trichloroacetimidates and to subsequently transfer it irreversibly in a catalyzed, sterically distinct reaction to a suitable acceptor, considerable strides have been taken to master the chemical synthesis of oligosaccharides (Paulsen, 1990; Boons, 1996; Schmidt, 1997). Owing to this progress even the synthesis of the core unit of N-linked glycosylation, depicted in Figure 2, and its naturally occurring extensions has been completed, as elegantly demonstrated for complex biantennary sugar chains (Unverzagt, 1994, 1996). As a cornerstone for the elaboration of the sugar antennae, the engineering of expression vectors for glycosyltransferases has recently ushered in the wholeheartedly welcomed era of chemo-enzymatic oligosaccharide synthesis (Gijsen *et al.*, 1996). Custom-made termini of the synthetic precursors can thus be generated. Together with a convenient chemical manipulation at the reducing end involving a spacer and an adapter for conjugation, as exemplified recently (André *et al.*, 1998), their preparation by a convenient scheme, shown in Figure 4, satisfies the demand for common, naturally present oligosaccharides. With the help of combinatorial libraries, mimetics for these natural ligands can even be sought in attempts to optimize the affinity of the investigated case of protein-carbohydrate binding (Sofia, 1996; Arya and Ben, 1997). As already indicated, in each case the necessity for a suitable adapter to effect high-efficiency coupling to the carrier must not be overlooked.

A diverse array of conjugation methods is compiled in the literature (Stowell and Lee, 1980; Aplin and Wriston, 1981; Lee and Lee, 1993, 1994a). Unless the integrity of the recognition sites on the ligand is disturbed, any tested method can well be applied, as verified by histochemical control studies (Gabius *et al.*, 1990b). A certain popularity has been gained by the diazonium and the phenylisothiocyanate reactions (McBroom *et al.*, 1972; Monsigny *et al.*, 1984; Reichert *et al.*, 1994). In contrast to the linkage region of reductive amination with its ring-opened product at the reducing end or an oxirane-activated sugar with its modification at the 6'-hydroxyl group the products of these reactions display a phenyl ring at the C1-position of the reducing end mimicking a further extension of a sugar chain, which does not

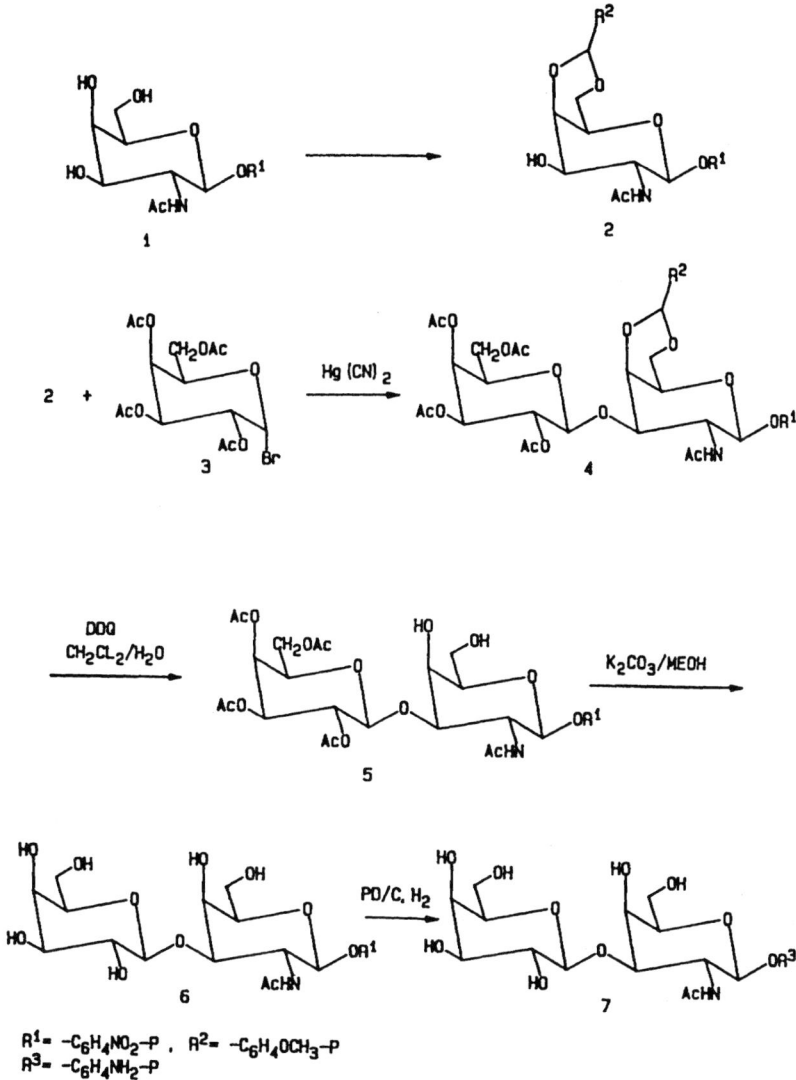

$R^1 = -C_6H_4NO_2-P$, $R^2 = -C_6H_4OCH_3-P$
$R^3 = -C_6H_4NH_2-P$

Figure 3 Schematic representation of the synthesis of the p-aminophenyl derivative of gal-β1,3-galNAc (compound 7), starting from the commercially available p-nitrophenyl derivative of 2-acetamido-2-deoxy-β-D-galactopyranoside (compound 1), which is converted to 2-acetamido-2-deoxy-4,6-O-(p-methoxybenzylidene)-β-D-galactopyranoside (compound 2) by reaction with p-methoxybenzaldehyde in the presence of zinc chloride to protect the 4' and 6' hydroxyl groups and reacted with 2, 3, 4, 6-tetra-O-acetyl-α-D-galactopyranosyl bromide (compound 3) in anhydrous nitromethane-benzene (1:1), sodium sulfate and mercuric cyanide.

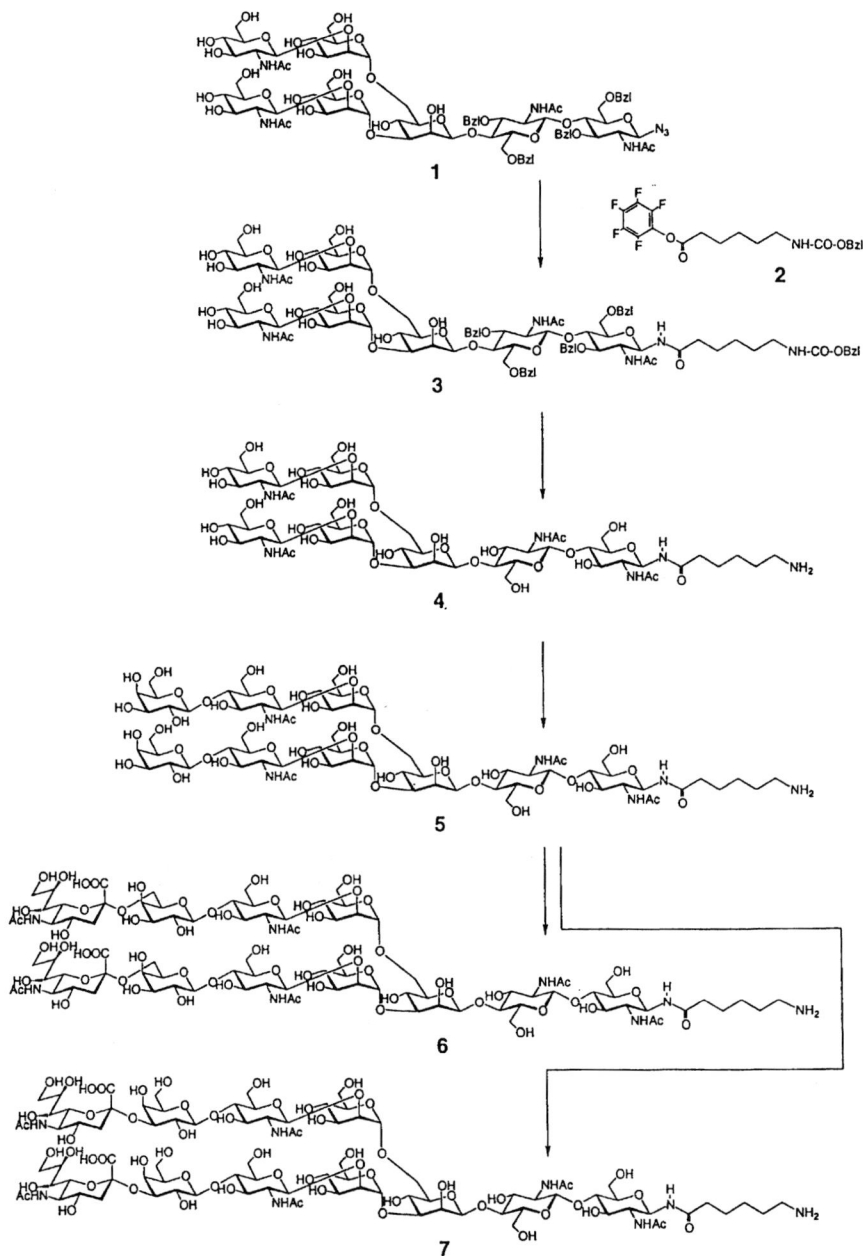

Figure 4 Schematic representation of the synthetic pathway to produce spacer-linked galactosylated and sialylated biantennary complex-type N-glycans. The chemically synthesized heptasaccharide-azide (compound 1) is converted to the amine, then to the 6-benzyl-oxycarbonyl-6-aminohexanoylamido-derivative (compound 3) and the 6-aminohexanoylamide form (compound 4). Subsequent application of β1,4-galactosyltransferase with UDP-gal as donor and either α2,3- or α2,6-sialyltransferases with CMP-N-acetylneuraminic acid as donor yields the galactosylated biantennary nonasaccharide sugar chain (compound 5) and its α2,3-/α2,6-sialylated undecasaccharide variants (compounds 6, 7).

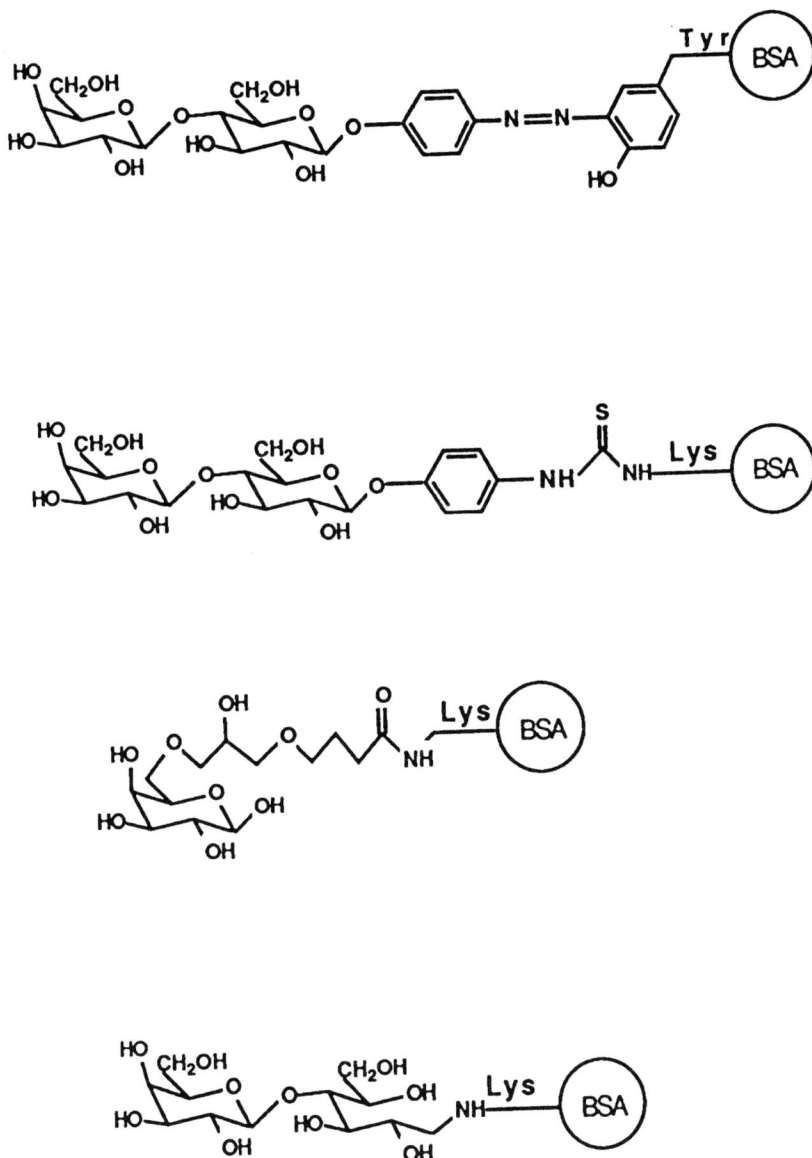

Figure 5 Illustration of the linkage of a sugar derivative or a disaccharide to the suitable functional group of an amino acid residue of the carrier (i.e. bovine serum albumin) for the product of coupling of a diazo derivative of p-aminophenyl lactoside (top), of a p-isothiocyanatophenyl lactoside derived from thiophosgene activation of p-aminophenyl derivatives, of a galactose derivative prepared by activation with (2,3-epoxypropane)-4-oxybutyric acid, and of reductive amination of lactose in the presence of NaCNBH$_3$ (bottom).

seem to impair the reactivity to a lectin (Figure 5). Inhibitory assays with lectins can prove the assumed potency of a sugar derivative prior to its conjugation. This reaction requires an entrance site on the carrier which is irreversibly modified in the course of the chemical reaction. Notably, the impact of conjugation to lysine groups of the carrier must not be neglected, as it shifts the isoelectric point of the protein and may cause undesired ionic interactions or binding to a scavenger receptor (Gabius and Bardosi, 1991; Jansen *et al.*, 1991; Danguy *et al.*, 1995). It also deserves emphasis that the maximum incorporation yield will not necessarily provide optimal results (Monsigny *et al.*, 1984; Gabius and Gabius, 1992; Bovin, 1993). Therefore, the density of the ligand has to be adapted experimentally.

Fine-tuning may also apply to the ligand structure, if key interaction points are composed of minor species of modifications in the glycan structure such as sulfation or O-acetylation (Varki, 1996; Sharon and Lis, 1997). Interestingly, the introduction of a single O-acetyl group into a ganglioside's sugar chain in the case of GD_{1a} has been shown to be without influence on the conformation of the oligosaccharide. But its presence markedly enhances the affinity to a receptor, what again highlights the already mentioned role of single-site substitutions in protein-carbohydrate interactions (Siebert *et al.*, 1996). Having assembled an oligosaccharide chain, additional refinements can thus be unavoidable. In these instances of site-directed modifications it is conducive to complement the synthetic strategy by the biochemical approach to purify a lectin-reactive oligosaccharide chain and to immobilize the natural product onto a carrier.

Biochemical Preparation of the Carbohydrate Ligand and its Conjugation to a Protein Carrier

Given the case that a glycoprotein is bound to a tissue receptor via a carbohydrate determinant which presently is not defined in detail, precise blueprint for the chemist is not available. The planned investigations can nonetheless be performed, because a probe can be prepared using the strategy given in Figure 6. Following proteolytic degradation or enzymatic deglycosylation a fractionation of the resulting glycopeptides or glycan chains can be added to the outlined protocol to obtain the most active compound. With appropriate manipulation of the glycopeptide or reducing end of the glycan (e.g. by 1-N-glycyl-β-derivatization after β-glycosylamine formation in ammonium bicarbonate whose lack of influence on the stability of any crucial modification must be confirmed) or use of a bifunctional crosslinker the covalent bond to the carrier can be established (Lee and Lee, 1989; Gabius *et al.*, 1991a; Wong *et al.*, 1993). A similar strategy works for gangliosides after their conversion to lysogangliosides. The opened amino function of the ceramide portion can then serve as attachment point for a crosslinking agent (Gabius *et al.*, 1990c; Mahoney and Schnaar, 1994; Hassid *et al.*, 1996). In this way, complex naturally occurring sugar chains can be covalently bound to a carrier. This part of a neoglycoconjugate ideally is histochemically inert. In view of any attempts of therapeutical evaluation based on such vehicles for drug targeting or tumor imaging (Gabius, 1988, 1997b), it would be preferable to work with a water-soluble and easily sterilizable polymer which is less immunogenic. These incentives have prompted to investigate alternatives for a protein as carrier substance in glycohistochemistry.

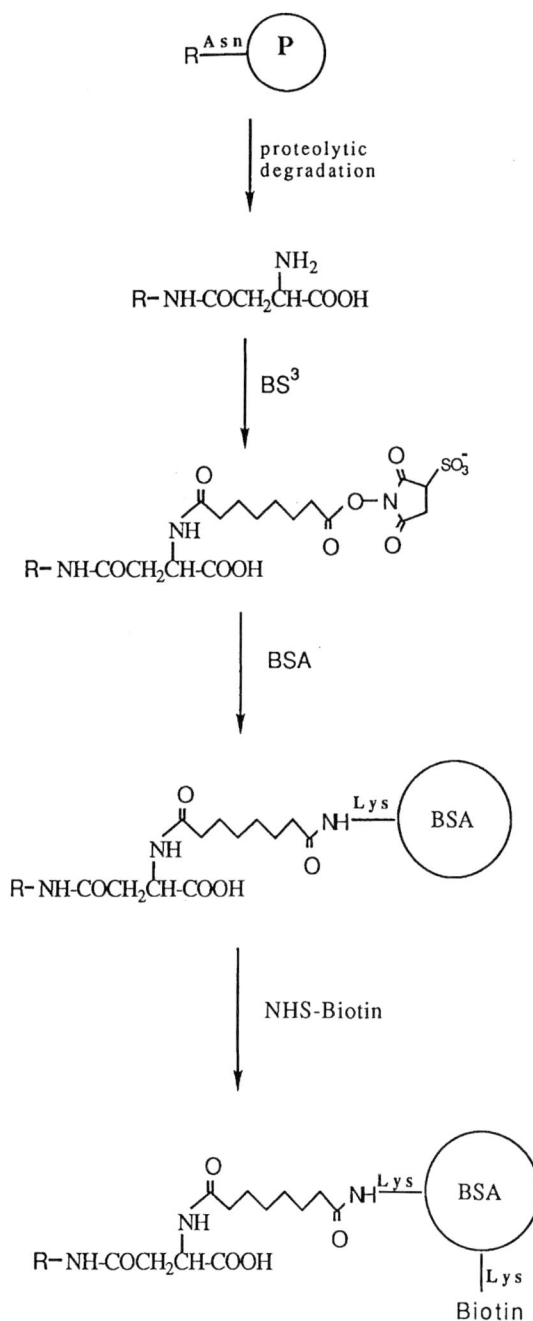

Figure 6 Illustration of the synthetic steps for the preparation of a glycopeptide-carrier (i.e. bovine serum albumin) conjugate, starting with the proteolytic degradation of a glycoprotein and the isolation of Asn-linked sugar chains (see Figure 1). By using a suitable bifunctional crosslinker (e.g. bis(sulfosuccinimidyl)suberate) the glycopeptide is activated at the α-amino group of the asparagine moiety and coupled to the carrier via its second functionality. Finally, this product can be labelled e.g. by biotinylation.

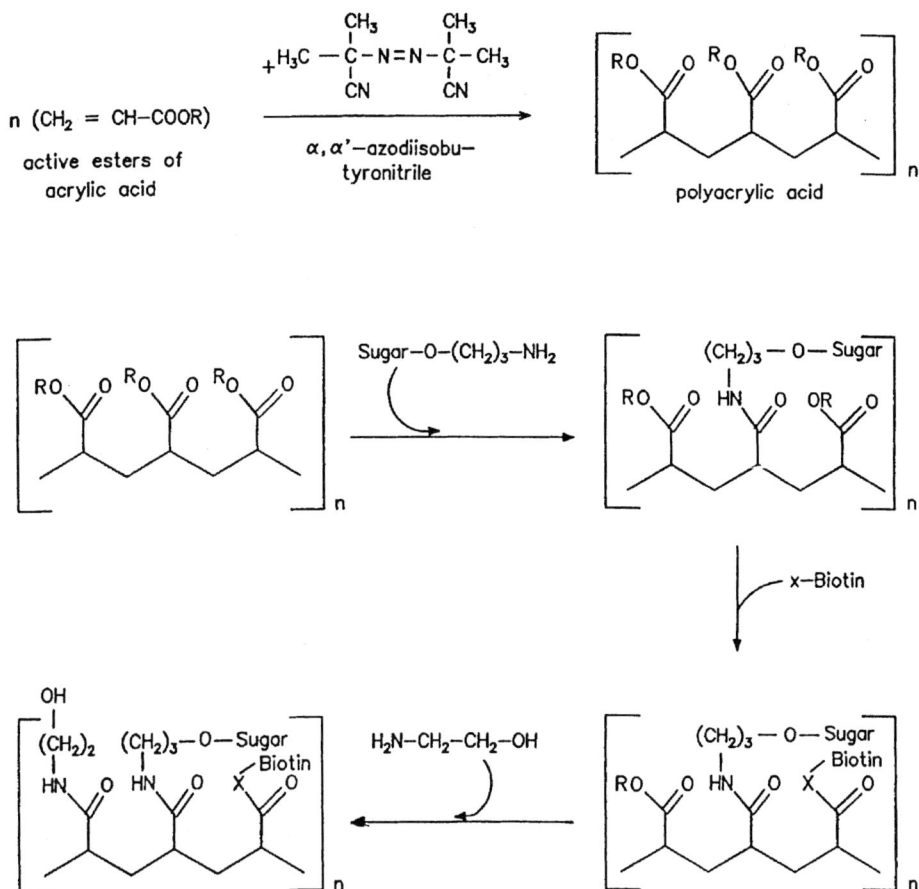

Figure 7 Schematic illustration of the chemical synthesis of a biotinylated, carbohydrate-bearing polymer [substituted poly(2-hydroxyethyl acrylamide)] starting with the catalyzed polymer formation from active ester monomers such as 4-nitrophenylacrylate and subsequent incorporation of 3-aminopropyl sugar derivatives (5-30% molar substitution), a suitable biotin derivative and finally ethanolamine to block residual active sites.

Synthetic Polymers as versatile Carrier Part of Neoglycoconjugates

Relative to the copolymerization of an olefinic group (allyl- or acryloyl)-containing sugar unit with acrylamide or the activation of polyacrylamide e.g. in the presence of carbodiimides the derivatization of activated polymers, i.e. poly(4-nitrophenylacrylate), presents favorable characteristics in terms of reproducibility and convenience of product features (Bovin 1993; Bovin *et al.* 1993). The detailed schematic representation of the reaction pathway is given in Figure 7. It illustrates polymer synthesis and the incorporation of the sugar ligand as nucleophilic compound which proceeds quantitatively to a degree of 25–30% substitution as well as of an appropriate label. Finally, the blocking of residual reactive esters with ethanolamine needs to be performed. Irrespective of the nature of the carrier

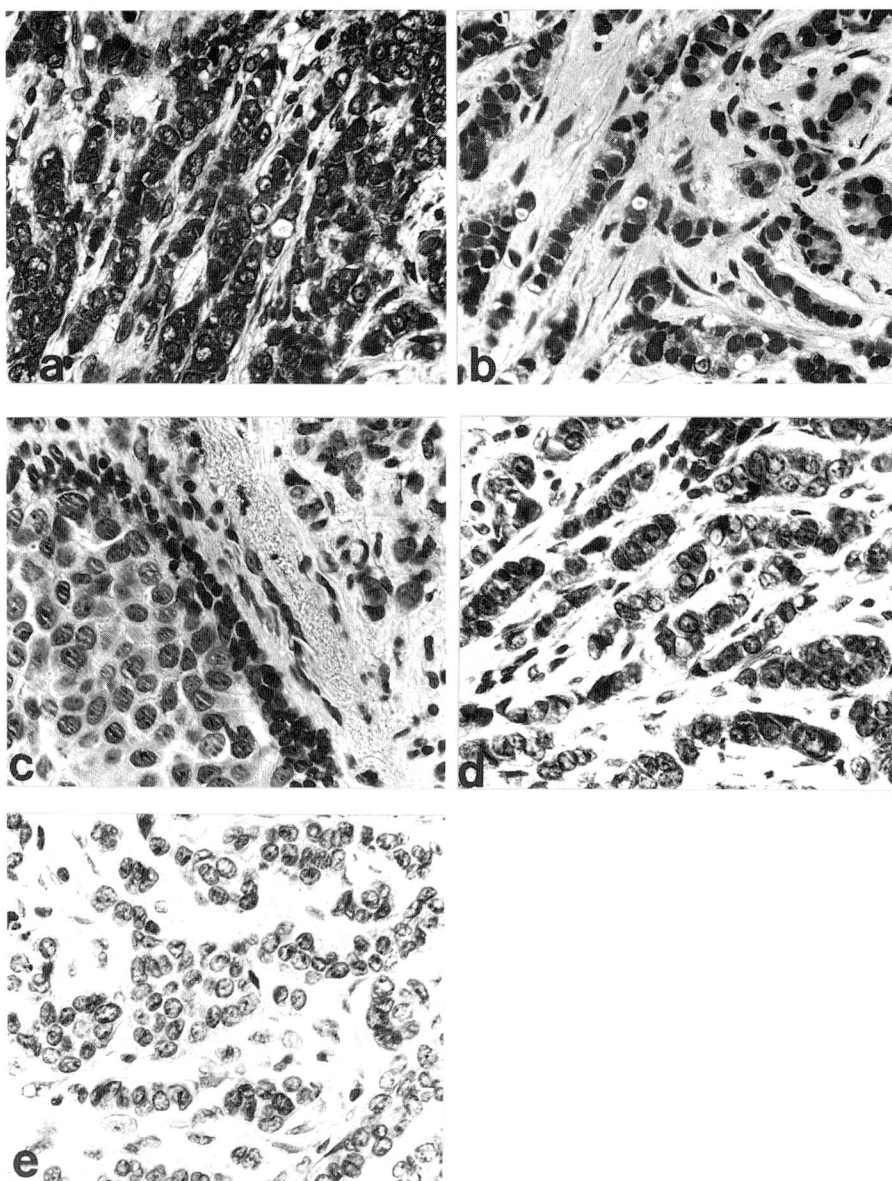

Figure 8 Light micrographs of formalin-fixed and paraffin-embedded sections of an infiltrating ductal mammary carcinoma after application of biotinylated neoglycoproteins exposing lactose or galactose moieties on their surface and of kit reagents to visualize specifically bound probe molecules as well as a hematoxylin counterstaining. The pictures show the staining reaction with lactosylated bovine serum albumin derived from diazo coupling (a; for structural details of the linkage between sugar part and bovine serum albumin, see Figure 5), from coupling of the p-isothiocyanatophenyl derivative (b) and from reductive amination (c) and with galactosylated albumin derived from covalent attachment of the aliphatic derivative of galactose, given in Figure 5 (d). An exemplary illustration of the result of a competitive inhibition of sugar-specific binding with label-free neoglycoprotein in the case of the first conjugate is given (e).

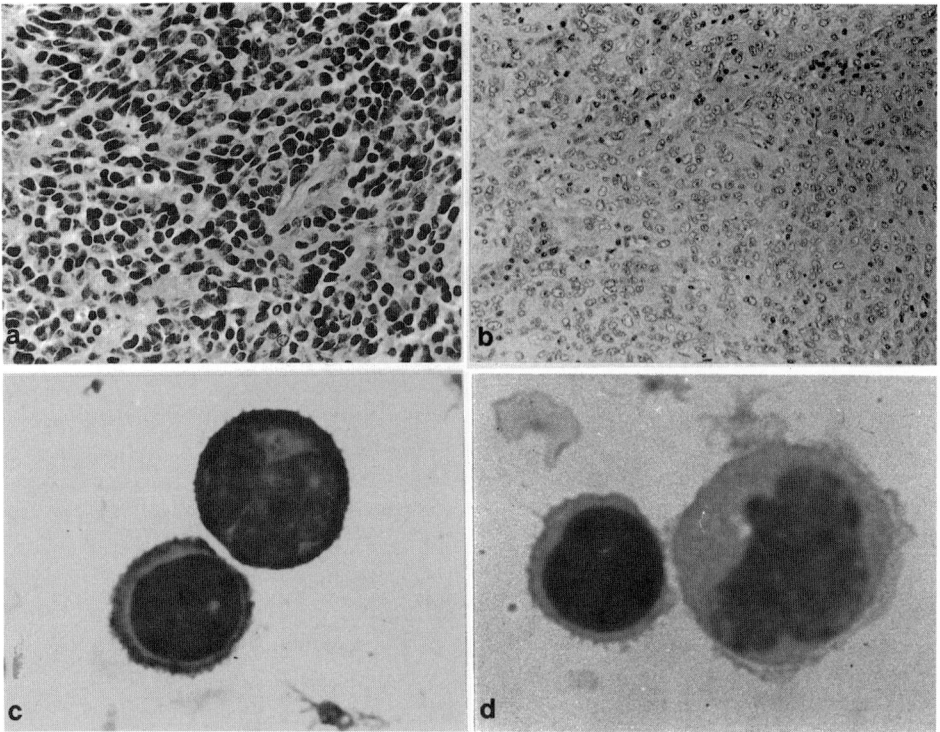

Figure 9   Light micrographs of formalin-fixed and paraffin-embedded sections of a small cell anaplastic carcinoma (a) and a large cell anaplastic carcinoma (b) of the human lung after application of a biotinylated neoglycoprotein (sugar part: purified lysoganglioside GM_1 (a) or a lysoganglioside mixture with GD_{1a}, GM_1, GT_{1b} and GD_{1b} (b)) and kit reagents as well as a hematoxylin counterstaining and of cytospin preparations of human peripheral blood lymphocytes and monocytes after subsequent incubation with the biotinylated lysoganglioside GM_1-bearing neoglycoprotein in the absence (c) or in the presence of competitive inhibitor (d) and kit reagents and after a hemalaun counterstaining.

adequate histochemical controls are indispensable to confirm that the staining reaction caused by the presence of the label is dependent on the protein (receptor in cells or tissues)-carbohydrate (sugar ligand of neoglycoconjugate) recognition, as already emphasized in the section on General Considerations.

Lectin Detection with Neoglycoconjugates

The glycohistochemical processing protocol offers various independent strategies to assure the assumed molecular interaction. Competitive inhibition studies can be performed to saturate ligand-binding sites with label-free glycocompound, as illustrated in Figures 8 and 9. Concomitantly, the labeled carrier without a histochemically reactive ligand structure should be employed to demonstrate the lack of presence of carrier backbone- and label-binding sites. Addition of carrier substance, which can even be modified to match the isoelectric point of the neoglycoconjugate, to the probe-containing solution is a further step to preclude false-positive interpretations. Internal controls with negative cases (Figure 9b) and with lectin-specific antibodies in instances where only one distinct lectin is responsible for binding (e.g. Bardosi *et al.*, 1989; Gabius *et al.*, 1991b) add to the reliability of the method. Moreover, slight structural changes of the crucial ligand part should affect the staining intensity and/or distribution, if no further chemical parameter of the panel of neoglycoconjugates is altered (Vidal-Vanaclocha *et al.*, 1990; Gabius and Bardosi, 1991; Kayser *et al.*, 1994a,b; Danguy *et al.*, 1995).

Similar to histological sections and cytological specimen native cells can be subjected to an evaluation of their lectin expression. To avoid the application of radioactive substances and to present a rather large ligand-exposing probe surface to the cells, the enzymatic activity of the tetrameric *E. coli* β-galactosidase was chosen. Glycosylation under activity-preserving conditions involving carbodiimide-mediated conjugation of p-aminophenyl glycosides has led to neoglycoenzymes (Gabius *et al.*, 1989). As illustrated in Figure 10, the assessment of extent of binding of these tools allows to determine receptor affinity and density in the case of native cells. The histogenetic origin of the cells and their state of differentiation are important factors which modulate the display of sugar-binding capacity on the surface (Gabius and Gabius, 1990; Gabius *et al.*, 1994a,b). Due to the assumed importance of recognitive glycobiological interactions, compiled in a recent collection of critical reviews (Gabius and Gabius, 1997), these results have encouraged to embark on the evaluation of the performance of this class of synthetic tool in histopathology.

GLYCOHISTOCHEMISTRY IN HISTOPATHOLOGY

As noted, the work with cultured normal and transformed cells has shown the impact of the cell type on the expression of neoglycoenzyme-binding sites. Monitoring of lectin presence with antibodies in the cases of certain C-type and I-type lectins are in full agreement with this result (Gabius, 1997a). This diagnostically potentially valuable conclusion dovetails with observations that this correlation between the binding pattern and the cell type is also valid for tumor cells in tissue sections. Remarkably, e.g. small cell and the non-small cell lung cancer types, mesotheliomas and metastatic adenocarcinomas to the pleura or subtypes of meningiomas can be distinguished glycohistochemically (Bardosi *et al.*, 1988; Kayser *et al.*, 1989, 1992). Relative to the complex biantennary nonasaccharide and its α2,3-/α2,6-sialylated derivatives, whose binding patterns are shown in Figure 11, carrier-immobilized mannose, fucose and maltose reach an enhanced level of discriminatory power, sensitivity and specificity in differential diagnosis of small cell versus non-small cell lung cancer being comparable to neuroendocrine markers (Kayser *et al.*, 1989; André *et al.*, 1998).

Not only the tumor cell type, but also the process of tumorigenesis alters the extent of binding to carbohydrate ligands quantitatively (Gabius *et al.*, 1991c; Brinck *et al.*, 1996). Such detectable parameter changes should not prematurely deceive into postulating physiolocigal consequences without solid experimental foundation. At any rate, caution should be exercized concerning the correlation of glycohistochemical properties to the metastatic phenotype. In this respect, it must not be ignored that a metastatic cell is exposed to the new organ microenvironment

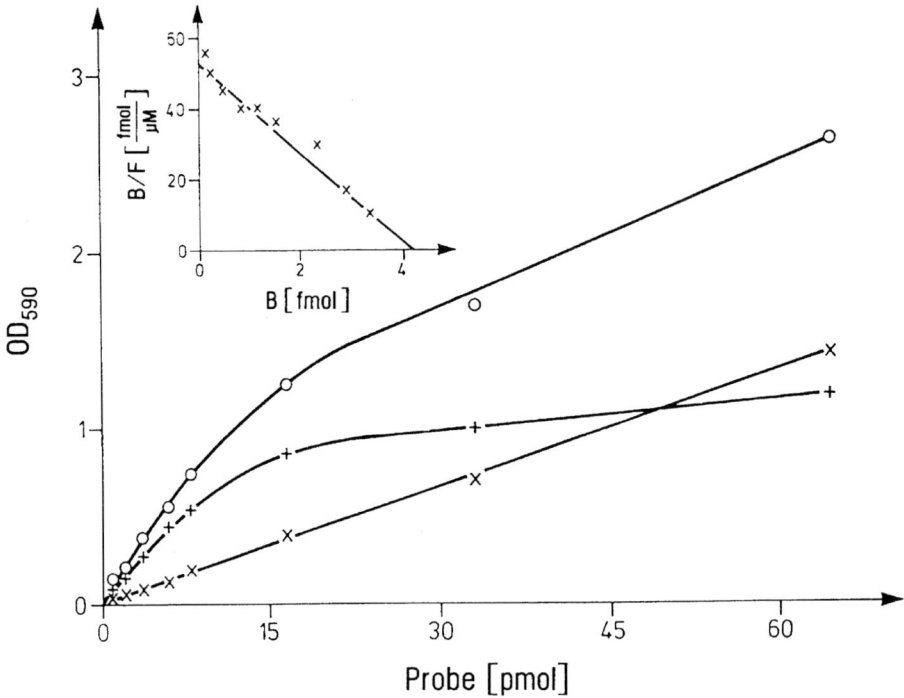

Figure 10 Determination of ligand-dependent binding (+) of N-acetyl-D-glucosaminylated
E. coli β-galactosidase (total binding: o; carrier-dependent binding: x) and Scatchard analysis
of the binding data (inset) for cells of the human monocytic leukemia cell line THP-1,
yielding a K_D value of 77 nM and an average of 7.1×10^4 bound neoglycoenzyme molecules
per cell at saturating probe concentration (B_{max}).

starting with its arrival and arrest in the parenchyma of the target organ. Owing
to the proven influence of diverse factors of the microenvironment on lectin
expression, as e.g. experimentally seen by tumor cell inoculation at different host
sites (Vidal-Vanaclocha *et al.*, 1990), the immediate assumption of participation of
any measurable correlation between lectin presence and metastasis establishment
in the homing mechanism which neglects other plausible explanations is not
warranted and may seriously lead astray.

  Another clinical parameter which can be set into correlation to glycohistochemical
features is the prognosis. For carcinoid tumors of the lung, the expression of the
heparin-binding lectin is positively correlated to the survival time (Kohnke-Godt
and Gabius, 1989, 1991; Kayser *et al.*, 1996a). Conversely, presence of binding sites
for β-N-acetyl-D-galactosamine is an indicator of comparatively poor survival in this
tumor class (Figure 12). These results graphically illustrate a phenomenological
relationship of the presence of these sugar-binding sites to the aggressiveness of the
tumors. Although protein-carbohydrate interactions are known to be involved in
the regulation of diverse cellular activities including proliferation and apoptosis
(Borrebaeck and Carlsson, 1989; Gabius *et al.*, 1995; Villalobo *et al.*, 1997), the actual

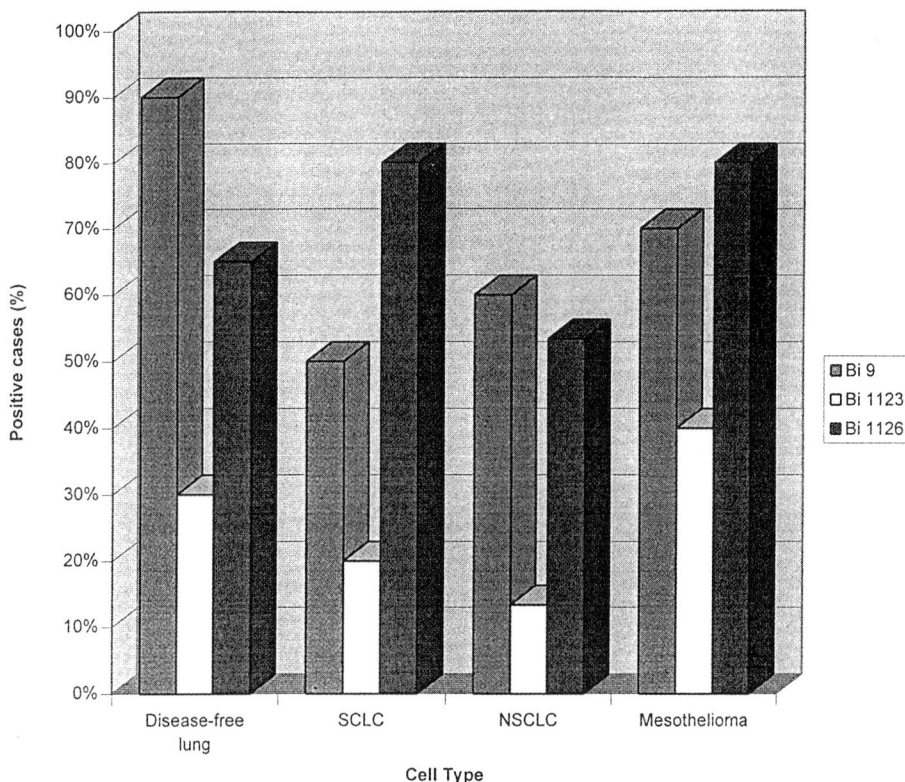

Figure 11 Illustration of the percentage of positive cases of disease-free lungs (N = 20), of small cell lung cancer (N = 10), of non-small cell lung cancer (N = 30) and mesotheliomas (N = 10), when formalin-fixed and paraffin-embedded sections were processed glycohistochemically under identical conditions with biotinylated neoglycoproteins containing biantennary complex-type sugar chains either with galactose termini (Bi 9) or with α2,3- or α2,6-sialylated termini (Bi1123, Bi1126). The structures of the nonasaccharide derivative and its two sialylated variants are given in Figure 4 (compounds 5-7).

pathway and effectors are presently biochemically not defined. This situation likewise applies to the still enigmatic question on the role of histo-blood group epitopes (Coon and Weinstein, 1986; King, 1994; Garratty, 1995; Dabelsteen, 1996; Greenwell, 1997).

Within the ABH-system the α1,3-directed introduction of a monosaccharide transforms the H(O)-core into the A- or B-determinants (Figure 13). Whereas monoclonal antibodies are common tools to assess the status of expression of these epitopes, neoglycoconjugates obviously open the way to address the question, whether suitable receptor sites for any blood group structure can be visualized in the tumor. They may e.g. act as modulators of tumor cell motility (Miyake and Hakomori, 1991; Ichikawa *et al.*, 1997), of tumor cell proliferation (Santos-Benito *et al.*, 1992; Coteron *et al.*, 1995) or of secretion of mediator substances by immune cells (Vellupillai and

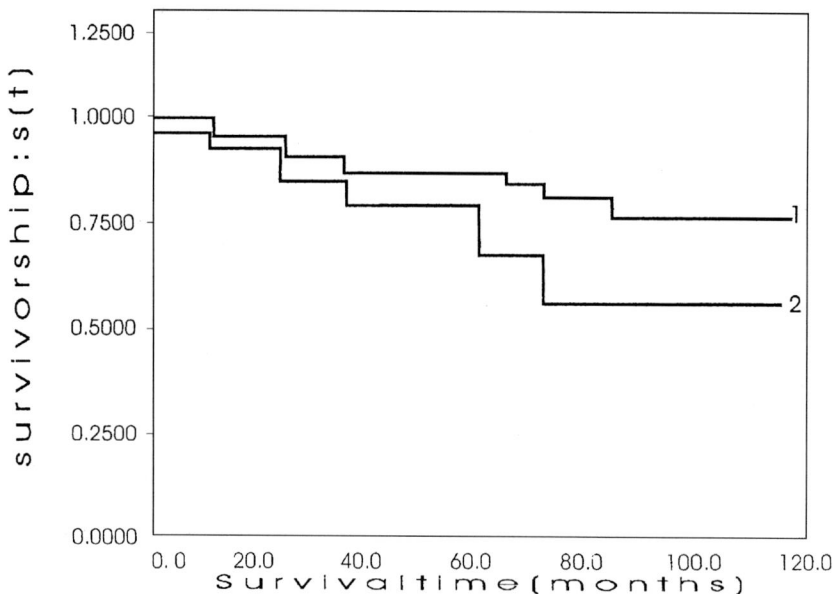

Figure 12 Illustration of the correlation between the survival time of surgically treated patients with carcinoid tumors of the lung (N = 82) and the absence (group 1) or presence (group 2) of binding sites for N-acetyl-D-galactosamine which had been detected glycohistochemically in formalin-fixed and paraffin-embedded tissue sections.

Blood group antigens

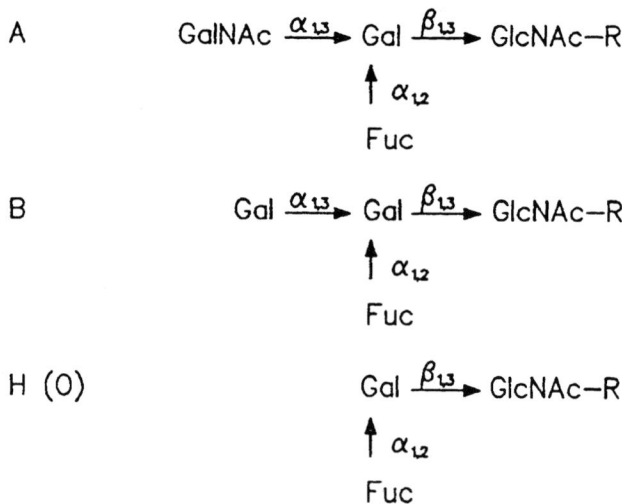

A   GalNAc $\xrightarrow{\alpha_{1,3}}$ Gal $\xrightarrow{\beta_{1,3}}$ GlcNAc–R

$\uparrow \alpha_{1,2}$

Fuc

B   Gal $\xrightarrow{\alpha_{1,3}}$ Gal $\xrightarrow{\beta_{1,3}}$ GlcNAc–R

$\uparrow \alpha_{1,2}$

Fuc

H (0)   Gal $\xrightarrow{\beta_{1,3}}$ GlcNAc–R

$\uparrow \alpha_{1,2}$

Fuc

Figure 13 Illustration of the structures of the ABH-histoblood group epitopes (type 1) with their establishment by an $\alpha 1,3$-extension at the galactose moiety of the common core (H(O) epitope) to yield the A and B determinants.

Harn, 1994). Starting with the monitoring of expression of sugar epitopes, their chemical synthesis and conjugation to produce the respective neoglycoconjugates is the second step in glycohistochemistry, as outlined in the section on General Considerations. When ABH-trisaccharide-exposing neoglycoconjugates had been tested in lung cancer in a model study in this field, a clear dependence of the result on the characteristics of the sugar structure was evident, presence of a 2'-N-acetyl group instead of a 2'-hydroxyl group in the terminal monosaccharide causing the significant difference (Kayser et al., 1994b). Whereas no correlation of tumor cell staining with prognosis was observable for B determinant-binding sites, presence of the capacity to recognize either the A- or H-epitopes intimates prolonged survival (Figure 14a-c). This biochemical reactivity may have a currently unknown functional bearing on other lines of evidence for prolonged survival which is now amenable to experimental consideration.

In a reasonable approach, these newly defined glycohistochemical features of the tumor cells can be correlated with further parameters which are gathered from the analysis of the tumor section, i.e. the distribution of the DNA content-related integrated optical density, the syntactic structure analysis and the transformation of the data into calculations of thermodynamic entities for the cellular organization such as the entropy or current of entropy (Kayser and Gabius, 1997). Although it is presently surely premature to attempt to raise mechanistic conclusions, it is noteworthy that the percentage of S-phases and of aneuploidic tumor cell nuclei is smaller in lung tumor cases with blood group H-trisaccharide-binding sites than in negative specimen (Kayser et al., 1994c). Interestingly, the correlation between decreased percentage of aneuploidic cells, shown in Table 3, is not confined to lung cancer. It is also seen for prostatic adenocarcinomas (Kayser et al., 1995). We thus consider this line of combined glycohistochemical and morphometric analysis potentially pertinent to delineate those biochemical pathways which are responsible for the generation of clinically favorable characteristics (Kayser and Gabius, 1997).

PERSPECTIVES

Communication is based on the transmission of physical signals. Defined sets of letters form words as messages. In the biological information transfer amino acids and nucleotides have been referred to as alphabets of life. In full appreciation of the proven premise of the functionality of protein-carbohydrate recognition it is justified to count the language of oligosaccharides among the natural code systems. As quoted by Montreuil (1995), Francois Jacob wrote in this respect that "together with nucleic acids and proteins, carbohydrates represent the third dimension of molecular biology". The ease of lectin purification and the abundance of these sugar receptors in plants were instrumental to take the first step in the monitoring of this type of molecular interaction by lectin histochemistry. These tools from plants and also invertebrates have underscored and continue to demonstrate that glycoconjugate expression is a multi-facetted and intriguingly precisely tuned process. The challenge to decipher the meaning of the — at first glimpse — phenomenological changes can be met by concerted efforts to which extensions of the panel of histochemical reagents can significantly contribute. Improvements of our knowledge in the realm of glycosciences are sure to be accrued by the deliberate application

(a)

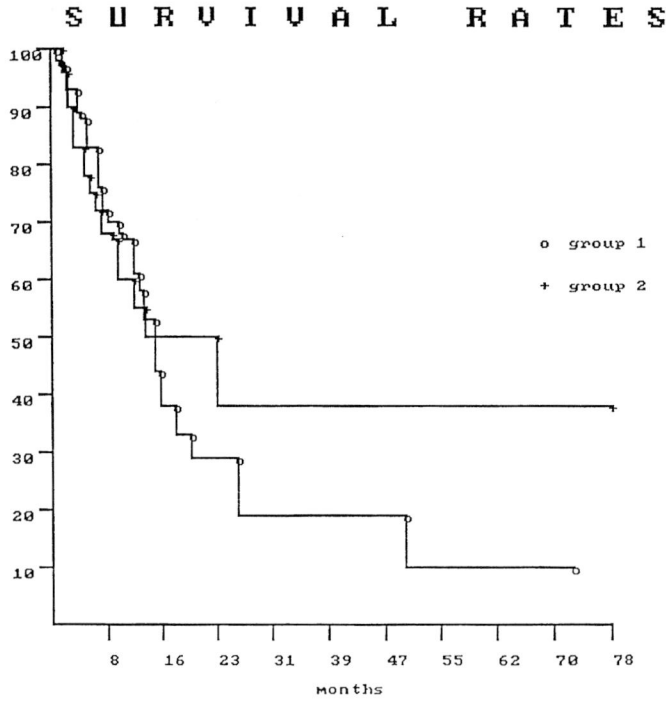

(b)

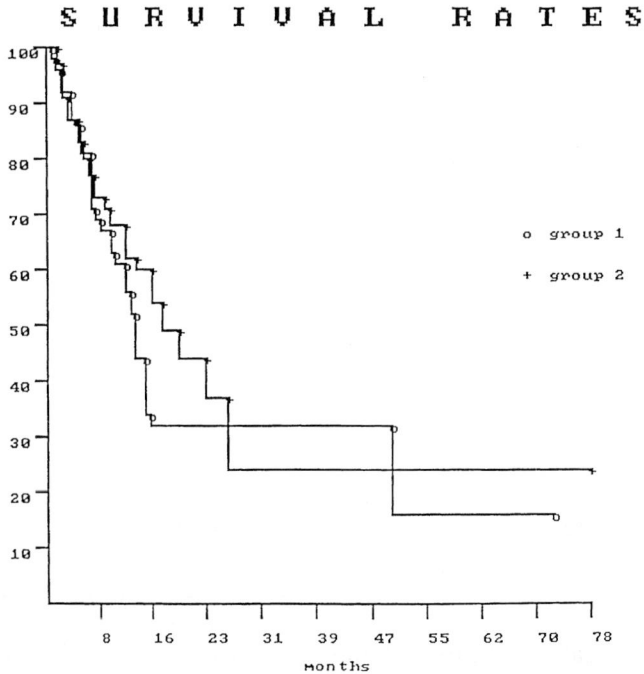

(c)

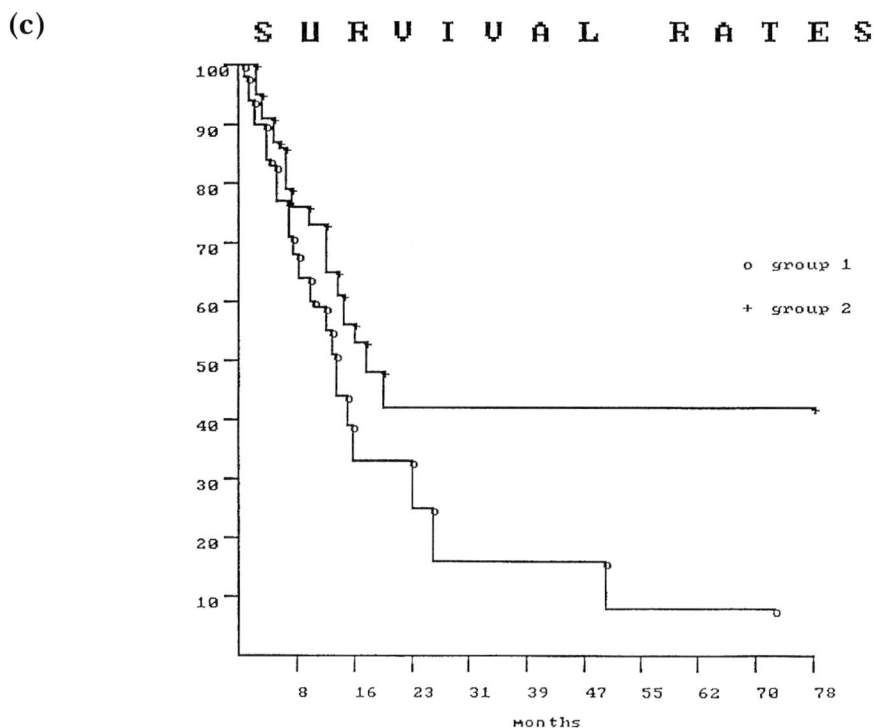

Figure 14 Illustration of the correlation between the survival time of surgically treated patients with lung cancer (N = 187) and the absence (group 1; o) or presence (group 2; +) of binding sites for histoblood group epitope A (a), B (b) and H (c).

of the custom-made neoglycoconjugates. The versatility of their synthesis by increasingly sophisticated chemoenzymatic process steps and the documented gain of information by the visualization of carbohydrate ligand-binding sites can even serve as a model to expand the reach of this technique. When following a similar strategic route to produce other carrier-immobilized ligands retaining their biological activity, the chemistry of derivatization and conjugation will only require slight adaptations, if any at all, to yield neoligandoconjugates (Gabius and Gabius, 1992; Kojima *et al.*, 1997).

The identification of endogenous lectins, to which neoglycoconjugates lend precious assistance (Table 1), and the long experience with plant and invertebrate lectins in histochemistry point to the next step to merge these two independent lines into a powerful line of investigation. The rapidly proceeding work on animal lectins will engender a wealth of projects to complement glyco- and immunohistochemical localization of tissue lectins with the visualization of the accessible complementary sites. As indicated by Figure 15a,b for T3/T4 lung tumors, the delineation of e.g. prognostic relevance by the measurement of extent of expression of galectin ligands already is a promising incentive to pursue such histopathological clues on the level of cell biology (Kayser *et al.*, 1996b). This

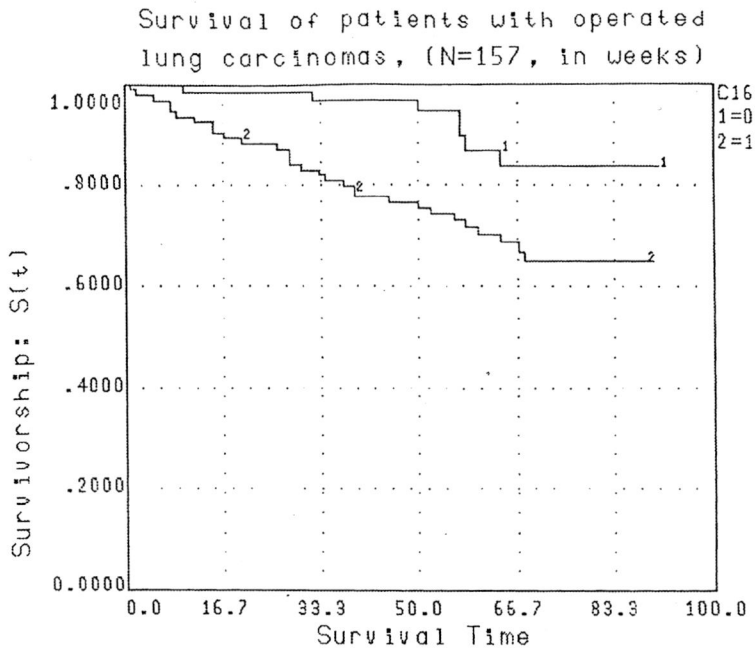

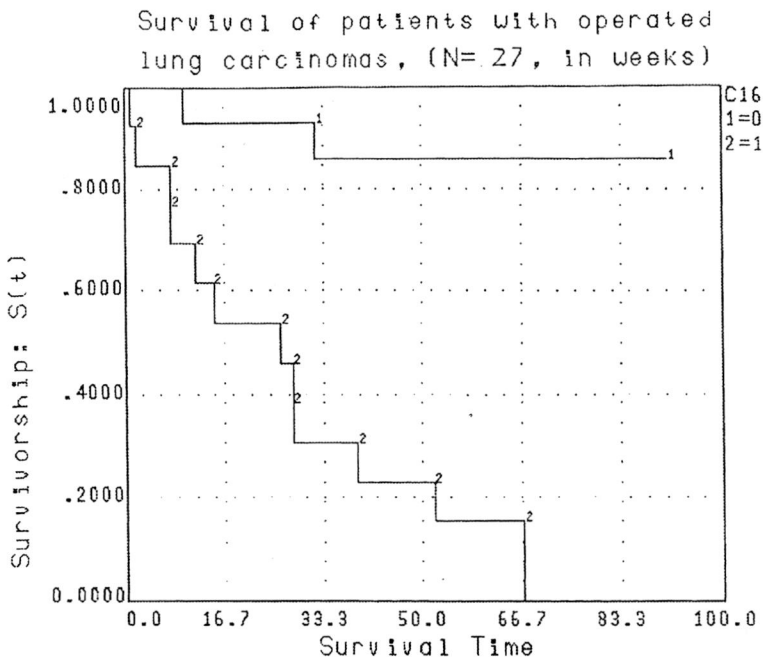

Figure 15 Illustration of the correlation between the survival time of surgically treated patients with lung cancer (a: N = 157; b: N = 27) and the absence (group 1) or presence (group 2) of ligands for the avian galectin from liver, termed C-16. The two groups differed in the pT stage, summarizing patients with pT1-pT4 (top) or with pT3/pT4 (bottom).

Table 3 IOD- and MST-Features in Relation to Expression of Binding Sites for A/H-Blood Group Trisaccharides for Prostatic Adenocarcinoma

Feature	Probe			
	A_{tri}		H_{tri}	
	−	+	−	+
SPRF	8.2	8.3	8.0	8.4
E(DNA)	2.9	2.7	2.6	2.8
2CV-index	2.4	2.7	3.6	2.1[*]
>5-Crate	8.8	9.7	12.6	7.8[*]
DI-Tu-TU	9.5	9.5	8.8	9.9
DI-TU-LY	9.1	9.2	8.3	9.8
DI-Pr-Pr	34.8	40.1	40.0	38.3
DI-5C-5C	24.0	30.7	25.1	31.1[*]
E(MST)	46.7	48.1	43.7	50.0

*statistically significant difference (p ≤ 0.05); abbreviations: A_{tri}, H-tri: synthetic blood group trisaccharides A and H, covalently attached to a labelled, histochemically inert carrier; SPRF: S-phase-related fraction; E(DNA): entropy of distribution of DNA; 2CV index: mean standard deviation of 2C index; 5C rate: aneuploidy-related fraction; Di-Tu-Tu: distance (in μm) betweeen nearest neighboring tumor cell nuclei; Pr, Ly, 5C: proliferating tumor cell (2.75 < IOD < 3.25), lymphocyte, tumor cell nuclei with IOD > 5; E(MST): entropy of the minimum spanning tree as a measure of deviation of the tumor structure from a biologic "non-proliferating" structure; from Kayser *et al.*, 1995.

indication flanked by the outlined chemical and biochemical schemes demonstrates the inevitably interdisciplinary nature of glycobiological investigations. The perspective of this work appears to be appropriately described by Alfred Gottschalk who wrote in 1973 on the outlook of glycoprotein research that "we are not at the end of all progress, but at the beginning. We have but reached the shores of a great unexplored continent", as quoted from Montreuil (1995). He legitimately added that "we are discovering, filled with wonder, its fascinating secrets".

REFERENCES

André, S., Unverzagt, C., Kojima, S., Dong, X., Fink, C., Kayser, K., and Gabius, H.-J. (1998) Neoglycoproteins with the synthetic complex biantennary nonasaccharide or its α2,3–/α2,6–sialylated derivatives: their preparation, the assessment of their ligand properties for purified lectins, for tumor cells *in vitro* and in tissue sections and their biodistribution in tumor-bearing mice. *Bioconjugate Chem.*, **8**, 845–855.

Aplin, J.D., and Wriston, J.C. (1981) Preparation, properties and applications of carbohydrate conjugates of proteins and lipids. *CRC Crit. Rev. Biochem.*, **10**, 259–306.

Arya, P., and Ben, R.N. (1997) Der Einsatz kombinatorischer Chemie zur Synthese von Kohlenhydratbibliotheken. *Angew. Chem.*, **109**, 1335–1337.

Baenziger, J.U. (1994) Protein-specific glycosyltransferases: how and why they do it! *FASEB J.*, **8**, 1019–1025.

Bardosi, A., Dimitri, T., and Gabius, H.-J. (1988) (Neo)glycoproteins as tools in neuropathology: histochemical patterns of the extent of expression of endogenous carbohydrate-binding receptors, like lectins, in meningiomas. *Virchows Arch. [Cell. Pathol.],* **56**, 35–43.

Bardosi, A., Dimitri, T., Wosgien, B., and Gabius, H.-J. (1989) Expression of endogenous receptors for neoglycoproteins, especially lectins, which allow fiber typing on formaldehyde-fixed, paraffin-embedded muscle biopsy specimens. A glycohistochemical, immunohistochemical and glycobiochemical study. *J. Histochem. Cytochem.,* **37**, 989–998.

Barondes, S.H. (1988) Bifunctional properties of lectins: lectins redefined. *Trends Biochem. Sci.,* **13**, 480–482.

Boons, G.-J. (1996) Synthetic oligosaccharides: recent advances. *Drug Discovery Today,* **1**, 331–342.

Borrebaeck, C.A.K., and Carlsson, R. (1989) Lectins as mitogens. *Adv. Lectin Res.,* **2**, 1–27.

Bovin, N.V. (1993) Sugar-polyacrylamide conjugates as probes for cell lectins. In H.-J. Gabius, and S. Gabius, (eds.), *Lectins and Glycobiology,* Springer Verlag, Heidelberg, pp. 23–28.

Bovin, N.V., and Gabius, H.-J. (1995) Polymer-immobilized carbohydrate ligands: versatile chemical tools for biochemistry and medical sciences. *Chem. Soc. Rev.,* **24**, 413–421.

Bovin, N.V., Korchagina, E.Y., Zemlyanukhina, T.V., Byramova, N.E., Galanina, O.E., Zemlyakov, A.E., Ivanov, A.E., Zubov, V.P., and Mochalova, L.V. (1993) Synthesis of polymeric neoglycoconjugates based on N-substituted polyacrylamides. *Glycoconjugate J.,* **10**, 133–141.

Brinck, U., Bosbach, R., Korabiowska, M., Schauer, A., and Gabius, H.-J. (1996) Histochemical study of expression of lectin-reactive carbohydrate epitopes and glycoligand-binding sites in normal human appendix vermiformis, colonic mucosa, acute appendicitis and colonic adenoma. *Histol. Histopathol.,* **11**, 919–930.

Brockhausen, I., and Schachter, H. (1997) Glycosyltransferases involved in N- and O-glycan biosynthesis. In H.-J. Gabius, and S. Gabius, (eds.), *Glycosciences: Status and Perspectives,* Chapman & Hall, London - Weinheim, pp. 79–113.

Bundle, D.R. (1997) Antibody-oligosaccharide interactions determined by crystallography. In H.-J. Gabius, and S. Gabius, (eds.), *Glycosciences: Status and Perspectives,* Chapman & Hall, Weinheim - London, pp. 311–331.

Cambillau, C. (1995) The structural features of protein-carbohydrate interactions revealed by X-ray crystallography. In J. Montreuil, J.F.G. Vliegenthart, and H. Schachter, (eds.), *Glycoproteins,* Elsevier, Amsterdam, pp. 29–66.

Colley, K.J. (1997) Golgi localization of glycosyltransferases: more questions than answers. *Glycobiology,* **7**, 1–13.

Cook, G.M.W. (1986) Cell surface carbohydrates: molecules in search of a function? *J. Cell Sci., Suppl.,* **4**, 45–70.

Coon, J.S., and Weinstein, R.S. (1986) Blood group-related antigens as markers of malignant potential and heterogeneity in human carcinomas. *Hum. Pathol.,* **17**, 1089–1106.

Coteron, J.M., Singh, K., Asensio, J.L., Dominguez-Dalda, M., Fernandez-Mayoralas, A., Jiménéz-Barbero, J., and Martin-Lomas, M. (1995) Oligosaccharides structurally related to E-selectin ligands are inhibitors of neural cell division: synthesis, conformational analysis and biological activity. *J. Org. Chem.,* **60**, 1502–1519.

Dabelsteen, E. (1996) Cell surface carbohydrates as prognostic markers in human carcinomas. *J. Pathol.,* **179**, 358–369.

Damjanov, I. (1987) Lectin cytochemistry and histochemistry. *Lab. Investig.,* **57**, 5–20.

Danguy, A. (1995) Perspectives in modern glycohistochemistry. *Eur. J. Histochem.,* **39**, 5–14.

Danguy, A., Kiss, R., and Pasteels, J.L. (1988) Lectins in histochemistry: a survey. *Biol. Struct. Morphol.,* **1**, 93–106.

Danguy, A., Kayser, K., Bovin, N.V., and Gabius, H.-J. (1995) The relevance of neoglycoconjugates for histology and pathology. *Trends Glycosci. Glycotechnol.,* **7**, 261–275.

Danguy, A., Camby, I., Salmon, I., and Kiss, R. (1997) Modern glycohistochemistry: a major contribution to morphological investigations. In H.-J. Gabius, and S. Gabius, (eds.), *Glycosciences: Status and Perspectives,* Chapman & Hall, London - Weinheim, pp. 547–562.

Drickamer, K. (1993) Evolution of Ca^{2+}-dependent animal lectins. *Progr. Nucl. Acid Res. Mol. Biol.,* **45**, 207–233.

Gabius, H.-J. (1988) Tumor lectinology: at the intersection of carbohydrate chemistry, biochemistry, cell biology and oncology. *Angew. Chem. Int. Ed. Engl.,* **27**, 1267–1276.

Gabius, H.-J. (1994) Non-carbohydrate binding partners/domains of animal lectins. *Int. J. Biochem.,* **26**, 469–477.

Gabius, H.-J. (1997a) Animal lectins. *Eur. J. Biochem.*, **243**, 543–576.

Gabius, H.-J. (1997b) Concepts of tumor lectinology. *Cancer Investig.*, **15**, 454–464.

Gabius, H.-J. (1998) The how and why of protein-carbohydrate interaction: a primer to the theoretical concept and a guide to application in drug design. *Pharm. Res.*, **15**, 23–30.

Gabius, H.-J., and Gabius, S. (1990) Tumorlektinologie: Status und Perspektiven klinischer Anwendung. *Naturwissenschaften*, **77**, 505–514.

Gabius, H.-J., and Bardosi, A. (1991) Neoglycoproteins as tools in glycohistochemistry. *Progr. Histochem. Cytochem.*, **22**, 1–66.

Gabius, H.-J., and Gabius, S. (1992) Chemical and biochemical strategies for the preparation of glycohistochemical probes and their application in lectinology. *Adv. Lectin Res.*, **5**, 123–157.

Gabius, H.-J., and Gabius, S. (eds.) (1993) *Lectins and Glycobiology*, Springer Verlag, Heidelberg.

Gabius, H.-J., and Gabius, S. (eds.) (1997) *Glycosciences: Status and Perspectives*, Chapman & Hall, London - Weinheim.

Gabius, S., Hellmann, K.P., Hellmann, T., Brinck, U., and Gabius, H.-J. (1989) Neoglycoenzymes: a versatile tool for lectin detection in solid-phase assays and glycohistochemistry. *Anal. Biochem.*, **182**, 447–451.

Gabius, H.-J., Schröter, C., Gabius, S., Brinck, U., and Tietze, L.-F. (1990a) Binding of T-antigen-bearing neoglycoprotein and peanut agglutinin to cultured tumor cells and breast carcinomas. *J. Histochem. Cytochem.*, **38**, 1625–1631.

Gabius, H.-J., Gabius, S., Brinck, U., and Schauer, A. (1990b) Endogenous lectins with specificity to b-galactosides and α- or β-N-acetylgalactosaminides in human breast cancer: their glycohistochemical detection in tissue sections by synthetically different types of neoglycoproteins, their quantitation on cultured cells by neoglycoenzymes and their usefulness as targets in lectin-mediated phototherapy *in vitro*. *Path. Res. Pract.*, **186**, 597–607.

Gabius, S., Kayser, K., Hellmann, K.P., Ciesiolka, T., Trittin, A., and Gabius, H.-J. (1990c) Carrier-immobilized derivatized lysoganglioside GM_1 is a ligand for specific binding sites in various human tumor cell types and peripheral blood lymphocytes and monocytes. *Biochem. Biophys. Res. Commun.*, **169**, 239–244.

Gabius, H.-J., Brinck, U., Lüsebrink, T., Ciesiolka, T., and Gabius, S. (1991a) Glycopeptide-albumin conjugate: its preparation and histochemical ligand properties. *Histochem. J.*, **23**, 303–311.

Gabius, H.-J., Wosgien, B., Brinck, U., and Schauer, A. (1991b) Localization of endogenous β-galactoside-specific lectins by neoglycoproteins, lectin-binding tissue glycoproteins and antibodies and of accessible lectin-specific ligands by a mammalian lectin in human breast cancer. *Path. Res. Pract.*, **187**, 839–847.

Gabius, H.-J., Gabius, S., Fritsche, M., and Brandner, G. (1991c) Transformation-associated decrease in cell surface binding of neoglycoenzymes in a temperature-sensitive, virally-transformed mouse model. *Naturwissenschaften*, **78**, 230–232.

Gabius, H.-J., Gabius, S., Zemlyanukhina, T.V., Bovin, N.V., Brinck, U., Danguy, A., Joshi, S.S., Kayser, K., Schottelius, J., Sinowatz, F., Tietze, L.F., Vidal-Vanaclocha, F., and Zanetta, J.-P. (1993) Reverse lectin histochemistry: design and application for detection of cell and tissue lectins. *Histol. Histopathol.*, **8**, 369–383.

Gabius, H.-J., André, S., Danguy, A., Kayser, K., and Gabius, S. (1994a) Detection and quantification of carbohydrate-binding sites on cell surfaces and in tissue sections by neoglycoproteins. *Meth. Enzymol.*, **242**, 37–46.

Gabius, S., Wawotzny, R., Martin, U., Wilholm, S., and Gabius, H.-J. (1994b) Carbohydrate-dependent binding of human myeloid leukemia cell lines to neoglycoenzymes, matrix-immobilized neoglycoproteins and bone marrow stromal cell layers. *Ann. Hematol.*, **68**, 125–132.

Gabius, H.-J., Kayser, K., and Gabius, S. (1995) Protein-Zucker-Erkennung. Grundlagen und medizinische Anwendung am Beispiel der Tumorlektinologie. *Naturwissenschaften*, **82**, 533–543.

Gahmberg, C.G., and Tolvanen, M. (1996) Why mammalian cell surface proteins are glycoproteins. *Trends Biochem. Sci.*, **21**, 308–311.

Gallagher, J.T., and Turnbull, J.E. (1992) Heparan sulphate in the binding and activation of basic fibroblast growth factor. *Glycobiology*, **2**, 523–528.

Garratty, G. (1995) Blood group antigens as tumor markers, parasitic/bacterial/viral receptors and their association with immunologically important proteins. *Immunol. Investig.*, **24**, 213–232.

Gijsen, H.J.M., Qiao, L., Fitz, W., and Wong, C.-H. (1996) Recent advances in the chemoenzymatic synthesis of carbohydrates and carbohydrate mimetics. *Chem. Rev.*, **96**, 443–473.

Greenwell, P. (1997) Blood group antigens: molecules seeking a function? *Glycoconjugate J.*, **14**, 159–173.

Hakomori, S., and Igarashi, Y. (1995) Functional role of glycosphingolipids in cell recognition and signaling. *J. Biochem.*, **118**, 1091–1103.

Hart, G.W. (1997) Dynamic O-linked glycosylation of nuclear and cytoskeletal proteins. *Annu. Rev. Biochem.*, **66**, 315–335.

Hassid, S., Salmon, I., Bovin, N.V., Kiss, R., Gabius, H.-J., and Danguy, A. (1996) Histochemical expression of binding sites for labelled hyaluronic acid and carrier-immobilized synthetic (histo-blood group trisaccharides) or biochemically purified (ganglioside GM$_1$) glycoligands in nasal polyps and other human lesions including neoplasms. *Histol. Histopathol.*, **11**, 985–992.

Hathaway, H.J., and Shur, B.D. (1997) Transgenic approaches to glycobiology. In H.-J. Gabius, and S. Gabius, (eds.), *Glycosciences: Status and Perspectives*, Chapman & Hall, London - Weinheim, pp. 507–517.

Helenius, A., Trombetta, E.S., Hebert, D.N., and Simons, J.F. (1997) Calnexin, calreticulin and the folding of glycoproteins. *Trends Cell Biol.*, **7**, 193–200.

Hooper, L.V., Manzella, S.M., and Baenziger, J.U. (1997) The biology of sulfated oligosaccharides. In H.-J. Gabius, and S. Gabius, (eds.), *Glycosciences: Status and Perspectives*, Chapman & Hall, London - Weinheim, pp. 261–276.

Hounsell, E.F. (1997) Methods of glycoconjugate analysis. In H.-J. Gabius, and S. Gabius, (eds.), *Glycosciences: Status and Perspectives*, Chapman & Hall, London - Weinheim, pp. 15–29.

Hounsell, E.F., Davies, M.J., and Renouf, D.V. (1996) O-Linked protein glycosylation structure and function. *Glycoconjugate J.*, **13**, 19–26.

Ichikawa, D., Handa, K., Withers, D.A., and Hakomori, S. (1997) Histoblood group A/B versus H status of human carcinoma cells as correlated with haptotactic cell motility: approach with A and B gene transfection. *Caner Res.*, **57**, 3092–3096.

Jansen, R.W., Molema, G., Ching, T.L., Oosting, R., Harms, G., Moolenaar, F., Hardonk, M.J., and Meijer, D.K.F. (1991) Hepatic endocytosis of various types of mannose-terminated albumins. What is important, sugar recognition, net charge or the combination of these features? *J. Biol. Chem.*, **266**, 3343–3348.

Kannan, S., and Nair, M.K. (1997) Lectins and neoglycoproteins in histopathology. In H.-J. Gabius, and S. Gabius, (eds.), *Glycosciences: Status and Perspectives*, Chapman & Hall, Weinheim, pp. 563–583.

Kayser, K., and Gabius, H.-J. (1997) Graph theory and the entropy concept in histochemistry. Theoretical considerations, application in histopathology and the combination with receptor-specific approaches. *Progr. Histochem. Cytochem.*, **32**(2), 1–106.

Kayser, K., Heil, M., and Gabius, H.-J. (1989) Is the profile of binding of a panel of neoglycoproteins useful as diagnostic marker in human lung cancer? *Path. Res. Pract.*, **184**, 621–629.

Kayser, K., Gabius, H.-J., Rahn, W., Martin, H., and Hagemeyer, O. (1992) Variations of binding of labelled tumor necrosis factor-a, epidermal growth factor, ganglioside GM$_1$, and N-acetylglucosamine, galactoside-specific mistletoe lectin and lectin-specific antibodies in mesothelioma and metastatic adenocarcinoma of the pleura. *Lung Cancer*, **8**, 185–192.

Kayser, K., Bovin, N.V., Zemlyanukhina, T.V., Donaldo-Jacinto, S., Koopmann, J., and Gabius, H.-J. (1994a) Cell type-dependent alterations of binding of synthetic blood group antigen-related oligosaccharides in lung cancer. *Glycoconjugate J.*, **11**, 339–344.

Kayser, K., Bovin, N.V., Korchagina, E.Y., Zeilinger, C., Zeng, F.-Y., and Gabius, H.-J. (1994b) Correlation of expression of binding sites for synthetic blood group A-, B-, and H-trisaccharides and for sarcolectin with survival of patients with bronchial carcinoma. *Eur. J. Cancer*, **30A**, 653–657.

Kayser, K., Bovin, N.V., Zeng, F.-Y., Zeilinger, C., and Gabius, H.-J. (1994c) Binding capacities to blood-group antigens A, B and H, DNA- and MST measurements, and survival in bronchial carcinoma. *Radiol. Oncol.*, **28**, 282–286.

Kayser, K., Bubenzer, J., Kayser, G., Eichhorn, S., Zemlyanukhina, T.V., Bovin, N.V., André, S., Koopmann, J., and Gabius, H.-J. (1995) Expression of lectin-, interleukin-2–, and histo-blood group-binding sites in prostate cancer and its correlation to integrated optical density and syntactic structure analysis. *Anal. Quant. Cytol. Histol.*, **17**, 135–142.

Kayser, K., Kayser, C., Rahn, W., Bovin, N.V., and Gabius, H.-J. (1996a) Carcinoid tumors of the lung: immuno- and ligandohistochemistry, analysis of integrated optical density, syntactic structure analysis, clinical data, and prognosis of patients treated surgically. *J. Surg. Oncol.*, **63**, 99–106.

Kayser, K., Kaltner, H., Dong, X., Knapp, M., Schmettow, H.K., Vlasova, E.V., Bovin, N.V., and Gabius, H.-J. (1996b) Prognostic relevance of detection of ligands for vertebrate galectins and a Lewis^y-specific monoclonal antibody: a prospective study of bronchial carcinoma patients treated surgically. *Int. J. Oncol.*, **9**, 893–900.

King, M.-J. (1994) Blood group antigens on human erythrocytes: distribution, structure and possible functions. *Biochim. Biophys. Acta,* **1197,** 14–44.

Kishore, U., Eggleton, P., and Reid, K.B.M. (1997) Modular organization of carbohydrate recognition domains in animal lectins. *Matrix Biol.,* **15,** 583–592.

Kohnke-Godt, B., and Gabius, H.-J. (1989) Heparin-binding lectin from human placenta: purification, partial molecular characterization and its relationship to basic fibroblast growth factors. *Biochemistry,* **28,** 6531–6538.

Kohnke-Godt, B., and Gabius, H.-J. (1991) Heparin-binding lectin from human placenta: further characterization of ligand binding and structural properties and its relationship to histones and heparin-binding growth factors. *Biochemistry,* **30,** 55–65.

Kojima, S., André, S., Korchagina, E.Y., Bovin, N.V., and Gabius, H.-J. (1997) Tyramine-containing poly(4–nitrophenylacrylate) as iodinatable ligand carrier in biodistribution analysis. *Pharm. Res.,* **14,** 879–886.

Kopitz, J. (1997) Glycolipids: structure and function. In H.-J. Gabius, and S. Gabius, (eds.), *Glycosciences: Status and Perspectives,* Chapman & Hall, London - Weinheim, pp. 163–189.

Kresse, H. (1997) Proteoglycans: structure and functions. In H.-J. Gabius, and S. Gabius, (eds.), *Glycosciences: Status and Perspectives,* Chapman & Hall, London - Weinheim, pp. 201–222.

Kuchler, S., Zanetta, J.-P., Vincendon, G., and Gabius, H.-J. (1990) Detection of binding sites for biotinylated neoglycoproteins and heparin (endogenous lectins) during cerebellar ontogenesis in the rat. *Eur. J. Cell Biol.,* **52,** 87–97.

Kuchler, S., Zanetta, J.-P., Vincendon, G., and Gabius, H.-J. (1992) Detection of binding sites for biotinylated neoglycoproteins and heparin (endogenous lectins) during cerebellar ontogenesis in the rat: an ultrastructural study. *Eur. J. Cell Biol.,* **59,** 373–381.

Laine, R.A. (1997) The information-storing potential of the sugar code. In H.-J. Gabius, and S. Gabius, (eds.), *Glycosciences: Status and Perspectives,* Chapman & Hall, Weinheim - London, pp. 1–14.

Lee, Y.C., and Lee, R.T. (1989) Conjugation of glycopeptides to proteins. *Meth. Enzymol.,* **179,** 253–257.

Lee, Y.C., and Lee, R.T. (1991) Neoglycoconjugates: fundamentals and recent progress. In H.-J. Gabius, and S. Gabius, (eds.), *Lectins and Cancer,* Springer Verlag, Heidelberg, pp. 53–70.

Lee, R.T., and Lee, Y.C. (1993) Synthetic ligands for lectins. In H.-J. Gabius, and S. Gabius, (eds.), *Lectins and Glycobiology,* Springer Verlag, Heidelberg, pp. 9–22.

Lee, Y.C., and Lee, R.T. (eds.) (1994a) *Neoglycoconjugates. Preparation and Applications,* Academic Press, San Diego.

Lee, R.T., and Lee, Y.C. (1994b) Enhanced biochemical affinities of multivalent neoglycoconjugates. In Y.C. Lee, and R.T. Lee, (eds.), *Neoglycoconjugates. Preparation and Applications,* Academic Press, San Diego, pp. 23–50.

Lee, R.T., and Lee, Y.C. (1997) Neoglycoconjugates. In H.-J. Gabius, and S. Gabius, (eds.), *Glycosciences: Status and Perspectives,* Chapman & Hall, London - Weinheim, pp. 55–77.

Mahoney, J.A., and Schnaar, R.L. (1994) Neoganglioproteins: probes for endogenous ganglioside receptors. In Y.C. Lee, and R.T. Lee, (eds.), *Neoglycoconjugates. Preparation and Applications,* Academic Press, San Diego, pp. 445–463.

McBroom, C.R., Samanen, C.H., and Goldstein, I.J. (1972) Carbohydrate antigens: coupling of carbohydrate to proteins by diazonium and phenylisothiocyanate reactions. *Meth. Enzymol.,* **28,** 212–219.

McDowell, G., and Gahl, W.A. (1997) Inherited disorders of glycoprotein synthesis: cell biological insights. *Proc. Soc. Exp. Biol. Med.,* **215,** 145–157.

Miyake, M., and Hakomori, S. (1991) A specific cell surface glycoconjugate controlling cell motility: evidence by functional monoclonal antibodies that inhibit cell motility and tumor cell metastasis. *Biochemistry,* **30,** 3328–3334.

Monsigny, M., Roche, A.-C., and Midoux, P. (1984) Uptake of neoglycoproteins via membrane lectin(s) of L1210 cells evidenced by quantitative flow cytofluorometry and drug targeting. *Biol. Cell,* **51,** 187–196.

Montreuil, J. (1995) The history of glycoprotein research, a personal view. In J. Montreuil, J.F.G. Vliegenthart, and H. Schachter, (eds.), *Glycoproteins,* Elsevier, Amsterdam, pp. 1–28.

Paulsen, H. (1990) Synthesen, Konformationen und Röntgenstrukturanalysen von Saccharidketten der Core-Region von Glycoproteinen. *Angew. Chem.,* **102,** 851–867.

Pavelka, M. (1997) Topology of glycosylation - a histochemist's view. In H.-J. Gabius, and S. Gabius, (eds.), *Glycosciences: Status and Perspectives,* Chapman & Hall, London - Weinheim, pp. 115–120.

Powell, L.D., and Varki, A. (1995) I-type lectins. *J. Biol. Chem.*, **270**, 14243–14246.

Reichert, C.M., Hayes, C.E., and Goldstein, I.J. (1994) Coupling of carbohydrates to proteins by diazonium and phenylisothiocyanate reactions. *Meth. Enzymol.*, **242**, 108–116.

Rini, J.M. (1995) Lectin structure. *Annu. Rev. Biophys. Biomol. Struct.*, **24**, 551–577.

Roche, A.-C., and Monsigny, M. (1996) Trafficking of endogenous glycoproteins mediated by intracellular lectins: facts and hypotheses. *Chemtracts*, **6**, 188–201.

Roth, J. (1987) Subcellular organization of glycosylation in mammalian cells. *Biochim. Biophys. Acta*, **906**, 405–436.

Sandhoff, K., and Kolter, T. (1995) Glykolipide der Zelloberfläche. *Naturwissenschaften*, **82**, 403–413.

Santos-Benito, F.F., Fernandez-Mayoralas, A., Martin-Lomas, M., and Nieto-Sampedro, M. (1992) Inhibition of proliferation of normal and transformed neural cells by blood group-related oligosaccharides. *J. Exp. Med.*, **176**, 915–918.

Schmidt, R.R. (1997) Strategies for chemical synthesis of carbohydrate structures. In H.-J. Gabius, and S. Gabius, (eds.), *Glycosciences: Status and Perspectives*, Chapman & Hall, London - Weinheim, pp. 31–53.

Sharon, N., and Lis, H. (1987) A century of lectin research (1888–1988). *Trends Biochem. Sci.*, **12**, 488–491.

Sharon, N., and Lis, H. (1997) Glycoproteins: structure and function. In H.-J. Gabius, and S. Gabius, (eds.), *Glycosciences: Status and Perspectives*, Chapman & Hall, Weinheim - London, pp. 133–162.

Siebert, H.-C., von der Lieth, C.-W., Dong, X., Reuter, G., Gabius, H.-J., and Vliegenthart, J.F.G. (1996) Molecular dynamics-derived conformation and intramolecular interaction analysis of the N-acetyl-9–O-acetylneuraminic acid-containing ganglioside GD_{1a} and NMR-based analysis of its binding to a human polyclonal immunoglobulin G fraction with selectivity for O-acetylated sialic acids. *Glycobiology*, **6**, 561–572.

Siebert, H.-C., von der Lieth, C.-W., Gilleron, M., Reuter, G., Wittmann, J., Vliegenthart, J.F.G., and Gabius, H.-J. (1997) Carbohydrate-protein interaction. In H.-J. Gabius, and S. Gabius, (eds.), *Glycosciences: Status and Perspectives*, Chapman & Hall, London - Weinheim, pp. 291–310.

Sofia, M.J. (1996) Generation of oligosaccharide and glycoconjugate libraries for drug discovery. *Drug Discovery Today*, **1**, 27–34.

Spicer, S.S., and Schulte, B.A. (1992) Diversity of cell glycoconjugates shown histochemically: a perspective. *J. Histochem. Cytochem.*, **40**, 1–38.

Stanley, P., and Ioffe, E. (1995) Glycosyltransferase mutants: key to new insights in glycobiology. *FASEB J.*, **9**, 1436–1444.

Stowell, C.P., and Lee, Y.C. (1980) Neoglycoproteins: the preparation and application of synthetic glycoproteins. *Adv. Carbohydr. Chem. Biochem.*, **37**, 225–281.

Unverzagt, C. (1994) Synthesis of a biantennary heptasaccharide by regioselective glycosylations. *Angew. Chem. Int. Ed.*, **33**, 1102–1104.

Unverzagt, C. (1996) Chemoenzymatic synthesis of a sialylated undecasaccharide-asparagine conjugate. *Angew. Chem. Int. Ed.*, **35**, 2350–2353.

Varki, A. (1993) Biological roles of oligosaccharides: all of the theories are correct. *Glycobiology*, **3**, 97–130.

Varki, A. (1996) "Unusual" modifications and variations of vertebrate oligosaccharides: are we missing the flowers for the trees? *Glycobiology*, **6**, 707–710.

Vellupillai, P., and Harn, D.A. (1994) Oligosaccharide-specific induction of interleukin 10 production by B220+ cells from schistosome-infected mice: a mechanism for regulation of CD4+ T-cell subsets. *Proc. Natl. Acad. Sci. USA*, **91**, 18–22.

Vidal-Vanaclocha, F., Barbera-Guillem, E., Weiss, L., Glaves, D., and Gabius, H.-J. (1990) Quantitation of endogenous lectin expression in 3LL tumors, growing subcutaneously and in the kidneys of mice. *Int. J. Cancer*, **46**, 908–912.

Villalobo, A., Horcajadas, J.A., André, S., and Gabius, H.-J. (1997) Glycobiology of signal transduction. In H.-J. Gabius, and S. Gabius, (eds.), *Glycosciences: Status and Perspectives*, Chapman & Hall, London - Weinheim, pp. 485–496.

Walker, R.A. (1989) The use of lectins in histopathology. *Path. Res. Pract.*, **185**, 826–835.

Wong, S.Y.C., Manger, I.D., Guile, G.R., Rademacher, T.W., and Dwek, R.A. (1993) Analysis of carbohydrate-protein interactions with synthetic N-linked neoglycoconjugate probes. *Biochem. J.*, **296**, 817–825.

3. LECTINS, MICROGLIA AND ALZHEIMER'S DISEASE

PAOLO ZATTA[1] and PAMELA ZAMBENEDETTI[2]

[1]*Centro CNR per lo Studio della Biochimica e della Fisiologia delle Metalloproteine, Via Trieste 75, 35131 Padova, Italy*
[2]*Dipartimento di Biologia, Università di Padova, Via Trieste 75, 35131 Padova, Italy*

ALZHEIMER'S DISEASE

Despite intense efforts, it has not yet been possible to clarify the etiopathogenesis, multigenic and multifactorial in character, of Alzheimer's disease (AD). AD is mainly characterized by the presence of two types of brain lesions consisting of the paired helical filaments (PHF) of proteinaceous material in neurons and amyloid deposits in the extracellular space. The main component of amyloid, called βA4 and formed by a chain of 39–43 amino acids, derives from a much larger transmembrane glycoprotein (Figure 1) which accumulates in diffuse and mature plaques associated with the neurodegenerative process.

Other biochemical characteristics of AD are the result of profound modifications in the tau and apolipoprotein E (ApoE) families of proteins. Tau, a family of axonal phosphoproteins that normally function as microtubule-associated proteins, are the major components of PHF (Kondo *et al.*, 1988). In AD, tau proteins are hyperphosphorylated and, with respect to normal tau, differ in terms of aggregation, proteolytic degradation, glycation, and ubiquitination. ApoE, a 229-amino acid protein encoded in a four exon gene located on the long arm of chromosome 19, is known to be synthesized and secreted by astrocytes and upregulated after neuronal injury (Poirier *et al.*, 1991). In 1991 a genetic linkage between ApoE and AD was reported (Pericak *et al.*, 1991) ApoE exists in three major allelic forms (ApoE2, ApoE3, ApoE4); ApoE3 is the most common one, and ApoE4 has recently been linked with late-onset AD (Saunders *et al.*, 1993). While it is unkonwn whether ApoE4 alters or accelerates the formation of extracellular amyloid accumulation in plaque formation, it has been demonstrated that it interacts *in vitro* with β-amyloid to form a stable complex. It is noteworthy that tau interacts with ApoE3, but not with ApoE4, suggesting a protective role of ApoE3 in tau hyperphosphorylation (Weisgraber *et al.*, 1994).

MICROGLIAL CELLS

The properties of migration and phagocytosis of microglial cells (MG) were first recognized by Nissl (*Stabchenzellen*) in 1899. However, their nature and identity

---

Abbreviations: CLA, leukocyte common antigen; CML, carboxymethyllysine; MHC, major histocompatibility complex; IL, interleukin; M-CSF, macrophage colony stimulating factor; TNF, tumor necrosis factor, HLA-DR, human leukocyte antigen

remained in doubt until the studies of Ramon y Cajal (1919), who identified a *tercer elemento* of the CNS distinct from neurons and astrocytes, and of Del Rio-Hortega (1919, 1932). In the normal adult brain MG functions are still ill-defined; however, a general consensus exists on the fact that MG play a major role in immuno-surveillance of the CNS (see Barron, 1995 and Kreutzberg, 1996). The first selective staining method for MG, using silver carbonate, was developed by Del Rio-Hortega (1919) and remained the sole method for almost half a century. At the present time several MG markers are in common use (Streit, 1995).

MG are ubiquitous throughout the brain in both gray and white matter and represent a significant and relevant cell component comprising between 5% and 12 % of all cells of the CNS. A study of mouse brain by Lawson *et al.* (1990) revealed that in the hippocampus, olfactory telencephalon, basal ganglia and substantia nigra contain the highest density of MG.

MG are pluripotent members of the macrophage/monocyte lineage, and respond in different ways to pathological changes in CNS. Their proliferation and activation are common events in the injured CNS, and lesions of the CNS result in invasion of peripheral phagocytes and/or *in situ* activation and proliferation of MG, depending on the direct or indirect nature of the injury (Riva *et al.*, 1994). MG activation is characterized by the appearance of rod-shaped cells in the area of cortical degeneration arising during chronic infections. Activated microglia may promote the progression of neuronal injury after a wide range of CNS insults, and their ability to release cytokines makes them candidates as modulators of secondary damage mechanisms. The presence of phosphotyrosine in ramified MG is consistent with a role for tyrosine phosphorylation in the activation of MG, and in the signaling events concomitant with the conversion of resting MG to the activated ameboid form (Karp *et al.*, 1994).

ALZHEIMER'S DISEASE AND GLYCOSYLATION

Glycation and glycosylation phenomena are largely present in the brains of AD subjects. The recent identification of advanced glycation end products (AGEs) of the Maillard reaction in the CNS has suggested the potential involvement of carbohydrate derivatives as a key component in aging and in pathological processes. AGEs are formed by an irreversible condensation reaction through the non-enzymatic long-term glycation of proteins, resulting in the formation of substances that are highly resistant to proteolytic cleavage and which induce protein cross linking (Table 1).

Glucose as well as other reducing sugars can react with proteins by a non-enzymatic, post-translational modification process called glycation, different from the enzymatic glycosylation. The sugar-derived carbonyl groups add to free $-NH_2$, and form, over time, an adduct which produces a class of AGEs (Smith *et al.*, 1994). These glycation products remain irreversibly bound to macromolecules and can covalently crosslink proximate amino groups. AGEs can activate cellular receptors, initiating a variety of physiological processes. In particular, AGEs represent structural components of amyloid plaques and neurofibrillary tangles (Yan *et al.*, 1994; Thome *et al.*, 1996). The AGEs pentosidine and carboxymethyllysine (CML) were identified in the neuronal perikarya and in the extraneuroperikaryal deposits of both AD and

Table 1   Maillard reaction: A condensation reaction between a carboxyl group of a reducing sugar with an amino group of a protein (e.g., Lysine e-NH_2) gives a Schiff Base that undergoes Amadori rearrangement. Further dehydration, rearrangement, fragmentation and condensation reactions produce a large variety of AGEs which can cause protein cross-links. In addition, Amadori products, generating a carboxyl group, can also react with other amino groups of proteins.

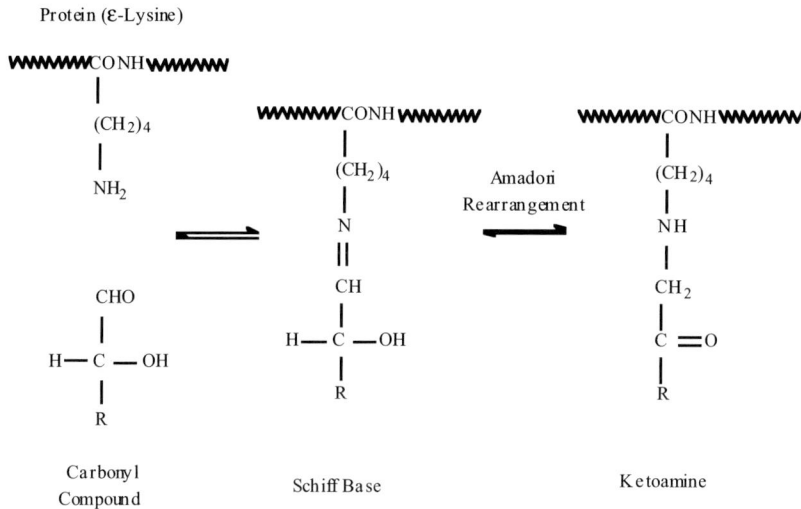

Protein (ε-Lysine)

〰〰〰CONH〰〰〰
|
$(CH_2)_4$
|
NH_2

⇌

〰〰〰CONH〰〰〰
|
$(CH_2)_4$
|
N
‖
CH
|
H—C—OH
|
R

Amadori
Rearrangement

⟶

〰〰〰CONH〰〰〰
|
$(CH_2)_4$
|
NH
|
CH_2
|
C=O
|
R

CHO
|
H—C—OH
|
R

Carbonyl
Compound

Schiff Base

Ketoamine

aged brain (Horie *et al.*, 1997). In CNS diseases pentosidine and CML co-localize with lipofuscin pigments. In contrast, control brains from young subjects not affected by neurodegenerative diseases contain little pentosidine and CML were only very faint (Horie *et al.*, 1997). Furthermore, Yan *et al.*, (1996) demonstrated that cell surface receptor sites for AGEs are highly expressed in AD. Such receptors mediate the biological effects of βA4 on neurons and microglial cells.

PHF tangles isolated from AD brain are modified by glycosylated, whereas no glycans are detected in normal tau (Wang *et al.*, 1996). Both phosphorylation and glycation increase the propensity of tau to form aggregates. In addition, glycation of tau stabilizes the assembled polymers and could facilitate formation of bundles from these polymers (Ledesma *et al.*, 1996). Recently, Dickson *et al.* (1996) have demonstrated that tau is multiply modified by Ser-(Thr)-O-linked N-acetylglucosamine, an abundant post-translational modification. O-GlcNAcylation of tau was demonstrated by immunoblotting and structural studies. It has been hypothesized that O-GlcNAcylation may not only modulate tau functions, but may also play a role in the formation of PHF.

βA4, which accumulates in the brain of AD subjects, represents a proteolytic product of a family of much larger amyloid glycosylated transmembrane precursors (Figure 1). The three major isoforms of amyloid are derived by alternative splicing and contain 695, 751 and 770 amino acids. They are heavily O-glycosylated and contain two N-linked glycosylation sites.

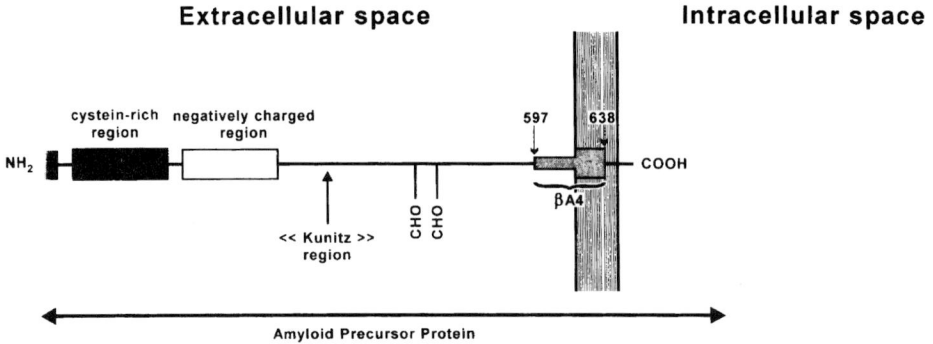

Extracellular space   **Intracellular space**

Figure 1 Scheme of amyloid precursor protein. CHO are the glycosylated sites of APP. βA4 represents the fragment that is abnormally secreted from APP and accumulated in the patological features in the brain of Alzheimer's disease subjects.

ApoE4, which is highly expressed in AD, exhibits a much greater AGE-binding activity than the ApoE3 isoform, suggesting that ApoE4 may participate in aggregate formation in AD brain by binding the AGE-modified plaque components, with consequent pathogenic effects (Li and Dickson, 1997).

In the brain of AD patients, a high level of acetylcholinesterase (AChE) activity has been detected in the senile plaque-rich fraction. AChE inhibitors such as tacrine and physostigmine have a weaker effect on AChE present in the senile plaque-fraction isolated from AD brain compared to the soluble fraction of AD brain or normal brain. Interestingly, AChE in AD brain seems to be highly glycosylated (Saez-Valero *et al.*, 1997), and the hydrophobic properties of anomalous AChE may explain its abnormal behavior with respect to the inhibitors, furthermore, it has been hypothesized that abnormally glycosylated AChE may serve as a seed of amyloid fibril deposition in senile plaques (Mimori *et al.*, 1997). This modification of enzymatic activity has been implicated in the biochemical events consequent to AD neurodegeneration.

While the role of glycation in the pathogenesis of AD is not yet unequivocally proven, it could be the only protein modification that may explain the formation of both of the characteristic pathological lesions first observed by Alois Alzheimer at the beginning of this century.

MICROGLIA, GLYCOSYLATION AND ALZHEIMER'S DISEASE

MG express macrophage markers such as elements of the complement cascade, complement receptors, and numerous surface markers, including LCA; MHC class-II antigen; IL 1, 3, and 6; TNF; and M-CSF (Barron, 1995). Lectins, which bind to carbohydrate residues of glycoconjugates, can also be used as histochemical markers for MG. It is known that the selectivity of lectin binding is achieved through a combination of hydrogen bonding to sugar hydroxyl groups and van de Waals interactions, and often involves packing of a hydrophobic sugar surface against aromatic amino acid side chains (Weis and Drickamer, 1996). Lectin histochemistry is considered nowadays as a consolidated technique utilized to identify cerebral structures and substructures as well as characteristic fine cellular and subcellular features associated with normal and neuropathological events, both in humans and experimental animals (Ko, 1987). Lectins can thus be used as markers of neurological diseases to better identify morphological modifications as well as molecular events involved in abnormal glycosylation that could occur with neuropathologies (Dickson et al., 1988; Kobayashi et al., 1989; De Gasperi et al., 1990; Esiri and Morris, 1991; Mann et al., 1992; Barcikowska et al., 1993; Sao et al., 1993; Nicolini and Zatta, 1994; Zatta et al. 1994; Barron, 1995; Knuckey et al., 1996). Thus, identification of both resting and activated microglia using lectin histochemistry may contribute to a better understanding of the mechanisms involved in neuropathological events. MG are highly plastic cells that exhibit heterogeneity in terms of distribution and morphology. In the normal brain resting MG are highly branched, with small amounts of perinuclear cytoplasm and a small, heterochromatic and dense nucleus, as can be visualized with lectin histochemistry in Figure 2(1). MG may become reactive and convert to different morphological shapes after receiving signals from a nearby microenvironment such as in the case of neuronal stress. Consequently, MG retract their branched ramifications, assume an ameboid form and move towards the zone from which the signal originated. Such morphological modifications can be related to pathological events, including viral infections, ischemia, trauma, neurodegeneration, etc., and also depend on the structure of the cerebral area where the MG are localized. In fact, become "bushy" in apparance if provided with enough volumetric space to move about (Figure 2(3)). Conversely, if they have to insert themselves in between neuronal axons or dendrites they become long and thick and rod-shaped in character (Figure 2(2)). In other situations MG may alight on damaged neuronal surfaces assuming a perinuclear shape. If the damaged neurons recover their integrity, MG can return to their resting status. On the other hand, when neurons die, MG assume a phagocytic shape (Figure 3(4)) and actively remove all cellular debris.

The mechanisms by which MG are eliminated after a pathological stimulation are still rather poorly understood. DNA fragmentation has been observed in activated MG, leading to the suspicion that programmed cell death (apoptosis) could be a mechanism by which activated MG are gradually eliminated following CNS injury (Gehrmann and Banati, 1995).

Table 3 reports a list of several lectins that recognize microglial cells; relevant literature is reported in Table 2.

The lectins most often utilized to recognize MG are *Bandeiraea simplicifolia* isolectin (I-B4), *Ricinus communis* agglutinin and *Viscum album* agglutinin. BSA I-B4 is able to detect both resting and reactive MG. RCA also recognizes both ramified and activated MG, and can also be used as a specific marker to study the development and differentiation of MG in human embryogenesis (Bobryshev and Ashwell, 1994). Both BSA I-B4 and RCA show similar abilities to recognize MG cells in sections from paraffin-embedded samples. However, BSA I-B4 performs better on rat samples, while RCA labelling appears to be superior on rabbit samples (Boya et al., 1991).

Other lectins have also been reported to be able to recognize MG, and the list is continuously growing as a reflection of the expanding popularity of this technique due to its simplicity, reliability, fast realization and low cost (Zatta and Cummings, 1992).

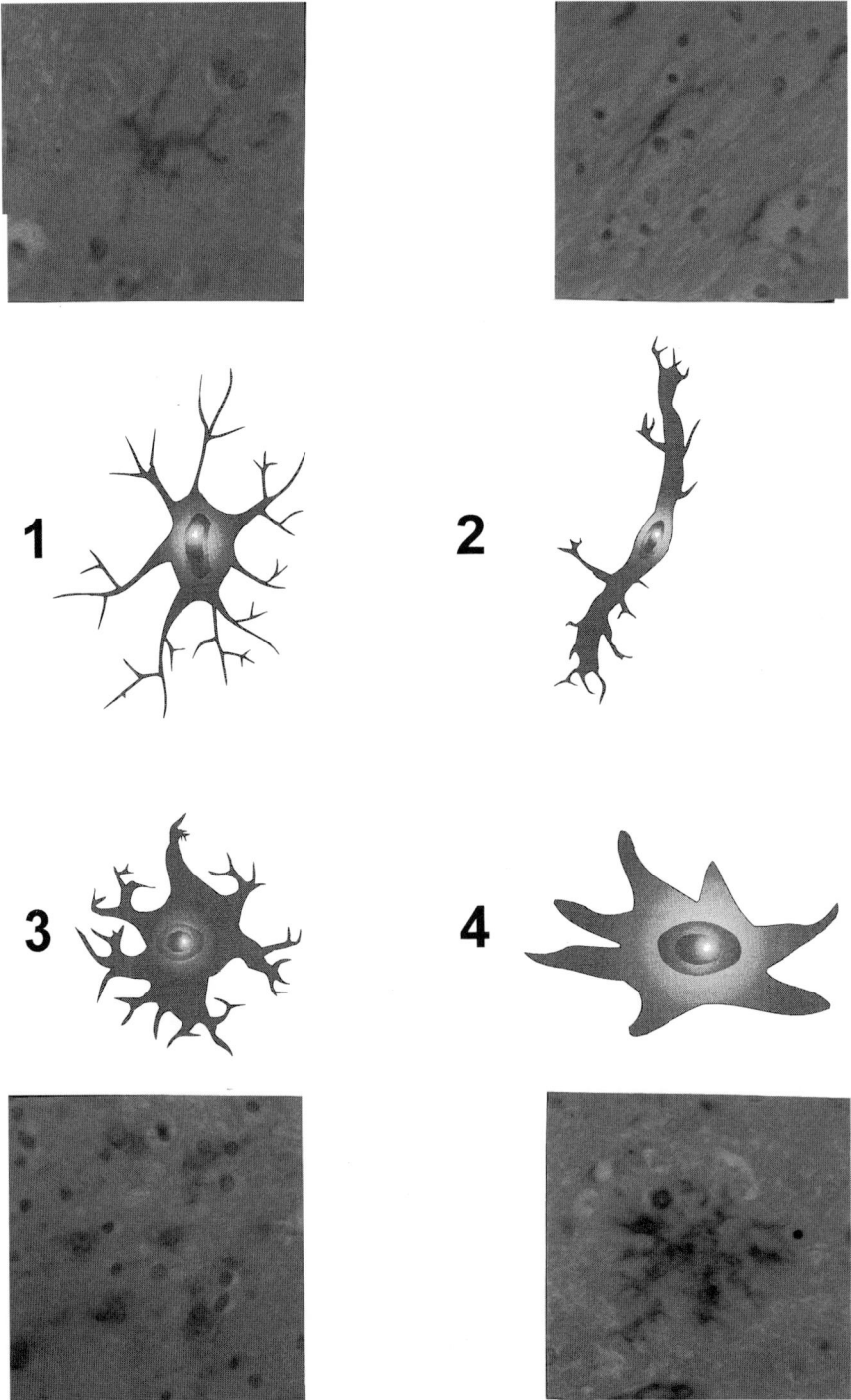

Figure 2   Different types of microglial cells stained with RCA: 1) Resting microglial cell; 2) Microglia cells of rod-shape type; 3) Microglial cells of "bushy" type; 4) Phagocytic microglial cells. Magnification × 40. (See Colour Plate I)

Table 2   Data from literature regarding the lectins that recognize microglia cells.

Lectin	References
Arachis hypogea	*Koval* et al. *(1994)*
Bandeiraea (Griffonia) simplicifolia I–B4	Ashwell (1990); Ashwell (1991); Boya *et al.* (1991a); Boya *et al.* (1991b); Breese *et al.* (1996); Bruckner *et al.* (1993); Buttini *et al.* (1996); Caggiano and Brunjes (1993); Cammer and Zhang (1993); Colton *et al.* (1992); Cuadros *et al.* (1992); Czlonkowska *et al.* (1996); Gehrmann *et al.* (1991); Glenn *et al.* (1993); Goodbrand and Gaze (1991); Guntinas-Lichius *et al.* (1994); Ivacko *et al.* (1996); Hewicher-Trautwein and Schultheis (1994); Hewicher-Trautwein *et al.* (1996); Humphrey and Moore (1995); Kaur and Ling (1991); Kaur *et al.* (1992); Kaur *et al.* (1993); Ludwin (1990); Morioka *et al.* (1991); Morioka *et al.* (1992); McNeill *et al.* (1994); Naujoks-Manteuffel and Niemann (1994); Nogradi (1993); Pennel *et al.* (1994); Pennisi *et al.* (1995); Roque and Caldwell (1993); Rossner *et al.* (1995); Streit (1990); Streit and Kreutzberg (1987); Wolswijk (1994); Wu *et al.* (1994)
Bahunia purpurea	*Colton* et al. *(1992)*
Canavalia ensiformis	Colton *et al.* (1992); Koval *et al.* (1994)
Dolichos biflorus	Colton *et al.* (1992)
Glycine max	Colton *et al.* (1992); Nagao *et al.* (1992); Ohshige-Hayashi and Kiyama (1997)
Lens culinaris	Koval *et al.* (1994)
Lycopersicon esculentum	Acarin *et al.* (1994); Acarin *et al.* (1996); Velasco *et al.* (1995); Moffet *et al.* (1997)
Ricinus communis	Bobryshev and Ashwell (1994); Boya *et al.* (1991a); Boya *et al.* (1991b); Cammer and Zhang (1993); Ciardi *et al.* (1990); Colton *et al.* (1992); Engel *et al.* (1996); Hauke and Korr (1993); Hewicher-Trautwein *et al.* (1995); Hewicher-Trautwein *et al.* (1996); Hulette (1996); Koeppen and Dentinger (1988); Ling *et al.* (1989a); Ling *et al.* (1989b); Lutsik *et al.* (1991); Mannoji *et al.* (1986); Pennisi *et al.* (1995); Perlmutter *et al.* (1992); Sangruchi and Sobel (1989); Suzuki *et al.* (1988); Weidenheim *et al.* (1993)
Sambucus nigra	Lutsik *et al.* (1991)
Triticum vulgaris	Colton *et al.* (1992)
Ulex europaeus	Colton *et al.* (1992); Steffan *et al.* (1994)
Vicia villosa	Bruckner *et al.* (1993)
Viscum album	Artigas *et al.* (1992); Hewicher-Trautwein *et al.* (1996); Konno *et al.* (1989); Schumacher *et al.* (1994a); Schumacher *et al.* (1994b); Suzuki *et al.* (1988)
Wisteria floribunda	Bruckner *et al.* (1993)

Several carbohydrate structures have thus been identified by lectins in microglial cells. The following lectins recognize Gal or GalNAc on MG:

(1)   **(DBA)** GalNAcα1-3GalNAc-specific agglutinin
(2)   **(WFA)** GalNAcα1-6Gal > GalNAcα1-3Gal
(3)   **(SBA)** GalNAcα or β1-linked, GalNAcα1-3Galβ1-3GlcNAc
(4)   **(VVA)** GalNAcα1-3Gal
(5)   **(BPA)** Galβ1-3GalNAc

Table 3 Lectins utilized for the identification of microglial cells

Lectin origin	Acronym	Specificity	Microglia
Anguilla anguilla	AAnA	α-L-Fuc	-
Agaricus bisporus	ABiA	β-D-Gal(1-3)D-GalNAc	-
Abrus precatorius	APA	D-Gal	+
Arachis hypogea	PNA	β-D-Gal(1-3)D-GalNAc	++
Artocarpus integrifolia	JCA	α-D-Gal; Gal-β(1-3)GalNAc	-
Bahunia purpurea	BPA	β-D-Gal(1-3)D-GalNAc	+
Bandeiraea simplicifolia I-B4	BSA I-B4	α-D-Gal; α-D-GalNAc	++
Canavalia ensiformis	ConA	α-D-Man; α-D-Glc	++
Caragana arborescens	CAA	GalNAc	-
Codium fragile	CFA	D-GalNAc	-
Cytisus scoparius	CSA	GalNAc; Gal	-
Datura stramonium	DSA	(D-GlcNAc)$_2$	-
Dolichos biflorus	DBA	α-D-GalNAc	+
Erythrina cristagalli	ECL	β-D-Gal(1-4)D-GlcNAc	-
Erythrina corallodendrum	ECo	β-D-Gal(1-4)D-GlcNAc	-
Euonymus europaeus	EEA	α-D-Gal(1-3)D-Gal	-
Galantus nivalis	GNA	Non-reduc. α-Man	-
Glycine max	SBA	α-D-GalNAc	+
Helix pomatia	HPA	α-D-GalNAc; D-GlcNAc	-
Lens culinaris	LCA	α-D-Man	+
Limulus polyphemus	LPA	NeuNAc; D-GlcNAc	-
Lotus tetragonolobus	PTA	L-Fuc-α(1-2)Gal-β(1-4)[L-Fuc-α(1-3)]GlcNAc	-
Lycopersicon esculentum	LEA	(D-GlcNAc)$_3$	++
Maackia amurensis	MAA	Neu-5-Ac-α(2-3)Gal	++
Maclura pomifera	MPA	Gal-β(1-3)GalNAc	-
Momordica charantia	MCA	Gal; GalNAc	++
Mycoplasma gallisepticum	MGA	Glycophorin	-
Phytolacca americana	PAA	(D-GlcNAc)$_3$	-
Phaseolus vulgaris	PHA-E	Bisected, bi-triantennary N-glycans	-
Pisum sativus	PSA	Man-α-1	-
Pseudomonas aeruginosa	PsAA	D-Gal	-
Psophocarpus tetragonolobus	PTA	D-GalNAc; D-Gal	-
Ptilota plumosa	PPA	α-Gal	-
Ricinus communis	RCA	D-GalNAc; D-Gal	++
Robinia pseudoacacia	RPA	———	-
Sambucus nigra	SNA	NeuNAc-α(2-6)Gal	+
Solanum tuberosus	STA	[GlcNAc-β(1-4)]$_4$	-
Sophora japonica	SJA	β-D-GalNAc	-
Triticum vulgaris	WGA	(D-GalNAcβ1-)$_2$; NeuNAc	++
Ulex europaeus	UEA-I	L-Fuc-α-Gal-β(1-4)GlcNAc	+
Vicia villosa	VVA	GalNAc	+
Vigna radiata	VRA	α-Gal	-
Viscum album	VAA	β-Gal	++
Wisteria floribunda	WFA	D-GalNAc-α(1-6)Gal-β1	+

– = negative; + = moderately positive; ++ = highly positive

(6)   (**RCA**) Biantennary oligosaccharides containing Galβ1-4GlcNAc-linked units
       at non-reducing end
(7)   (**BSA I-B4**) Galα1-3Gal
(8)   (**PNA**) Galβ1-3GalNAc

Other carbohydrates such as mannose, glucose and fucose are also recognizable
in MG:

(9)   (**ConA**) Manα1-linked
(10)   (**LCA**) Manα1-linked
(11)   (**LEA**) (GlcNAcβ1-4)$_4$
(12)   (**WGA**) (GlcNAcβ1-4)$_5$
(13)   (**UEA**) LFucα1-2Galβ1-4GlcNAcβ1-6R

Sialic acid on MG can also be detected:

(14)   (**SNA**) Neu5NAcα2-6Galβ1-4, GlcNAc-R. In AD, a considerable decrease in
       the activity of both soluble and membrane-bound forms of sialyltransferase
       enzyme was observed in the frontal and temporal cortical lobes, but not in
       the hippocampus (Maguire and Breen, 1995).

The predominant carbohydrate residues observed in MG are Gal and GalNAc
residues especially common in O-linked carbohydrates.

In Alzheimer's disease a correlation between MG cell activation and pathological
features has been observed. According to Di Parte *et al.* (1997), MG activation was
moderately higher in non-demented elderly subjects, while in AD a more striking
activation was observed in all sectors of the hippocampus. MG glycosylation in AD
can be readily and specifically detected in most fixed, paraffin-embedded autoptsy
samples (Zatta *et al.*, 1994).

In our study, herein briefly described, we have utilized samples from 10 AD
brains. Diagnosis was carried out according to CERAD-defined criteria (Mirra *et al.*,
1993). Using fortyfour lectins, we were able to demonstrate, that MG cells stain
intensely and reproducibly with BSA-I B4, RCA, WGA, ConA, PNA, LEA, and VAA.
In contrast, MG cells were only moderately stained by BPA, DBA, SBA, UEA-I, VVA,
SNA and WFA. Some lectins are not entirely specific for MG, in that they also stain
other cell types or cerebral structures (Zatta *et al.*, 1994). Some examples of MG
lectin staining are shown in Figures 3 and 4.

In addition to the above reported lectins, we have recently found, for the first
time, that MG are recognized by three other lectins from *Abrus precatorius, Maackia
amurensis* and *Momordica charantia* (Figure 4) (Zambenedetti *et al.*, 1998). In terms
of specificity, hemagglutination/inhibition data show that βDGal is the best
monosaccharide inhibitor for binding of MCA, while MAA has a high binding
affinity for sialic acid α2, 3 galactose β1, 4N-acetylglucosamine, but not for 2, 6-
linked isomers (Wang and Cummings, 1988). Finally, APA has a specific affinity for
DGal (Figure 4).

The glycation reaction has been observed in AD suggesting a possible role for
AGEs as etiological factors in disease development. Recently Dickson *et al.* (1996)
raised the hypothesis that glycation of ApoE may serve as one of the signals for
activation of MG associated with amyloid deposits and extracellular NFT in AD.

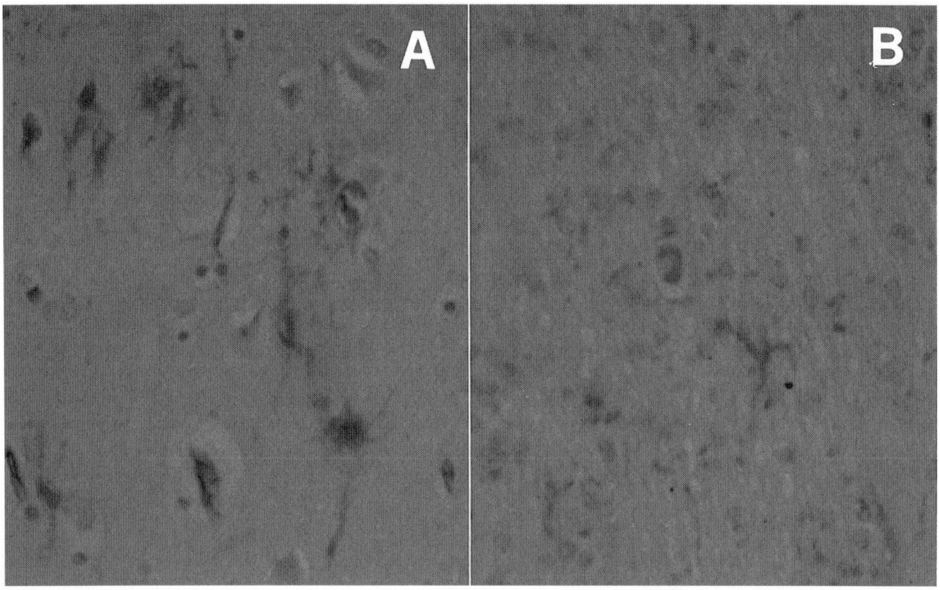

Figure 3 (A) *Sambucus nigra* agglutinin. Cerebral cortex from AD subjects shows staining of the microglial cells and vessel walls. MG are prevalently of branched type with the tendency to be crowded in some restricted areas. Magnification × 40. (B) *Lycopersicon esculentum* agglutinin. White matter from an AD brain. Diffuse staining can be seen in the neuropil, corresponding to the thin microglial prolongations. Magnification × 40. (See Colour Plate II)

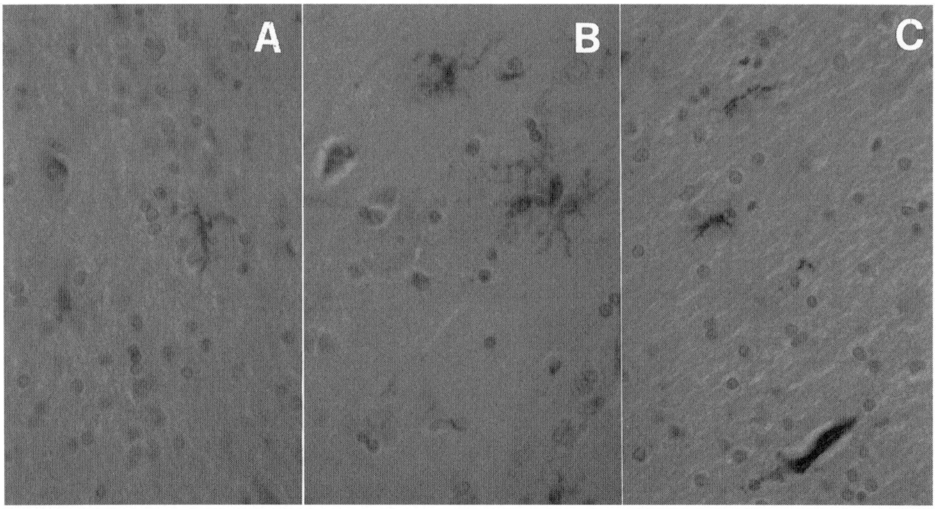

Figure 4 (A) *Maackia amurensis* agglutinin. White matter from an AD brain. Only a small number of microglial elements of rod-shaped type can be seen which are found among oligodendroglial cells. This lectin probably stains only some MG cell subtypes. (B) *Abrus precatorius* agglutinin. Cerebral cortex from an AD brain with positive MG cells. MG cells of rod-shaped and branched type. The latter type appears to be more dense in an area close to the neuron and in between the neuron and perineuronal oligodendroglia. Magnification × 40. (C) *Momordica charantia* agglutinin. White matter from an AD brain. Only a few microglial cells are positive, and of rod-shaped type. Magnification × 40. (See Colour Plate III)

From a therapeutic point of view it is interesting to note that both indomethacin and acetylsalicylic acid are able to inhibit the Maillard reaction (Colaco *et al.*, 1996). In addition, indomethacin, a specific inhibitor of prostaglandin (PG) synthesis, has been observed to inhibit the production PG synthesized by MG (DuBois *et al*, 1986).

In conclusion, lectin histochemistry, is very useful for the study of pathological features of AD such in the case of MG cells as well as of other neuropathologies. The use of lectin histochemistry, either alone or in combination with other MG markers, could represent a valid tool for the study of MG and other types of cells in normal brain as well as in brains affected by neurodegenerative diseases to better understand both etiopathogenetic, morphological, physiological and pathological aspects related to reactivity with lectins and thus to normal or altered glycosylation.

ACKNOWLEDGEMENTS

Authors are deeply grateful to the Brain Bank facility of Dolo General Hospitel–Venice, Italy, for brain samples supplied to carry out experiments described in this paper and Regione Venets for partial financial support.

REFERENCES

Acarin, L., Vela, J.M., Gonzales, B. and Castellano, B. (1994) Demonstration of poly-N-acetyl Lactosamine residues in ameboid and ramified microglial cells in rat brain by tomato lectin binding. *J. Histochem. Cytochem.*, **42**, 1033–1041.

Artigas, J., Bachler, B., Habedank, S., Taube, F., Franz, H. and Niedobitek, F. (1992) Comparative lectin histochemical studies on paraffin- and glycolmethacrylate-embedded CNS tissue specimens from AIDS autopsies. Mistletoe lectin I (ML I) as cell marker. *Zentralbl. Pathol.*, **138**, 272–277.

Ashwell, K. (1990) Microglia and cell death in the developing mouse cerebellum. *Brain Res. Dev. Brain Res.*, **55**, 219–230.

Ashwell, K. (1991) The distribution of microglia and cell death in the fetal rat forebrain. *Brain Res.Dev.Brain Res.*, **15**, 1–12.

Barcikowska, M., Liberski, P.P., Boellaard, J.W., Brown, P., Gajdusek, D.C. and Budka, H. (1993) Microglia is a component of the prion protein amyloid plaques in the Gerstmann-Straussler-Scheinker syndrome. *Acta Neuropathol.*, **85**, 623–627.

Barron, K.D. (1995) The microglia cell. A historical review. *J. Neurol. Sc.*, **143** Suppl, 57–68.

Bobryshev, Y. and Ashwell, K. (1994) Ultrastructural identification of Ricinus communis agglutinin-1 positive cells in primary dissociated cell cultures of human embryonic brain. *Arch. Histol. Cytol.*, **57**, 481–491.

Boya, J., Carbonell, A.L., Calvo, J.L. and Borregon, A.(1991a) Microglial cells in the central nervous system of the rabbit and rat: cytochemical identification using two different lectins. *Acta Anat.* (Basel.) **140**, 250–253.

Boya, J., Calvo, J.L., Carbonell, A.L. and Borregon, A. (1991) A lectin histochemistry study on the development of the rat microglial cells. *J. Anat.*, **175**, 229–236.

Breese, C.R., D'Costa, A., Rollins, Y.D., Adams, C., Booze, R.M., Sonntag, W.E. and Leonard, S. (1996) Expression of insulin-like growth factor-1 (IGF-1) and IGF-binding protein 2 (IGF-BP2) in the hippocampus following cytotoxic lesion of the dentate gyrus. *J. Comp. Neurol._* **369**, 388–404.

Bruckner, G., Bauer, K., Hartig, W., Wolff, J.R., Rickmann, M.J., Derouiche, A., Delpech, B., Girard, N., Oertel, W.H., and Reichenbach, A. (1993) Perineuronal nets provide a polyanionic, glia-associated for of microenvironment around certain neurons in many parts of rat brain. *Glia*, **8**, 183–200.

Cammer, W. and Zhang, H. (1993) Lectin binding and microglia cells in the brain of young normal and myelin-deficient mutant rats. *Glycobiology*, **3**, 627–631.

Caggiano, A.O. and Brunjes, P.C. (1993) Microglia and the developing of olfactory bulb. *Neuroscience*, **52**, 717–724.

Ciardi, A., Sinclair, E., Scaravilli, F., Harcourt-Webster, N.J. and Lucas, S. (1990) The involvement of the cerebral cortex in human immunodeficiency virus encephalopathy: a morphological and immunohistochemical study. *Acta Neuropathol.* (Berl.), **81**, 51–59.

Colaco, C.A., Ledesma, M.D., Harrington, C.R. and Avila, J. (1996) The role of Maillard reaction in other pathologies: Alzheimer's disease. *Nephrol. Dial. Transpl.*, Suppl., **5**, 7–12.

Colton, C.A., Abel, C., Patchett, J., keri, J. and Yao, J. (1992) Lectin staining of cultured CNS microglia. *J. Histochem. Cytochem.*, **40**, 505–512.

Czlonkowska, A., Kohutnicka, M., Kurkuwska-Jastrzebska, I and Czlonkowski, A. (1996) Microglial reaction in MPTP (1-methyl-4-phenyl-1, 2, 3, 6-tetrahydropyridine) induced Parkinson's disease mice model. *Neurodegeneration*, **5**, 137–143.

De Gasperi, R., Alroy, J., Richard, R., Goyal, V., Orgad, U., Lee, R.R. and Warren, C.D. (1990) Glycoprotein storage in Gaucher disease: lectin histochemistry and biochemical studies. *Lab. Inv.*, **63**, 385–392.

Del Rio-Hortega, P. (1919) El tercer elemento de los centros nerviosus . I. La microglia en estado normal. II. Intervencion de la microglia en los procesos patologicos (Celulas en bastoncino y cuerpos granulo-adiposos). III. Naturaleza probable de la microglia. *Bol. Soc. Espan. Biol.*, **9**, 68–120.

Del Rio-Hortega P. (1932) Microglia. In W. Penfield (ed.), *Cytology and Cellular Pathology of the Nervous System*, vol. III, Paul B.Hoeber, New York, pp. 481–534.

Dickson, D.W., Farlo, J., Davies, P., Crystal, H., Fuld, P. and Yen, S.H. (1988) Alzheimer's disease. A double labelling immunohistochemical study of senile plaques. *Am. J. Pathol.*, **132**, 86–101.

Dickson, D.W., Sinicropi, S., Yen, S.H., Ko, L.W., Mattiace, L.A., Bucala, R. and Vlassara, H. (1996) Glycation and microglia reaction in lesions of Alzheimer's disease. *Neurobiol. Aging.*, **17**, 733–743.

DiParte, P.L. and Gelman, B.B. (1997) Microglial cell activation in aging and Alzheimer's disease: Partial linkage with neurofibrillary tangle burden in the hippocampus. *J. Neuropathol. Exp. Neurol.*, **56**, 143–149.

Engel, S., Wehner, H.D. and Meyermann, R. (1996) Expression of microglia markers in the human CNS after closed head injury. *Acta Neuroch. Suppl. Wien*, **66**, 89–95.

Esiri, M.M. and Morris, C.S. (1991) Immunocytochemical study of macrophages and microglial cells and the extracellular matrix components in human CNS disease. 2. Non-neoplastic diseases. *J. Neurol. Sci.*, **101**, 59–72.

Gehrmann, J., Schoen, S.W. and Kreutzberg, G.W. (1991) Lesion of the rat entorhinal cortex leads to a rapid microglial reaction in the dentate gyrus. A light and electron microscopical study. *Acta Neuropathol.* (Berl.), **82**, 442–455.

Glenn, J.A., Sonceau, J.B., Wynder, H.J. and Thomas, W.E. (1993) Histochemical evidence for microglia-like macrophages in the rat trigeminal ganglion. *J. Anat.*, **183**, 475–481.

Gehrmann, J. and Banati, R.B. (1995) Microglia turn-over in the injured CNS: activated microglia undergo delayed DNA fragmentation following peripheral nerve injury. *J. Neuropathol. Exp. Neurol.*, **54**, 680–688.

Guntinas-Lichius, A., Neiss, W.F., Gunkel-A. and Stennert, E. (1994) Differences in glia, synaptic and motoneuron responses in the facial nucleus of the rat brainstem following facial nerve resection and nerve suture reanastomosis. *Eur. Arch. Otorhinolaryngol.*, **251**, 410–417.

Hauke, C. and Korr, H. (1993) RCA-I lectin histochemistry after trypsinisation anables the identification of microglia cells in thin paraffin sections of the mouse brain. *J. Neurosci. Meth.*, **50**, 273–277.

Hewicker-Trautwein, M., Schultheis, G. and Trautwein, G. (1995) Effects of trypsinization and microwave treatment on lectin labelling of microglial cells in paraffin-embedded sections from pre- and postnatal bovine brains. *Acta Histochem.*, **97**, 455–461.

Hewicker-Trautwein, M. and Schultheis, G. (1994) Lectin labelling of ameboid and ramified microglial cells in the telencephalon of ovine fetuses with the B4 isolectin from Griffonia simplicifolia. *J. Comp. Pathol.*, **111**, 21–31.

Hewicker-Trautwein, M., Schultheis, G. and Trautwein, G. (1996). Demonstration of ameboid and ramified microglial cells in pre- and postnatal bovine brain by lectin histochemistry. *Anat. Anz*, **178**, 25–31.

Horie, K., Miyata, T., Yasuda, T., Takeda, A., Yasuda, Y., Maeda, K., Sobue, G. and Kurokawa, K. (1997) Immunohistochemical localization of advanced glycation end products, pentosidine, and carboxylmethyllysine in lipofuscin pigments of Alzheimer's disease and aged neurons. *Biochim. Biophys. Res. Comm.*, **236**, 327–332.

Hulette, C.M. (1996) Microglioma, a histiocytic neoplasm of the central nervous system. *Mod. Pathol.*, **9**, 316–319.

Humphrey, M.F. and Moore, S.R. (1995) Strain differences in the distribution of NDP-ase labelled microglia in the normal rabbit retina. *Glia*, **15**, 367–376.

Ivacko, J.A., Sun, R. and Silverstein, F.S. (1996) Hypoxic-ischemic brain injury induces an acute microglial reaction in perinatal rats. *Pediatr. Res.*, **39**, 39–47.

Karp, H.L., Tillotson, M.L., Soria, J., Reich, C. and Wood, J.G. (1994) Microglial tyrosine phosphorylation systems in normal and degeneration brain. *Glia*, **11**, 284–290.

Kaur, C. and Ling, E.A. (1991) Study of the transformation of amoeboid microglial cells into microglia labelled with isolectin Griffonia simplicifolia in postnatal rats. *Acta Anat.* (Basel), **142**, 118–125.

Kaur, C., Chan, Y.G., Ling, E.A. (1992) Ultrastructural and immunocytochemical studies of macrophages in an excitotoxin induced lesion in the rat brain. *J. Hirnforsch.*, **33**, 645–652.

Kaur, C., Singh, J. and Ling, E.A. (1993) Immunohistochemical and lectin-labelling studies of the distribution and development of microglia in the spinal cord of postnatal rats. *Arch. Histol. Cytol.*, **56**, 475–484.

Knuckey, N.W., Finch, P., Palm, D.E., Primiano, M.J., Johanson, C.E., Flanders, K.C. and Thompson, N.L. (1996) Differential neuronal and astrocytic expression of transforming growing factor beta isoform in rat hippocampus following transient forebrain ischemia. *Brain Res. Mol. Brain Res.*, **40**, 1–14.

Ko, C.P. (1987) A lectin, peanut agglutinin, as a probe for the extracellular matrix in living neuromuscolar junction. *J. Neurocytol.*, **16**, 567–576.

Kobayashi, K., Emson, P.C., and Mountjoy, CX.Q. (1989) Vicia villosa lectin-positive in human cerebral cortex. Loss in Alzheimer's-type dementia. *Brain Res.*, **498**, 170–174.

Koeppen, A.H. and Dentinger, M.P. (1988) Brain hemosiderin and superficial siderosis of the central nervous system. *J. Neuropathol. Exp. Neurol.* **47**, 249–270.

Kondo, J., Honda, T., Mori, H., Hamada, Y., Miura, R., Ogawara, H. and Ihara, Y. (1988) The carboxyl third of tau is tightly bound to paired helical filaments. *Neuron*, **1**, 827–834.

Konno, H., Yamamoto, T., Iwasaki, Y., Suzuki, H., Saito, T., Terunuma, H. (1989) Wallerian degeneration induces Ia-antigen expression in the rat brain. *J. Neuroimmunol.*, **25**, 151–159.

Koval, L.M., Kononenko, N.I., Lutsik, M.D. and Yavorskaya, E.N. (1994) Electron cytochemical study of carbohydrate components in different types of cultured glial cells of snail Helix pomatia. *Comp. Biochem. Physiol., Comp. Physiol.*, **108**, 195–212.

Kreutzberg, G.W. (1996) Microglia sensor for pathological events in the CNS. *TINS*, **19**, 312–318.

Lawson, L.J., Perry, V.H., Cri P. and Gordon, S. (1990) Heterogeneity in the distribution and morphology of microglia in the normal adult mouse brain. *Neuroscience*, **39**, 151–170.

Ledesma, M.D., Medina, M. and Avila, J. (1996) The *in vitro* formation of recombinant tau polymers: effect of phosphorylation and glycation. *Mol. Chem. Neuropathol.*, **27**, 249–258.

Li, Y.M. and Dickson, D.W. (1997) Enhanced binding of advanced glycan endproducts (AGE) by the ApoE4 isoform links to the mechanism of plaque deposition in Alzheimer's disease. *Neurosci. Lett.*, **226**, 155–158.

Ling, E.A., Shieh, J.Y., Wen, C.Y., Yick, T.Y. and Wong, W.C. (1989a) Neuronal degeneration and non-neuronal cellular reactions in the hypoglossal nucleus following intraneuronal injection of toxic ricin. *Arch. Histol. Cytol.*, **52**, 345–354.

Ling, E.A., Wen, C.Y., Shieh, J.Y., Yick, T.Y. and Leong, S.K. (1989b) Neuroglial response to neuron injury. A study using intraneural injection of ricinus communis agglutinin-60. *J. Anat.*, **164**, 201–213.

Ludwin, S.K. (1990) Phagocytosis in the rat optic nerve following Wallerian degeneration. *Acta Neuropathol.*, (Berl.), **80**, 266–273.

Lutsik, B.D., Lashchewnko, A.M. and Lutsik, A.D. (1991) Lectin peroxidase markers of the microglia in paraffin sections. *Arkh. Patol.*, **53**, 60–63.

Maguire, T.M. and Breen, K.C. (1995) A decrease in neuronal sialyltransferase activity in Alzheimer's disease. *Dementia*, **6**, 185–190.

Mann, D.M., Purkiss, M.S., Bonshek, R.E., Jones, D., Brown A.M. and Stoddart, R.W. (1992) Lectin histochemistry of cerebral microvessels in ageing, Alzheimer's disease and Down syndrome. *Neurobiol. Aging*, **13**, 137–143.

Mannoji, H., Yeger, H. and Becker, L.E. (1986) A specific histochemical marker (lectin Ricinus communis agglutinin-1) for normal human microglia, and application to routine histopathology. *Acta Neuropathol.*, (Berl.), **71**, 341–343.

Marioka, T., Kalehua, A.N. and Streit, W.J. (1991) The microglial reaction in the rat dorsal hippocampus following transient forebrain ischemia. *J. Cereb. Blood Flow Metab.*, **11**, 966–973.

Marioka, T., Baba, T., Black, K.L. and Streit, W.J. (1992) Response of microglial cells to experimental glioma. *Glia*, **6**, 75–79.

McNeill, C., Guan, J., Dragunow, M., Lawlor, P., Sirimanne, E., Nikolics, K. and Gluckman, P. (1994) Neuronal rescue with transforming growth factor-beta 1 after hypoxic-ischemic brain injury. *Neuroreport*, **5**, 901–904.

Mimori, Y., Nakamura, S. and Yukawa, M. (1997) Abnormalities of acetylcholinesterase in Alzheimer's disease with special reference to effect of acetylcholinesterase inhibitor. *Behav. Brain Res.*, **83**, 25–30.

Mirra, S.S., Hart, M.N., and Terry, R.D. (1993) Making the diagnosis of Alzheimer's disease. Aprimer for practicing pathologists. *Arch. Pathol.Lab.Med.*, **117**, 132–134.

Nagao, M., Kamo, H., Akiguchi, I. and Kimura, J. (1992) Soybean agglutinin binds commonly to a subpopulation of a small-diameter neurons in dorsal root ganglion, vascular endothelium and microglia in human spinal cord. *Neurosci. Lett.*, **142**, 131–134.

Naujoks-Manteuffel, C. and Niemann, U. (1994) Microglial cells in the brain of Pleurodeles watl (Urodela, Salamandridae) after wallerian degeneration in the primary visual system using Bandeiraea simplicifolia isolectin B4-cytochemistry. *Glia*, **10**, 101–113.

Nicolini, M. and Zatta, P. (1994) *Glycobiology and the Brain*. Pergamon Press, Oxford, U.K.

Nissl, F. (1899) Ueber einige Beziehungen zwischen Nerven-zellerkrankungen und gliosen Erscheinungen bei verschiedenen Psychosen. *Arch. Psychiatr.*, **32**, 1–21.

Nogradi, A. (1993) Differential expression of carbonic anhydrase isozymes in microglia cell types. *Glia*, **8**, 133–142.

Pennisi, E.M., Palladini, G., Buttinelli, C., Cusimano, G., Galgani, S., Lauro, G., and Fieschi, C. (1995) Immunohistological study of a case of cerebral Langerhans cell histiocytosis in brain biopsy. *Clin. Neuropathol.*, **14**, 25–28.

Pericak, V.M., Bebout, J.L., Gaskell, P.J. (1991) Linkage studies in familial Alzheimer's disease: Evidence for chromosome 19 linkage. *Am. J. Human Genet.*, **48**, 1034–1050.

Perlmutter, L.S., Scott, S.A., Barron, E. and Chui, H.C. (1992) MHC class II- positive microglia in human brain: association with Alzheimer's lesions, *J. Neurosci. Res.*, **35**, 346–349.

Poirier, J.M., Hess, M., May, P.C. and Finch, E. (1991) Astrocytic apolipoprotein E mRNA and GFAP mRNA in hippocampus after entorhinal cortex lesions. *Brain Res. Mol. Brain Res.*, **11**, 97–106.

Riva-Depaty, I., Fardeau, C., Mariani, J., Bouchaud, C., Delhaye-Bouchaud, N. (1994) Contribution of peripheral macrophages and microglia to the cellular reaction after mechanical or neurotoxin-induced lesions of the rat brain. *Exp. Neurol.*, **128**, 77–87.

Roques, R.S. and Caldwell, R.B. (1993) Isolation and culture of retinal microglia. *Curr. Eye Res.*, **12**, 285–290.

Rossner, S., Hartig, W., Schliebs, R., Bruckner, G., Brauer, K., Perez-Polo, J.R., Wiley, R.G. and Bigl, V. (1995) 192IgG-saporin immunotoxin-induced loss of cholinergic cells differentially acitivates microglia in rat basal forebrain nuclei. *J. Neurosci. Res.*, **41**, 335–346.

Saez-Valero, J., Sberna, G., McLean, C.A., Masters, C. and Small, D.H. (1997) Glycosylation of acetylcholinesterase as diagnostic marker for Alzheimer's disease. *Lancet*, **350**, 929.

Sangruchi, T. and Sobel, R.A. (1989) Microglial and neural differentiation in human teratoma. *Acta neuropathol.* (Berl.), **78**, 258–263.

Saso, L., Silvestrini B., Guglielmotti , A., Lahita, R. and Cheng, C.Y. (1993) Abnormal glycosylation of alpha 2-macroglobulin, a non-acute-phase protein in patients with autoimmune diseases. *Inflammation*, **17**, 465–479.

Saunders, A.M, Strittmatter, W.J., Schmechel, D., St-George Hyslop, P.H. , Pericak-Vance, M.A., Joo, S.H., Rosi, B.L., Gusella, J.F., Crapper-McLachlan, D.B., Alberts, M.J., Hulette, C., Crain, B., Goldgaber, D. and Roses, D. (1993) Association of apolipoprotein E allele e4 with late-onset familial and sporadic Alzheimer's disease. *Neurology*, **43**, 1467–1472.

Schumacher, U., Kretzschamar, H. and Pfuller, U. (1994a) Staining of cerebral amyloid plaque glycoproteins in patients with Alzheimer's disease with microglia-specific lectin from mistletoe. *Acta Neuropathol.* (Berl.), **87**, 422–424.

Schumacher, U., Adam, E., Kretzschmar, H. and Pfuller, U. (1994b) Binding pattern of mistletoe lectins I, II and III to microglia and Alzheimer plaque glycoprotein in human brains. *Acta Histochem.*, **96**, 399–403.

Smith, M.A., Richey, P.L., Taneda, S., Kutty, K., Sayre, L. M., Monnier, V.M. and Perry, G. (1994) Advanced Maillard reaction end products, free radicals, and protein oxidation in Alzheimer's disease. *Ann. N.Y. Acad. Sci.*, **738**, 447–454.

Steffan, A.M., Lafon, M.E., Gendrault, J.L., Koehren, F., De Monte, M., Royer, C., Kirn, A. and Gut, J.P. (1994) Feline immunodeficiency virus can productively infect cultured endothelial cells from cat brain microvessels. *J. Gen. Virol.*, **75**, 3547–3653.

Streit, W.J. and Kreutzberg, G.W. (1987) Lectin binding by resting and reactive microglia. *J. Neurocytol.*, **16**, 249–260.

Streit, W.J. (1990) An improved staining method for rat microglial cells using the lectin Griffonia simplicifolia (GSA-IB4). *J. Histochem. Cytochem.*, **38**, 1683–1686.

Streit, W.G. (1995) Microglial cells. In H. Kettenmann, and B.R. Ransom, (eds), *Neuroglia*, Oxford University Press, Oxford, U.K., pp. 85–96.

Suzuki, H., Franz, H., Yamamoto, T., Iwasaki, Y. and Konno, H. (1988) Identification of the normal microglial population in human and rodent nervous tissue using lectin-histochemistry. *Neuropathol. Appl. Neurobiol.*, **14**, 221–227.

Thome, J., Munch, G., Muller, R., Schinzel, R., Kornhuber, J., Blum-Degen, D., Sitzmann, L., Rosler, M., Heidland, A. and Riederer, P. (1996) Advanced glycation end product-associated parameters in the peripheral blood of patient with Alzheimer's disease. *Life Sci.*, **59**, 679–685.

Velasco, A., Caminos, E., Vecino, E., Lara, J.M. and Aijon, J. (1995) Microglia in normal and regenerating visual pathways of the tench (*Tinca tinca* L., 1758; Teleost): a study with tomato lectin. *Brain Res.*, **705**, 315–324.

Wang, W.C. and Cummings, R.D. (1988) The immobilized leukoaggutinin from the seed *Maackia amurensis* binds with high affinity to complex-type Asn-linked oligosaccharides containing terminal sialic acid-linked a2, 3 to penultimate galactose residue. *J. Biol. Chem.*, **263**, 4576–4585.

Wang, J.Z., Grundke-Iqbal, I. and Iqbal, K. (1996) Glycosylation of microtubule-associated protein tau: an abnormal postranslational modification in Alzheimer's disease. *Nature-Medicine*, **2**, 850–852.

Weidenheim, K.M., Epshteyn, I and Lyman, W.D. (1993) Immunocytochemical identification of T-cells in HIV-1 encephalitis: implications for pathogenesis of CNS disease. *Mod. Pathol.*, **6**, 167–174.

Weis, W.I. and Drickamer, K. (1996) Structural basis of lectin-carbohydrate recognition. *Ann. Rev. Biochem.*, **65**, 441–473.

Weisgraber, K.H., Roses, A.D., and Strittmatter, W.J. (1994) The role of apolipoprotein E in the nervous system. *Curr. Opin. Lipidol.*, **5**, 110–116.

Wolswijk, G. (1994) CD3+ cells in the adult rat optic nerve are ramified microglia rather than O-2A adult progenitor cells. *Glia*, **10**, 244–249.

Wu, C.H., Wen, C.Y., Shieh, J.Y. and Ling, E.A. (1994) Down-regulation of membrane glycoprotein in ameboid microglia transforming into ramified microglia in postnatal rat brain. *J. Neurocytol.*, **23**, 258–269.

Yan, S.D., Chen, X., Schmidt, A.M., Brett, J., Godman, G., Zou, Y.S., Scott, C.W., Caputo, C., Frappier, T., Smith, M.A., Perry, G., Yen, S-H., and Stern, D. (1994) Glycated tau protein in Alzheimer's disease: a mechanism for induction of oxidant stress. *Proc. Natl. Acad. Sci. USA*, **91**, 7787–7791.

Yan, S.D., Chen, X., Fu, J., Chen, M., Zhu, H., Roher, A., Slattery, T., Zhao, L., Nagashima, M., Morser, J., Migheli, A., Nawroth, P., Stern, D. and Schmidt, A.M. (1996) Rage and amyloid-beta peptide neurotoxicity in Alzheimer's disease. *Nature*, **382**, 685–691.

Zambenedetti, P., Giordano, R. and Zatta, P. (1998) Histochemical localization of glycocoyugates on microglial cells in Alzheimer's disease brain samples by using *Abrus pecatorius, Maackie amureusis, Momordice charautie* and *Sauluas nigre* lectins. *Exp. Neurol.*, **153**, 167–171.

Zatta, P. and Cummings, R.D. (1992) Lectins and their uses as biotechnological tools. *Biochem. Educ.*, **20**, 2–9.

Zatta, P., Zanoni, S. and Favarato, M. (1994) Lectin histochemistry of the brain cortex in Alzheimer's disease. In M. Nicolini and P. Zatta, (eds), *Glycobiology and the Brain*, Pergamon Press, Oxford, U.K., pp. 141–153.

4. ROLE OF GALECTIN-3 IN INFLAMMATION

FU-TONG LIU

Division of Allergy, La Jolla Institute for Allergy and Immunology, 10355 Science Center Drive, San Diego, CA 92121, USA

INTRODUCTION

Galectins are members of a recently identified family of β-galactoside-binding animal lectins (Barondes *et al.*, 1994a). Presently, ten members have been identified (Barondes *et al.*, 1994b; Kasai and Hirabayashi, 1996) and more are likely to be discovered. Members are defined by shared consensus amino acid sequences and affinity for β-galactose-containing oligosaccharides. The family can be subdivided into prototypical type (galectin-1, -2, -5, -7, and -10), existing as monomers or homodimers made of one lectin domain; chimeric type (galectin-3), containing a nonlectin part connected to a lectin domain; and tandem repeat type (galectin-4, -6, -8, and -9), composed of two distinct but homologous lectin domains in a single polypeptide chain. The exact functions of most of the family members remain to be elucidated. Nevertheless, some of the members, especially galectin-3, have been extensively characterized and experimental results suggest that these lectins may have diverse functions and participate in a variety of physiological and pathological processes. This chapter focuses on galectin-3; some characteristics of this protein that are relevant to its function in inflammation are first presented, and information on various biological activities of this protein that support its role in inflammation is then reviewed.

BIOCHEMICAL PROPERTIES OF GALECTIN-3

Galectin-3 was initially designated with various different names by a number of research groups. It was identified as IgE-binding protein (εBP) for its affinity for IgE (Liu, 1990) and was also designated as Mac-2, CBP35, CBP30, L-29, and L-34. A great deal of structural information on this lectin has accumulated through studies by a number of laboratories. cDNAs for several animal species have been cloned, including human (Robertson *et al.*, 1990; Cherayil *et al.*, 1990; Raz *et al.*, 1991; Oda *et al.*, 1991), rat (Liu *et al.*, 1985; Albrandt *et al.*, 1987), mouse (Jia and Wang, 1988; Cherayil *et al.*, 1989; Raz *et al.*, 1989), hamster (Mehul *et al.*, 1994), and rabbit (Gaudin *et al.*, 1995), and the primary structure of dog galectin-3 has been determined by protein sequencing (Herrmann *et al.*, 1993).

Galectin-3 is composed of a small amino-terminal domain, a domain consisting of proline-, glycine-rich tandem repeats, and a carboxyl-terminal domain that shares sequence similarity with other galectin family members and represents the carbohydrate-recognition domain. The protein can be phosphorylated both *in vivo* and *in vitro* at two different serine residues in the amino-terminal region (Huflejt *et al.*, 1993). The repetitive sequences share some similarities with collagens in that it has a high proline, glycine and alanine content. However, they do not have a glycine residue at every third position, as is found in collagens and shown to be necessary for the formation of a triple helix. Nevertheless, galectin-3 can be digested by collagenase at the amino-terminal region (Hsu *et al.*, 1992; Herrmann *et al.*, 1993; Ochieng *et al.*, 1993; Mehul *et al.*, 1994) and is also a substrate for metalloproteinases (Mehul *et al.*, 1994; Ochieng *et al.*, 1994). Human (Kadrofske *et al.*, 1998) and mouse (Gritzmacher *et al.*, 1992; Rosenberg *et al.*, 1993) genomic DNAs have been cloned and sequenced. The entire gene is composed of six exons: Exon I contains part of the 5'-untranslated region, exon II contains the remainder of the 5'-untranslated region plus 18 bp encoding the first six amino acids of the protein. The entire tandem repeats are encoded by exon III and the carboxyl-terminal domain is encoded by exons IV-VI.

TISSUE DISTRIBUTION OF GALECTIN-3 AND ITS EXPRESSION IN INFLAMMATORY CELLS

Galectin-3 has a wide tissue distribution (Flotte *et al.*, 1983; Crittenden *et al.*, 1984); it is abundantly present in the epithelia of several organs (Flotte *et al.*, 1983), including intestine (Brassart *et al.*, 1992) and skin (Wollenberg *et al.*, 1993; Konstantinov *et al.*, 1994). Other cell types that express galectin-3 include osteoblasts, osteoclasts (Aubin *et al.*, 1996), cultured microglial cells (Pesheva *et al.*, 1998), activated Schwann cells (Reichert *et al.*, 1994), and myelopoietic cells and surrounding stroma in the bone marrow (Krugluger *et al.*, 1997). Most relevant to this review is the fact that galectin-3 is expressed by various inflammatory cells, including mast cells (Frigeri and Liu, 1992), neutrophils (Truong *et al.*, 1993b), monocytes/macrophages (Liu *et al.*, 1995) and eosinophils (Truong *et al.*, 1993a). The expression is dependent on cell differentiation. For example, the level of galectin-3 increases over 10-15 fold as human monocytes differentiate into macrophages *in vitro* (Liu *et al.*, 1995). The differentiation-dependent expression in the monocyte lineage is also observed with cell lines (Nangia-Makker *et al.*, 1993). In addition, there is enhanced expression of this lectin when monocytes adhere to synthetic polymers *in vitro* (Smetana *et al.*, 1997). The levels of galectin-3 in some of the inflammatory cells in human appear to vary from individual to individual. Neutrophils from some allergic patients contain much higher levels of galectin-3 mRNA as compared to those from healthy donors (Truong *et al.*, 1993b), and the level of galectin-3 expression in eosinophils also exhibits individual variation (Truong *et al.*, 1993a). Normal resting lymphocytes do not express galectin-3; however, the lectin is abundantly expressed in T cell lines or thymocytes infected by human T lymphotropic virus-I (HTLV-I) (Hsu *et al.*, 1996).

SUBCELLULAR LOCALIZATION AND SECRETION

Earlier work showed that galectin-3 is localized primarily in the cytoplasm and is not detectable in the plasma membrane fraction (Moutsatsos *et al.*, 1986; Gritzmacher *et al.*, 1988), and that the protein is also detectable in the nucleus (Moutsatsos *et al.*, 1986; Gritzmacher *et al.*, 1988). An immunoelectron microscopic study of intracellular localization of galectin-3 in mast cells has been reported (Craig *et al.*, 1995). Some mast cells show a preferential distribution of the label over the nucleus and others exhibit predominantly cytoplasmic labeling. In the cytoplasm, the labeling is over or adjacent to secretory granules and concentrated over electron dense components of secretory granules. The cellular distribution pattern of galectin-3 in a human colonic adenocarcinoma cell line T84 was studied and found to vary with culturing conditions (Huflejt *et al.*, 1997). In confluent cells, galectin-3 is concentrated in large granular inclusions, whereas in subconfluent cells, it is found primarily in lamellipodia. The distribution is distinctly different from that of galectin-4.

None of the galectin family members contains a classical signal sequence in the protein. However, a number of studies have demonstrated the secretion of these lectins. By pulse-chase experiments, galectin-3 has been shown to be secreted by thioglycollate-stimulated murine macrophages (Cherayil *et al.*, 1989). Other investigators demonstrated the secretion of galectin-3 from murine macrophage cell lines that are activated by a calcium ionophore; some cell lines secrete substantial amounts of the newly synthesized lectin (Sato and Hughes, 1994b; Sato and Hughes, 1994a). The secretion of this lectin has also been demonstrated in kidney cell lines, which selectively secrete the protein from the apical side of the cell (Sato *et al.*, 1993; Lindstedt *et al.*, 1993).

The mechanism underlying the secretion of galectin-3 is not well understood. Recent data suggest that plasma membrane targeting and vesicular budding are responsible for the release of galectin-3 from the cytoplasm of mammalian cells during secretion (Mehul and Hughes, 1997). While galectin-3 secretion from cells is not a very efficient process, a chimera of galectin-3 fused to the amino-terminal acylation sequence of p56lck protein tyrosine kinase (Nt-p56lck-galectin-3) is efficiently released from the transfected cells. The sequence from p56lck contains sites for myristoylation and palmitoylation, which are responsible for the rapid transportation of the fusion protein to plasma membrane domains. In contrast, chimeras lacking palmitoylation sites show no or minimal transport to plasma membranes. Thus, the movement of cytoplasmic galectin-3 to plasma membrane domains appears to be a rate-limiting step in lectin secretion. Immunofluorescence studies confirm that Nt-p56lck-galectin-3 aggregates underneath the plasma membrane and is exocytosed by membrane blebbing. The authors suggested that plasma membrane targeting and vesicular budding are critically involved in galectin-3 secretion and that galectin-3 is eventually released from the vesicles into the extracellular space.

Consistent with the fact that galectin-3 is secreted by cells is the detection of this lectin on cell surfaces. Thioglycollate-elicited murine peritoneal macrophages contain $1.7–3.4 \times 10^5$ surface galectin-3 molecules per cell (Ho and Springer, 1982). Galectin-3 is present in variable amounts in several mast cell lines and the surface expression in these cell lines is variable and does not correlate with the total amounts of galectin-3 in the cells (Frigeri and Liu, 1992). When rat basophil leukemia cells are activated by IgE immune complexes, there is greater than 7.5-fold increase in cell surface galectin-3 (Frigeri and Liu, 1992). This finding is consistent with the presence of this lectin in the secretory granules of mast cells (Craig *et al.*, 1995), as mentioned above. Cells which do not express galectin-3 mRNA can also display this lectin on the cell surface by acquiring the protein from other cells. For example, skin Langerhans cells have been noted to acquire galectin-3 released from neighboring epidermal keratinocytes (Wollenberg et al., 1993).

How the galectin-3 molecules are expressed on the cell surface is not established. Based on the fact that exogenous galectin-3 can bind to surfaces of various cell types through lectin-carbohydrate interactions, it is not difficult to envision that galectin-3 is displayed on the cell surface through binding to glycoconjugates. Indeed, when the number of cell surface galectin-3 was determined by using anti-galectin-3 antibodies, it was found that the antibody binding to the cell surface was significantly reduced in the presence of lactose (Frigeri and Liu, 1992). The data suggest that galectin-3 is eluted off the cell surface by lactose. However, when the cells were treated with lactose and then washed to remove lactose, there was only a partial reduction in the number of cell surface galectin-3, as determined by the antibody-binding assay (Frigeri and Liu, 1992). While the data might suggest the existence of galectin-3 molecules on cell surfaces that are not elutable by lactose, one possible explanation is that during the washing step after the cells are exposed to lactose, there is additional release of galectin-3 from the cells, either due to active secretion or leakage from the damaged cells. Nevertheless, based on these findings, the possibility can not be excluded that galectin-3 exists as a peripheral membrane protein. It has been suggested that the two distinct hydrophobic domains in the molecule (amino acids 134-152 and 214-235) may interact with the membrane lipid bilayer (Ochieng et al., 1993).

GLYCOCONJUGATE LIGANDS OF GALECTIN-3

Each member of the galectin family exhibits selectivity for certain galactose-containing oligosaccharide structures (Leffler and Barondes, 1986; Sparrow et al., 1987). The binding of galectin-3 to oligosaccharide ligands has been studied extensively by several laboratories (Leffler and Barondes, 1986; Sparrow et al., 1987; Sato and Hughes, 1992; Knibbs et al., 1993; Feizi et al., 1994). Studies of the binding of radiolabeled galectin-3 to a series of lipid-linked oligosaccharide sequences of the lacto-neolacto family showed that the minimum lipid-linked oligosaccharides that can support galectin-3 binding are pentasaccharides (Feizi et al., 1994). A unique and interesting feature of this lectin is that it binds more strongly to the blood group A, B, or B-like determinants than to those bearing group H. This preferential binding of galectin-3 is also manifested with whole cells, as erythrocytes of blood groups A and B are more strongly bound by galectin-3 than those of blood group O. These results raise the possibility that the in vivo activity of galectin-3 may be influenced by the blood group status of the host (Feizi et al., 1994). Another key feature is that recognition of saccharides by galectin-3 is significantly modulated by sialylation. For example, sialylation of the outer galactose of an oligosaccharide at position 6 renders it poorly recognizable by galectin-3 (Leffler and Barondes, 1986; Sparrow et al., 1987; Knibbs et al., 1993; Feizi et al., 1994).

A large number of glycoproteins has been reported to be recognized by galectin-3. As mentioned above, this lectin was initially identified as an IgE-binding protein (Liu et al., 1985; Albrandt et al., 1987; Robertson et al., 1990). The binding was found to be dependent on the source of IgE and the degree of sialylation. Although a murine monoclonal IgE is recognized by both human and rat galectin-3, three different human IgE myeloma proteins are not, unless they are first treated with neuraminidase (Robertson *et al.*, 1990), suggesting that the determinants in the untreated IgE are masked by sialylation. A similar observation was made with polyclonal IgE in that IgE from some individuals is well recognized by galectin-3, while that from others is poorly recognized unless pre-treated with neuraminidase (Robertson and Liu, 1991). The saccharide structures recognized by galectin-3 appear to be IgE specific and are not displayed by IgG or IgM (Cherayil *et al.*, 1989). They are not abolished by treating the IgE with N-glycanase, suggesting that they may be O-linked (Cherayil *et al.*, 1989).

Galectin-3 also binds laminin, a major basement membrane protein (Woo *et al.*, 1990; Lee *et al.*; 1991; Castronovo *et al.*, 1992; Sato and Hughes, 1992; Massa *et al.*, 1993). Galectin-3 interacts with bacterial lipopolysaccharide (LPS) via two independent sites (Mey *et al.*, 1996). The amino-terminal region of galectin-3 interacts with part of LPS that contains lipid A and the inner core, while the carboxyl-terminal domain binds β-galactoside-containing polysaccharide. The first interaction can confer binding to a wide range of LPS and lipooligosaccharide (LOS) that have similar structures in their lipid A and core regions, while the second interaction can confer binding to a subset of LPS and LOS that has properly displayed β-galactoside moieties.

It is now evident that galectin-3 recognizes a number of different glycoproteins, but it does exhibit some selectivity. For example, galectin-3 binds only limited numbers of glycoprotein species present on the surfaces of rat basophilic leukemia cells (Frigeri *et al.*, 1993), neutrophils (Yamaoka *et al.*, 1995), and macrophages (Dong and Hughes, 1997). Some of the glycoproteins on different cells have been identified, and these include: 1) IgE receptor (FcεRI) on mast cells (Frigeri *et al.*, 1993); 2) CD66 (NCA160) on human neutrophils (Yamaoka *et al.*, 1995); 3) the α-subunit (CD11b) of the CD11b/CD18 integrin, the lysosomal membrane glycoproteins LAMPs 1 and 2, the Mac-3 antigen, the heavy chain of CD98, and Mac-2-binding protein (a new member of the superfamily defined by the macrophage scavenger receptor cysteine-rich domain), on macrophages (Rosenberg *et al.*, 1991; Koths *et al.*, 1993; Dong and Hughes, 1997); 4) LAMPs 1 and 2, and Mac-2-binding protein on melanoma cells (Inohara and Raz, 1994); 5) carcinoembryonic antigen in human colon carcinoma cells (Ohannesian *et al.*, 1995); and 6) neural tissue-derived glycoproteins L1 (a cell recognition molecule), MAG (myelin-associated glycoprotein), N-CAM (neural cell adhesion molecule), tenascin-C and tenascin-R, on neuronal cells (Probstmeier *et al.*, 1995).

MOLECULAR BASIS FOR BIVALENCY (OR MULTIVALENCY) OF GALECTIN-3

Like many other lectins, galectin-3 behaves in at least a bivalent fashion, as best demonstrated by its ability to agglutinate erythrocytes (Frigeri *et al.*, 1990). Both recombinant galectin-3 and the carboxyl-terminal domain fragment of galectin-3 (galectin-3C) contain only one lactose-binding site (Hsu *et al.*, 1992). The fact that galectin-3C does not have hemagglutination activity suggests that the amino-terminal region is critical for the multivalent property of this protein. Based on the finding of cooperative binding of galectin-3 to IgE, as well as results from chemical cross-linking experiments, it has been proposed that galectin-3 forms dimers or oligomers through noncovalent inter-molecular interactions involving the amino-terminal region (Hsu et al., 1992). This proposition is also supported by a similar positive cooperativity found in the binding of galectin-3 to laminin (Massa et al., 1993) and the finding that recombinant amino-terminal region fragment has a tendency to self-associate (Mehul et al., 1994). Monoclonal antibodies recognizing the amino-terminal region have been shown to either potentiate or inhibit the self-association process and thus modulate the functional properties of galectin-3 (Liu et al., 1996).

There are other data supporting the existence of dimeric galectin-3 (Ochieng et al., 1993). When recombinant galectin-3 fractionated by Sephacryl-100 gel filtration was analyzed, the peak fractions (protein >20 ug/ml) exhibited Mr 31,000 and 62,000 polypeptides, corresponding to monomeric and dimeric galectin-3, respectively, by SDS-PAGE under non-reducing conditions. Y-shaped structures, with the overall dimension and structure similar to other collagen-like molecules, such as C1q, SP-A, and MBP, are detectable by electron microscopy (Ochieng et al., 1993). These observations led the authors to conclude that galectin-3 dimers exist, which are favored by high protein concentrations.

Galectin-3 purified from thioglycollate-elicited peritoneal macrophages by laminin affinity chromatography also contains dimers (Woo et al., 1991). The ratio of galectin-3 monomers to dimers in the initial preparation and in the material that does not bind laminin-Sepharose indicates that the dimer has a higher affinity for laminin than the monomer. Site-directed mutagenesis indicates that Cys186 is required for dimerization. However, as pointed out by other authors (Hirabayashi, 1992), there remains a possibility that disulfide bound-linked dimers are formed during or after the chromatography. In fact, experimental evidence suggests that this cysteine residue is probably not exposed in native galectin-3 (Frigeri et al., 1990).

Another possible way for galectin-3 dimerization is cross-linkage by transglutaminase (Mehul et al., 1995). Incubation of recombinant hamster galectin-3 with guinea pig liver transglutaminase results in the formation of a Mr 60,000 species, as detected by SDS-PAGE and immunoblotting. A single lysine residue present in the amino-terminal domain and two glutamine residues present in the tandem repeats are probably involved in the cross-linkage.

EXTRACELLULAR FUNCTIONS OF GALECTIN-3: ACTIVATION OF CELLS

Because of the bivalent or multivalent property of galectin-3 and the fact that it recognizes cell surface glycoproteins, this protein has potential to cross-link cell surface glycoproteins of effector cells. Indeed, galectin-3 can activate rat basophilic leukemia (RBL) cells, as demonstrated by mediator release, in a manner that is inhibitable by saccharide ligands of the lectin (Frigeri et al., 1993). In another series of experiments, both IgE-sensitized and unsensitized RBL cells were found to be activated upon exposure to microtiter wells coated with galectin-3, again in a manner involving galectin-3-carbohydrate interactions (Zuberi et al., 1994). Furthermore, galectin-3 augments antigen-induced activation of RBL cells and the effects by the two stimuli appear to be synergistic (Zuberi et al., 1994).

Another example is provided by the galectin-3's activation of human neutrophils (Yamaoka *et al.*, 1995). Galectin-3 binds neutrophils in a manner that is dependent on the lectin activity. In addition, there is positive cooperativity in the binding of galectin-3, analogous to the situation of the lectin binding to IgE, suggesting that galectin-3 has a tendency to self-associate upon binding to the neutrophil surfaces. Recombinant human galectin-3 activates human neutrophils in a dose-dependent manner as demonstrated by superoxide production. Furthermore, the observed activity is dependent on the galectin-3's lectin property, since it can be inhibited by lactose. The amino-terminal domain fragment is also necessary for the observed activity, since the carboxyl-terminal domain fragment is devoid of the neutrophil-activating activity, even though it retains the carbohydrate-binding activity. This suggests that neutrophil activation by galectin-3 requires oligomerization of this protein.

Galectin-3 has been found to potentiate IL-1 production by human monocytes (Jeng *et al.*, 1994). A subsequent study showed that galectin-3 induces superoxide production by human monocytes (Liu *et al.*, 1995). Galectin-3 also stimulates uptake of extracellular Ca^{2+} in human Jurkat T-cells and the response is completely blocked by lactose or thiodigalactoside (Dong and Hughes, 1996). Jurkat cells do not express galectin-3, but express surface glycoproteins recognized by the lectin, one of which has been shown to be the glycosylated heavy chain of CD98.

Galectin-3 expressed on human neutrophils has been suggested to mediate neutrophil activation induced by IgE-anti-IgE immune complexes, because the cell activation can be inhibited by an anti-galectin-3 monoclonal antibody (Truong *et al.*, 1993b). It has also been shown that neutrophils from patients allergic to specific allergens, but not from healthy donors, are sensitive to the same allergens that produce the clinical symptoms. These authors proposed that the restricted recognition of allergens by neutrophils is probably associated with the binding of the allergens to its corresponding IgE molecule, bound to the cell surface via galectin-3 (Monteseirin *et al.*, 1996).

EXTRACELLULAR FUNCTIONS OF GALECTIN-3: PROMOTION OF CELL ADHESION

The properties of galectin-3 described above also suggest that this lectin may contribute to cell-cell and cell-extracellular matrix adhesion. In fact, the ability of this lectin to mediate cell-cell adhesion is already suggested by its hemagglutination activity (Frigeri *et al.*, 1990). It is possible that galectin-3 can form dimers or oligomers which serve as a bridge to bind cells together or cells to the extracellular matrix. Galectin-3 promotes adhesion of human neutrophils to laminin coated on microtiter plates (Kuwabara and Liu, 1996). The effect is dependent on the lectin's carbohydrate-binding function as well as the amino-terminal region. At lower concentrations, galectin-3 is effective in inducing cell adhesion via a Ca^{2+}, Mg^{2+}-dependent process, whereas at higher concentrations, it can induce cell adhesion in a divalent cation-independent manner. Galectin-3 at higher concentrations induces neutrophil adhesion to fibronectin, even though galectin-3 does not bind fibronectin, and the process is entirely dependent on Ca^{2+} and Mg^{2+}. Galectin-3-induced neutrophil adhesion to fibronectin is at least partially β2 integrin dependent, whereas that to laminin is independent of β2 integrins. These findings suggest a combination of two independent mechanisms: 1) the lectin bridges neutrophils to laminin in a Ca^{2+}, Mg^{2+}-independent manner, and 2) the lectin induces activation of neutrophils in the presence of the divalent cations, resulting in upregulation of other cell adhesion molecules, leading to enhanced adhesiveness. At inflammatory sites, increased amounts of galectin-3 released by epithelial cells and inflammatory cells may play an important role in promoting neutrophils to traverse through the basement membrane, in conjunction with other cell adhesion molecules.

The role of galectin-3 in cell adhesion is also suggested by its expression and effects on cyst enlargement and tubulogenesis in kidney epithelial MDCK cells cultured in three-dimensional matrices *in vitro* (Bao and Hughes, 1995). MDCK cells grown in a three-dimensional collagen gel shows a sequence of morphologic events and dynamic galectin-3 expression patterns. In single cells, galectin-3 is expressed uniformly within the cytosol, but as cells multiply and form loose aggregates, the cytoplasmic staining decreases and prominent cell surface staining becomes apparent. When cells progress to form cysts with a single layer of cells lining the lumen, the distribution of galectin-3 is confined to the basolateral domains of the cell surface and the staining tapers off towards the apical (luminal) surface. In addition, the lectin is co-localized with laminin on the basal surface. In tubule-forming cysts, galectin-3 is excluded from the initial spikes and the progressing tips of the tubules, although its basolateral expression in the cyst body remains. Galectin-3 added exogenously to cultures exerts an inhibitory effect on cyst enlargement, while galectin-3-specific antibodies promote this process. The authors proposed that galectin-3 has an enhancing role in intercellular adhesion and that galectin-3 at the basal surface may be involved in cell interactions with the surrounding matrix.

A similar effect of exogenously added galectin-3 on cell adhesion has also been shown with another cell line, baby hamster kidney (BHK) cells. Exogenously added hamster galectin-3 blocks BHK cell attachment to laminin and cell spreading in a dose-dependent fashion (Sato and Hughes, 1992). At a lower dose, galectin-3 does not affect cell attachment but inhibits the spreading of attached cells, whereas at a higher dose, it inhibits significantly cell attachment and the attached cells show round or poorly spread morphology. Exogenously added galectin-3 has also been shown to markedly enhance the migration of human breast carcinoma cells through a Matrigel barrier, an effect which likely reflects a role of this lectin in cell adhesion (Le Marer and Hughes, 1996).

Whether the endogenous cell surface galectin-3 indeed plays a role in cell adhesion remains to be clarified. BHK cell attachment to laminin is not inhibited by anti-galectin-3 (Sato and Hughes, 1992), although galectin-3 is expressed on the cell surface. While SCM-153 (human breast epithelial) cells express galectin-3 and adhere to laminin, addition of lactose or anti-galectin-3 fails to inhibit the adhesion, suggesting that galectin-3 in these cells may not be functioning in cell adhesion (Ochieng *et al.*, 1992). Likewise, A2058 and A375 melanoma cells express galectin-3 on their surfaces and attach to laminin *in vitro*, but anti-galectin-3 antiserum does not alter the adhesion (van den Brûle *et al.*, 1995). Also, unlike the other cell adhesion systems described above, adhesion of human melanoma cells to laminin is not modulated by exogenously added galectin-3 (van den Brûle *et al.*, 1995).

Data have also been provided to support that cell surface galectin-3 can mediate homotypic cell adhesion (Inohara and Raz, 1995). A recombinant strain of baculovirus encoding galectin-3 is used to infect Sf9 insect cells and galectin-3 was found to be localized on the cell surface as well as cytoplasm of infected cells. Sf9 cells infected with recombinant virus undergo homotypic aggregation in the presence of an exogenous glycoprotein (i.e., asialofetuin) recognized by galectin-3, whereas control cells uninfected or infected with wild-type virus do not. Furthermore, the aggregation is inhibitable by lactose and anti-galectin-3. Co-suspension of Sf9 cells infected with the recombinant virus with uninfected cells, in the presence of asialofetuin, results in preferential cell-cell adhesion of the galectin-3-expressing cells. Similarly, galectin-3 expressed on the surface of A375 melanoma cells was shown to mediate homotypic cell adhesion in the presence of a glycoprotein recognized by galectin-3, Mac-2-binding protein (Inohara *et al.*, 1996).

INTRACELLULAR FUNCTIONS OF GALECTIN-3: REGULATION OF CELL GROWTH AND APOPTOSIS

The expression of galectin-3 has been shown to be upregulated in proliferating cells. In addition, while the protein is present diffusely in the cytoplasm in quiescent cells, it is localized primarily in the nucleus in proliferating cells (Moutsatsos *et al.*, 1987; Agrwal *et al.*, 1989). The mechanisms responsible for transport of the lectin from the cytoplasm to the nucleus is unknown. Recently, it has been reported that galectin-3 recognizes cytokeratins by binding to αGalNAc-containing glycans present on cytokeratins (Goletz *et al.*, 1997). It has been suggested that cytokeratins may serve as a cytoplasmic anchor for galectin-3 and be involved in regulating transport of this lectin between the cytoplasm and the nucleus.

Based on the upregulated expression of galectin-3 in proliferating cells and the fact that this lectin can be localized in the nucleus in proliferating cells, it has been suggested that galectin-3 may be a component of growth regulatory systems that are elicited in stimulated and transformed cells (Moutsatsos *et al.*, 1987). Consistent with the possible intracellular role of galectin-3, some intracellular proteins recognized by galectin-3 have been identified, including nuclear lectins such as CBP67 and CBP70 (Lauc *et al.*, 1993; Sève *et al.*, 1993). Significantly, galectin-3 has been identified as a factor in pre-mRNA splicing (Dagher *et al.*, 1995), as demonstrated in a cell-free system involving the splicing of pre-mRNA by HeLa cell nuclear extracts (NEs). When the NE is preincubated with various saccharides prior to the addition of the pre-mRNA substrate, saccharide ligands of galectin-3 are among those which inhibit product formation. Depleting galectin-3 from NEs with lactosyl-agarose results in a complete loss of the splicing activity. Galectin-3-depleted extracts reconstituted with recombinant galectin-3 regain the ability to form splicing complexes.

The role of galectin-3 in regulation of cell growth has been supported by studies of human leukemia T cells transfected with galectin-3 cDNA (Yang *et al.*, 1996). In medium containing 1% fetal bovine serum, Jurkat transfectants expressing galectin-3 grow significantly faster than control transfectants. In addition, Jurkat transfectants expressing galectin-3 are significantly more resistant to apoptosis induced by anti-Fas antibody and staurosporine, respectively. The effects of galectin-3 are not due to the lectin exerting its function extracellularly in an autocrine or paracrine fashion, as they are not reversed by addition of lactose. Evidence for a role of galectin-3 in regulation of cell growth has also been provided by the study in which galectin-3 expression is suppressed by using an antisense approach (van den Brule *et al.*, 1997). Stable transfectants from human breast cells expressing antisense galectin-3 cDNA have reduced levels of galectin-3 protein and grow significantly more slowly than wild-type transfectants.

Galectin-3 has significant sequence similarity with Bcl-2 (Yang *et al.*, 1996), a well established suppressor of apoptosis. The amino-terminal portions of both proteins are rich in proline (P), glycine (G), and alanine (A): Amino acid residues 36 to 85 of human Bcl-2 have a P + G + A content of 62%, while residues 30-113 of human galectin-3 contain 75% P + G + A. Significant sequence similarity is also found in the carboxyl-terminal regions of the two proteins. In particular, the highly conserved NWGR motif within the BH1 domain of the Bcl-2 family members, which is critical for Bcl-2's apoptosis-suppressing activity, is present in galectin-3. This motif is highly conserved among galectin-3 from different species.

Because both galectin-3 and Bcl-2 have a tendency to self-associate and Bcl-2 is known to heterodimerize with its homologues, sequence homology between the two proteins suggests the possibility that they may interact with each other. Indeed, Bcl-2 in Jurkat cell lysates can be specifically adsorbed by galectin-3-Sepharose 4B. Interestingly, lactose, a ligand for galectin-3, but not sucrose, inhibits the binding of Bcl-2 to galectin-3-Sepharose-4B, even though Bcl-2 is not a glycoprotein, suggesting the relevance of the carbohydrate-recognition domain in the galectin-3-Bcl-2 association (Yang *et al.*, 1996). The anti-apoptotic activity of galectin-3 has been confirmed in another cell type by other investigators (Akahani *et al.*, 1997).

CONCLUSION

Based on the information summarized in this chapter, it should be convincing that galectin-3 may play an important role in inflammation and help amplify inflammatory responses. This is because i) various inflammatory cells as well as epithelial cells express galectin-3, ii) this protein can be secreted by activated leukocytes and epithelial cells, especially under inflammatory conditions, iii) released protein can bind to various cells and cause activation of these cells, through recognition (and cross-linking) of appropriately glycosylated cell surface glycoproteins, and iv) it can participate in inflammation by promoting cell-cell interaction and cell adhesion. The high cytosolic concentrations of this lectin in many cells make it possible that when these cells are activated by inflammatory stimuli, there may be a burst release of large amounts of the protein, resulting in a profound modulatory effect. The exocytosed galectin-3 may function in either a paracrine or juxtacrine fashion. In the latter case, the cells producing galectin-3 are in contact with cells responding to the lectin, and in this way, galectin-3 would be less likely to bind to, and be neutralized by, other glycoconjugates which are present in the extracellular space. Galectin-3 may also contribute to inflammatory responses through its intracellular functions, specifically, its activity in modulating apoptosis. Some of the inflammatory cells express galectin-3 or exhibit upregulated expression of this lectin under inflammatory conditions. Because galectin-3 has been shown to suppress programmed cell death in some cells, it is possible that its expression in inflammatory cells can result in prolonged survival of these cells, and this may contribute to persistence of inflammation. Because of the potential role of galectin-3 in inflammation, it is conceivable that inhibitors of galectin-3 may prove to be useful therapeutic agents for suppressing inflammatory responses.

ACKNOWLEDGMENTS

The author thanks Dr. Dan Hsu for reviewing the manuscript. The work from this laboratory has been supported in part by NIH grants AI-20958 and AI-39620.

REFERENCES

Agrwal, N., Wang, J.L. and Voss, P.G. (1989). Carbohydrate-binding protein 35. Levels of transcription and mRNA accumulation in quiescent and proliferating cells. *J. Biol. Chem.*, **264**, 17236–17242.

Akahani, S., Nangia-Makker, P., Inohara, H., Kim, H.R.C. and Raz, A. (1997). Galectin-3: A novel antiapoptotic molecule with a functional BH1 (NWGR) domain of Bcl-2 family. *Cancer Res.*, **57**, 5272–5276.

Albrandt, K., Orida, N.K. and Liu, F.-T. (1987). An IgE-binding protein with a distinctive repetitive sequence and homology with an IgG receptor. *Proc. Natl. Acad. Sci. USA*, **84**, 6859–6863.

Aubin, J.E., Gupta, A.K., Bhargava, U. and Turksen, K. (1996). Expression and regulation of galectin 3 in rat osteoblastic cells. *J. Cell. Physiol.*, **169**, 468–480.

Bao, Q. and Hughes, R.C. (1995). Galectin-3 expression and effects on cyst enlargement and tubulogenesis in kidney epithelial MDCK cells cultured in three-dimensional matrices in vitro. *J. Cell Sci.*, **108**, 2791–2800.

Barondes, S.H., Castronovo, V., Cooper, D.N.W., Cummings, R.D., Drickamer, K., Feizi, T., Gitt, M.A., Hirabayashi, J., Hughes, C., Kasai, K., Leffler, H., Liu, F.-T., Lotan, R., Mercurio, A.M., Monsigny, M., Pillai, S., Poirer, F., Raz, A., Rigby, P.W.J., Rini, J.M. and Wang, J.L. (1994a). Galectins: A family of animal b-galactoside-binding lectins. [Letter to the Editor]. *Cell*, **76**, 597–598.

Barondes, S.H., Cooper, D.N.W., Gitt, M.A. and Leffler, H. (1994b). Galectins. Structure and function of a large family of animal lectins. *J. Biol. Chem.*, **269**, 20807–20810.

Brassart, D., Kolodziejczyk, E., Granato, D., Woltz, A., Pavillard, M., Perotti, F., Frigeri, L.G., Liu, F.-T., Borel, Y. and Neeser, J.-R. (1992). An intestinal galactose-specific lectin mediates the binding of murine IgE to mouse intestinal epithelial cells. *Eur. J. Biochem.*, **203**, 393–399.

Castronovo, V., Campo, E., van den Brûle, F.A., Claysmith, A.P., Cioce, V., Liu, F.-T., Fernandez, P.L. and Sobel, M.E. (1992). Inverse modulation of steady state mRNA levels of two non-integrin laminin binding proteins in human colon carcinoma. *J. Natl. Cancer Inst.*, **84**, 1161–1167.

Cherayil, B.J., Weiner, S.J. and Pillai, S. (1989). The Mac-2 antigen is a galactose-specific lectin that binds IgE. *J. Exp. Med.*, **170**, 1959–1972.

Cherayil, B.J., Chaitovitz, S., Wong, C. and Pillai, S. (1990). Molecular cloning of a human macrophage lectin specific for galactose. *Proc. Natl. Acad. Sci. USA*, **87**, 7324–7329.

Craig, S.S., Krishnaswamy, P., Irani, A.-M.A., Kepley, C.L., Liu, F.-T. and Schwartz, L.B. (1995). Immunoelectron microscopic localization of galectin-3, an IgE binding protein, in human mast cells and basophils. *Anat. Rec.*, **242**, 211–216.

Crittenden, S.L., Roff, C.F. and Wang, J.L. (1984). Carbohydrate-binding protein 35: Identification of the galactose- specific lectin in various tissues of mice. *Mol. Cell Biol.*, **4**, 1252–1259.

Dagher, S.F., Wang, J.L. and Patterson, R.J. (1995). Identification of galectin-3 as a factor in pre-mRNA splicing. *Proc. Natl. Acad. Sci. USA*, **92**, 1213–1217.

Dong, S. and Hughes, R.C. (1996). Galectin-3 stimulates uptake of extracellular Ca^{2+} in human Jurkat T-cells. *FEBS Lett.*, **395**, 165–169.

Dong, S. and Hughes, R.C. (1997). Macrophage surface glycoproteins binding to galectin-3 (Mac-2–antigen). *Glycoconjugate J.*, **14**, 267–274.

Feizi, T., Solomon, J.C., Yuen, C.-T., Jeng, K.C.G., Frigeri, L.G., Hsu, D.K. and Liu, F.-T. (1994). Adhesive specificity of the soluble human lectin, IgE-binding protein (eBP), towards lipid-linked oligosaccharides. Presence of the blood group A, B, B-like and H monosaccharides confers a binding activity to tetrasaccharide (lacto-<u>N</u>- and lacto-<u>N</u>-neo-tetraose) backbones. *Biochemistry*, **33**, 6342–6349.

Flotte, T.J., Springer, T.A. and Thorbecke, G.J. (1983). Dendritic cell and macrophage staining by monoclonal antibodies in tissue sections epidermal sheets. *Am. J. Pathol.*, **111**, 112–124.

Frigeri, L.G., Robertson, M.W. and Liu, F.-T. (1990). Expression of biologically active recombinant rat IgE-binding protein in *Escherichia coli*. *J. Biol. Chem.*, **265**, 20763–20769.

Frigeri, L.G. and Liu, F.-T. (1992). Surface expresson of functional IgE binding protein, an endogenous lectin, on mast cells and macrophages. *J. Immunol.*, **148**, 861–869.

Frigeri, L.G., Zuberi, R.I. and Liu, F.-T. (1993). eBP, a b-galactoside-binding animal lectin, recognizes IgE receptor (FceRI) and activates mast cells. *Biochemistry*, **32**, 7644–7649.

Gaudin, J.C., Monsigny, M. and Legrand, A. (1995). Cloning of the cDNA encoding rabbit galectin-3. *Gene*, **163**, 249–252.

Goletz, S., Hanisch, F.-G. and Karsten, U. (1997). Novel aGalNAc containing glycans on cytokeratins are recognized in vitro by galectins with type II carbohydrate recognition domains. *J. Cell Sci.*, **110**, 1585–1596.

Gritzmacher, C.A., Robertson, M.W. and Liu, F.-T. (1988). IgE-binding protein: subcellular location and gene expression in many murine tissues and cells. *J. Immunol.*, **141**, 2801–2806.

Gritzmacher, C.A., Mehl, V.S. and Liu, F.-T. (1992). Genomic cloning of the gene for an IgE-binding lectin reveals unusual utilization of 5'untranslated regions. *Biochemistry*, **31**, 9533–9538.

Herrmann, J., Turck, C.W., Atchinson, R.E., Huflejt, M.E., Poulter, L., Gitt, M.A., Burlingame, A.L., Barondes, S.H. and Leffler, H. (1993). Primary structure of the soluble lactose binding lectin L-29 from rat and dog and interaction of its non-collagenous proline-, glycine-, tyrosine-rich sequence with bacterial and tissue collagenase. *J. Biol. Chem.*, **268**, 26704–26711.

Hirabayashi, J. (1992). Carbohydrate-binding protein 35 forms functional dimers using a conservative cysteine residue? *Trends in Glycoscience and Glycotechnology*, **4**, 218–220.

Ho, M.-K. and Springer, T.A. (1982). Mac-2, a novel 32,000 Mr mouse macrophage subpopulation-specific antigen defined by monoclonal antibodies. *J. Immunol.*, **128**, 1221–1228.

Hsu, D.K., Zuberi, R. and Liu, F.-T. (1992). Biochemical and biophysical characterization of human recombinant IgE-binding protein, an S-type animal lectin. *J. Biol. Chem.*, **267**, 14167–14174.

Hsu, D.K., Hammes, S.R., Kuwabara, I., Greene, W.C. and Liu, F.-T. (1996). Human T lymphotropic virus-1 infection of human T lymphocytes induces expression of the b-galactose-binding lectin, galectin-3. *Am. J. Pathol.*, **148**, 1661–1670.

Huflejt, M.E., Turck, C.W., Lindstedt, R., Barondes, S.H. and Leffler, H. (1993). L-29, a soluble lactose-binding lectin, is phosphorylated on serine 6 and serine 12 *in vivo* and by casein kinase I. *J. Biol. Chem.*, **268**, 26712–26718.

Huflejt, M.E., Jordan, E.T.,Gitt; M.A., Barondes, S.H. and Leffler, H. (1997). Strinkingly different localization of galectin-3 and galectin-4 in human colon adenocarcinoma T84 cells-galectin-4 is localized at sites of cell adhesion. *J. Biol. Chem.*, **272**, 14294–14303.

Inohara, H. and Raz, A. (1994). Identification of human melanoma cellular and secreted ligands for galectin-3. *Biochem. Biophys. Res. Commun.*, **201**, 1366–1375.

Inohara, H. and Raz, A. (1995). Functional evidence that cell surface galectin-3 mediates homotypic cell adhesion. *Cancer Res.*, **55**, 3267–3271.

Inohara, H., Akahani, S., Koths, K. and Raz, A. (1996). Interactions between galectin-3 and Mac-2–binding protein mediate cell-cell adhesion. *Cancer Res.*, **56**, 4530–4534.

Jeng, K.C.G., Frigeri, L.G. and Liu, F.-T. (1994). An endogenous lectin, galectin-3 (eBP/Mac-2), potentiates IL-1 production by human monocytes. *Immunol. Lett.*, **42**, 113–116.

Jia, S. and Wang, J.L. (1988). Carbohydrate binding protein 35. Complementary DNA sequence reveals homology with proteins of the heterogeneous nuclear RNP. *J. Biol. Chem.*, **263**, 6009–6011.

Kadrofske, M.M., Openo, K.P. and Wang, J.L. (1998). The human *LGALS3* (galectin-3) gene: Determination of the gene structure and functional characterization of the promoter. *Arch. Biochem. Biophys.*, **349**, 7–20.

Kasai, K. and Hirabayashi, J. (1996). Galectins: A family of animal lectins that decipher glycocodes. *J. Biochem.* (Tokyo) **119**, 1–8.

Knibbs, R.N., Agrwal, N., Wang, J.L. and Goldstein, I.J. (1993). Carbohydrate-binding protein 35. II. Analysis of the interaction of the recombinant polypeptide with saccharides. *J. Biol. Chem.*, **268**, 14940–14947.

Konstantinov, K.N., Shames, B., Izuno, G. and Liu, F.-T. (1994). Expression of eBP, a b-galactoside-binding soluble lectin, in normal and neoplastic epidermis. *Exp. Dermatol.*, **3**, 9–16.

Koths, K., Taylor, E., Halenbeck, R., Casipit, C. and Wang, A. (1993). Cloning and characterization of a human Mac-2–binding protein, a new member of the superfamily defined by the macrophage scavenger receptor cysteine-rich domain. *J. Biol. Chem.*, **268**, 14245–14249.

Krugluger, W., Frigeri, L.G., Lucas, T., Schmer, M., Forster, O., Liu, F.T. and Boltz-Nitulescu, G. (1997). Galectin-3 inhibits granulocyte-macrophage colony-stimulating factor (GM-CSF)-driven rat bone marrow cell proliferation and GM-CSF-induced gene transcription. *Immunobiology*, **197**, 97–109.

Kuwabara, I. and Liu, F.-T. (1996). Galectin-3 promotes adhesion of human neutrophils to laminin. *J. Immunol.*, **156**, 3939–3944.

Lauc, G., Seve, A.-P., Hubert, J., Flögel-Mrsic, M., Müller, W.E.G. and Schröder, H.C. (1993). HnRNP CBP35_CBP67 interaction during stress response and ageing. *Mech. Ageing Dev.*, **70**, 227–237.

Le Marer, N. and Hughes, R.C. (1996). Effects of the carbohydrate-binding protein galectin-3 on the invasiveness of human breast carcinoma cells. *J. Cell. Physiol.*, **168**, 51–58.

Lee, E.C., Woo, H.-J., Korzelius, C.A., Steele, G.D. and Mercurio, A.M. (1991). Carbohydrate-binding protein 35 is the major cell-surface laminin-binding protein in colon carcinoma. *Arch. Surg.*, **126**, 1498.

Leffler, H. and Barondes, S.H. (1986). Specificity of binding of three soluble rat lung lectins to substituted and unsubstituted mammalian b-galactosides. *J. Biol. Chem.*, **261**, 10119–10126.

Lindstedt, R., Apodaca, G., Barondes, S.H., Mostov, K.E. and Leffler, H. (1993). Apical secretion of a cytosolic protein by Madin-Darby canine kidney cells. Evidence for polarized release of an endogenous lectin by a nonclassical secretory pathway. *J. Biol. Chem.*, **268**, 11750–11757.

Liu, F.-T., Albrandt, K., Mendel, E., Kulczycki Jr, A. and Orida, N.K. (1985). Identification of an IgE-binding protein by molecular cloning. *Proc. Natl. Acad. Sci. USA*, **82**, 4100–4104.

Liu, F.-T. (1990). Molecular biology of IgE-binding protein, IgE-binding factors and IgE receptors. *CRC Crit. Rev. Immunol.*, **10**, 289–306.

Liu, F.-T., Hsu, D.K., Zuberi, R.I., Kuwabara, I., Chi, E.Y. and Henderson, W.R.,Jr. (1995). Expression and function of galectin-3, a b-galactoside-binding lectin, in human monocytes and macrophages. *Am. J. Pathol.*, **147**, 1016–1029.

Liu, F.-T., Hsu, D.K., Zuberi, R.I., Shenhav, A., Hill, P.N., Kuwabara, I. and Chen, S.-S. (1996). Modulation of functional properties of galectin-3 by monoclonal antibodies binding to the non-lectin domain. *Biochemistry* **35**, 6073–6079.

Massa, S.M., Cooper, D.N.W., Leffler, H. and Barondes, S.H. (1993). L-29, an endogenous lectin, binds to glycoconjugate ligands with positive cooperativity. *Biochemistry*, **32**, 260–267.

Mehul, B., Bawumia, S., Martin, S.R. and Hughes, R.C. (1994). Structure of baby hamster kidney carbohydrate-binding protein CBP30, an S-type animal lectin. *J. Biol. Chem.*, **269**, 18250–18258.

Mehul, B., Bawumia, S. and Hughes, R.C. (1995). Cross-linking of galectin 3, a galactose-binding protein of mammalian cells, by tissue-type transglutaminase. *FEBS Lett.*, **360**, 160–164.

Mehul, B. and Hughes, R.C. (1997). Plasma membrane targetting, vesicular budding and release of galectin 3 from the cytoplasm of mammalian cells during secretion. *J. Cell Sci.*, **110**, 1169–1178.

Mey, A., Leffler, H., Hmama, Z., Normier, G. and Revillard, J.P. (1996). The animal lectin galectin-3 interacts with bacterial lipopolysaccharides via two independent sites. *J. Immunol.*, **156**, 1572–1577.

Monteseirin, J., Camacho, M.J., Montano, R., Llamas, E., Conde, M., Carballo, M., Guardia, P., Conde, J. and Sobrino, F. (1996). Enhancement of antigen-specific functional responses by neutrophils form allergic patients. *J. Exp. Med.*, **183**, 2571–2579.

Moutsatsos, I.K., Davis, J.M. and Wang, J.L. (1986). Endogenous lectins from cultured cells: Subcellular localization of carbohydrate-binding protein 35 in 3T3 fibroblasts. *J. Cell Biol.*, **102**, 477–483.

Moutsatsos, I.K., Wade, M., Schindler, M. and Wang, J.L. (1987). Endogenous lectins from cultured cells: Nuclear localization of carbohydrate-binding protein 35 in proliferating 3T3 fibroblasts. *Proc. Natl. Acad. Sci. USA*, **84**, 6452–6456.

Nangia-Makker, P., Ochieng, J., Christman, J.K. and Raz, A. (1993). Regulation of the expression of galactoside-binding lectin during human monocytic differentiation. *Cancer Res.*, **53**, 1–5.

Ochieng, J., Gerold, M. and Raz, A. (1992). Dichotomy in the laminin-binding properties of soluble and membrane-bound human galactoside-binding protein. *Biochem. Biophys. Res. Commun.*, **186**, 1674–1680.

Ochieng, J., Platt, D., Tait, L., Hogan, V., Raz, T., Carmi, P. and Raz, A. (1993). Structure-function relationship of a recombinant human galactoside- binding protein. *Biochemistry*, **32**, 4455–4460.

Ochieng, J., Fridman, R., Nangia-Makker, P., Kleiner, D.E., Liotta, L.A., Stetler-Stevenson, W.G. and Raz, A. (1994). Galectin-3 is a novel substrate for human matrix metalloproteinases- 2 and -9. *Biochemistry*, **33**, 14109–14114.

Oda, Y., Leffler, H., Sakakura, Y., Kasai, K.-I. and Barondes, S.H. (1991). Human breast carcinoma cDNA encoding a galactoside-binding lectin homologous to mouse Mac-2 antigen. *Gene*, **99**, 279–283.

Ohannesian, D.W., Lotan, D., Thomas, P., Jessup, J.M., Fukuda, M., Gabius, H.-J. and Lotan, R. (1995). Carcinoembryonic antigen and other glycoconjugates act as ligands for galectin-3 in human colon carcinoma cells. *Cancer Res.*, **55**, 2191–2199.

Pesheva, P., Urschel, S., Frei, K. and Probstmeier, R. (1998). Murine microglial cells express functionally active galectin-3 in vitro. *J. Neurosci. Res.*, **51**, 49–57.

Probstmeier, R., Montag, D. and Schachner, M. (1995). Galectin-3, a b-galactoside-binding animal lectin, binds to neural recognition molecules. *J. Neurochem.*, **64**, 2465–2472.

Raz, A., Pazerini, G. and Carmi, P. (1989). Identification of the metastasis-associated, galactoside-binding lectin as a chimeric gene product with homology to an IgE-binding protein. *Cancer Res.*, **49**, 3489–3493.

Raz, A., Carmi, P., Raz, T., Hogan, V., Mohamed, A. and Wolman, S.R. (1991). Molecular cloning and chromosomal mapping of a human galactoside-binding protein. *Cancer Res.*, **51**, 2173–2178.

Reichert, F., Saada, A. and Rotshenker, S. (1994). Peripheral nerve injury induces Schwann cells to express two macrophage phenotypes: Phagocytosis and the galactose- specific lectin MAC-2. J. *Neurosci.*, **14**, 3231–3245.

Robertson, M.W., Albrandt, K., Keller, D. and Liu, F.-T. (1990). Human IgE-binding protein: A soluble lectin exhibiting a highly conserved interspecies sequence and differential recognition of IgE glycoforms. *Biochemistry*, **29**, 8093–8100.

Robertson, M.W. and Liu, F.-T. (1991). Heterogeneous IgE glycoforms characterized by differential recognition of the IgE-binding protein lectin. *J. Immunol.*, **147**, 3024–3030.

Rosenberg, I., Cherayil, B.J., Isselbacher, K.J. and Pillai, S. (1991). Mac-2–binding glycoproteins. Putative ligands for a cytosolic b-galactoside lectin. J. Biol. Chem. *26*, 18731–18736.

Rosenberg, I.M., Iyer, R., Cherayil, B., Chiodino, C. and Pillai, S. (1993). Structure of the murine *Mac-2* gene. Splice variants encode proteins lacking functional signal peptides. *J. Biol. Chem.*, **268**, 12393–12400.

Sato, S. and Hughes, R.C. (1992). Binding specificity of a baby hamster kidney lectin for H type I and II chains, polylactosamine glycans and appropriately glycosylated forms of laminin and fibronectin. *J. Biol. Chem.*, **267**, 6983–6990.

Sato, S., Burdett, I. and Hughes, R.C. (1993). Secretion of the baby hamster kidney 30–kDa galactose-binding lectin from polarized and nonpolarized cells: A pathway independent of the endoplasmic reticulum-Golgi complex. *Exp. Cell Res.*, **207**, 8–18.

Sato, S. and Hughes, R.C. (1994a). Regulation of secretion and surface expression of Mac-2, a galactoside-binding protein of macrophages. *J. Biol. Chem.*, **269**, 4424–4430.

Sato, S. and Hughes, R.C. (1994b). Control of Mac-2 surface expression on murine macrophage cell lines. *Eur. J. Immunol.*, **24**, 216–221.

Smetana, K.Jr., Lukás, J., Palecková, V., Bartunková, J., Liu, F.T., Vacík, J. and Gabius, H.J. (1997). Effect of chemical structure of hydrogels on the adhesion and phenotypic characteristics of human monocytes such as expression of galectins and other carbohydrate-binding sites. *Biomaterials*, **18**, 1009–1014.

Sparrow, C.P., Leffler, H. and Barondes, S.H. (1987). Multiple soluble b-galactoside-binding lectins from human lung. *J. Biol. Chem.*, **262**, 7383–7390.

Sève, A.-P., Felin, M., Doyennette-Moyne, M.-A., Sahraoui, T., Aubery, M. and Hubert, J. (1993). Evidence for a lactose-mediated association between two nuclear carbohydrate-binding proteins. *Glycobiology*, **3**, 23–30.

Truong, M.-J., Gruart, V., Liu, F.-T., Prin, L., Capron, A. and Capron, M. (1993a). IgE-binding molecules (Mac-2/eBP) expressed by human eosinophils. Implication in IgE-dependent eosinophil cytotoxicity. *Eur. J. Immunol.*, **23**, 3230–3235.

Truong, M.-T., Gruart, V., Kusnierz, J.-P., Papin, J.-P., Loiseau, S., Capron, A. and Capron, M. (1993b). Human neutrophils express immunoglobulin E (IgE)-binding proteins (Mac-2/eBP) of the S-type lectin family: Role in IgE-dependent activation. *J. Exp. Med.*, **177**, 243–248.

van den Brule, F.A., Bellahcene, A., Jackers, F., Liu, F.-T., Sobel, M.E. and Castronovo, V. (1997). Antisense galectin-3 alters thymidine incorporation in human MDA-MB435 breast cancer cells. Int. *J. Oncology*, **11**, 261–264.

van den Brûle, F.A., Buicu, C., Sobel, M.E., Liu, F.-T. and Castronovo, V. (1995). Galectin-3, a laminin binding protein, fails to modulate adhesion of human melanoma cells to laminin. *Neoplasma*, **42**, 215–219.

Wollenberg, A., De la Salle, H., Hanau, D., Liu, F-T. and Bieber, T. (1993). Human keratinocytes release the endogenous b-galactoside-binding soluble lectin eBP (IgE-binding protein) which binds to Langerhans cells where it modulates their binding capacity for IgE glycoform. *J. Exp. Med.*, **178**, 777–785.

Woo, H.-J., Lotz, M.M., Jung, J.U. and Mercurio, A.M. (1991). Carbohydrate-binding protein 35 (Mac-2), a laminin-binding lectin, forms functional dimers using cysteine 186. *J. Biol. Chem.*, **266**, 18419–18422.

Woo, H.J., Shaw, L.M., Messier, J.M. and Mercurio, A.M. (1990). The major non-integrin laminin binding protein of macrophages is identical to carbohydrate binding protein 35 (Mac-2). *J. Biol. Chem.*, **265**, 7097–7099.

Yamaoka, A., Kuwabara, I., Frigeri, L.G. and Liu, F.-T. (1995). A human lectin, galectin-3 (eBP/Mac-2), stimulates superoxide production by neutrophils. *J. Immunol.*, **154**, 3479–3487.

Yang, R.-Y., Hsu, D.K. and Liu, F.-T. (1996). Expression of galectin-3 modulates T cell growth and apoptosis. *Proc. Natl. Acad. Sci. USA*, **93**, 6737–42.

Zuberi, R.I., Frigeri, L.G. and Liu, F.-T. (1994). Activation of rat basophilic leukemia cells by eBP, an IgE-binding endogenous lectin. *Cell. Immunol.*, **156**, 1–12.

5. THE ROLE OF GALECTIN-3 IN TUMOR METASTASIS[*]

PRATIMA NANGIA-MAKKER[1], SHIRO AKAHANI[2], ROBERT BRESALIER[3] and
AVRAHAM RAZ[1]

[1]*Metastasis Research Program, Karmanos Cancer Institute, Department of Pathology and Radiation Oncology, Wayne State University, School of Medicine, 110 E. Warren Avenue, Detroit, MI 48201, USA*

[2]*Present Address: Department of Otolaryngology, Osaka Teishin Hospital, 2-6-40 Karasugatsuji, Tennouji-ku, Osaka City, Osaka, 543 JAPAN,*

[3]*Cancer Research Laboratory, Henry Ford Health Sciences Center, Detroit, MI 48202, USA*

INTRODUCTION

Background

Lectins, (formerly found as hemagglutinins), are carbohydrate-binding proteins, which recognise specific carbohydrate structures on mammalian cells. They are classified into four small families according to their particular recognitions of glycoconjugates: C-type lectins (calcium-dependent); P-type lectins; pentraxins; and S-type lectins which are also known as galectins (Drickamer and Taylor, 1993).

Galectins are soluble proteins of 14-36 kDa molecular weight which exhibit a specific affinity for β-galactoside derivatives in a calcium-independent manner. They are distributed in a wide range of animals from invertebrates such as sponges to higher vertebrates, and at least 10 mammalian galectins have been identified (Barondes *et al.*, 1994).

The Structure of Galectin-3

Galectin-3, originally identified as the macrophage differentiation marker Mac-2 (Ho and Springer, 1982), is a member of this growing family of proteins with approximate molecular weights ranging between 26 kDa and 31 kDa and is also known as CBP-35, IgE-binding protein, L-29, L-31, L-34 and other names (Barondes *et al.*, 1994). As shown in Figure 1, it consists of three domains: an N-terminal domain; repeating elements; and a carbohydrate recognition domain (CRD) (Herrmann *et al.*, 1993). The N-terminal domain has less than 20 amino acid residues and the repeating elements are composed of proline, glycine-rich tandem repeats (9 repeats in human) which are susceptible to collagenases. The CRD, cleavable with trypsin, contains a carbohydrate-binding site and consists of 140

[*]This work was supported in part by the NIH grants R01-CA46120 and R01-CA69480 and the Paul Zuckerman Support Foundation for Cancer Research and Henry Ford Health Sciences Center Research Foundation.

Galectin's Sequence

N-Terminal Domain

[Carbohydrate Recognition Domain]

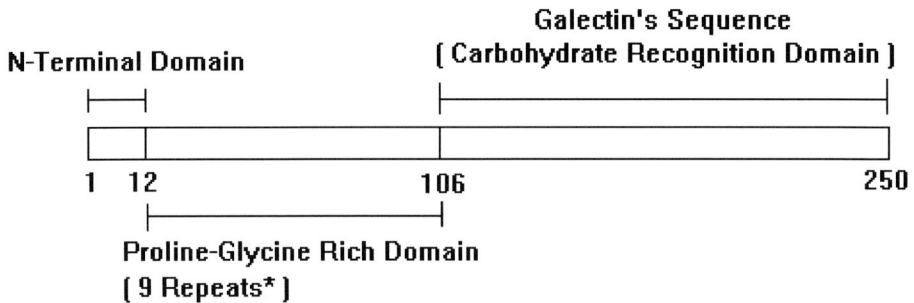

1   12   106   250

Proline-Glycine Rich Domain
[9 Repeats*]

* The number of repeats varies in species.

Figure 1   Scheme of Galectin-3 structure.

amino acid residues. Galectin-3, isolated as a monomer, can form multivalent aggregates in the presence of glycoconjugate ligands which leads to cell aggregation (Hsu *et al.*, 1992; Inohara and Raz, 1995). The N-terminal domain is thought to be responsible for this oligomerization property. In addition, since galectin-3 can agglutinate erythrocytes like many other lectins, it must behave at least bivalently (Frigeri *et al.*, 1990).   Moreover, the acetylated N-terminal domain can be phosphorylated on Ser_6 and Ser_{12} by casein kinase 1 (Huflejt *et al.*, 1993). The CRD shares significant sequence similarity with those of other galectins, and hydrophobic and hydrophilic regions in the CRD lead to globular formation of this protein (Barondes *et al.*, 1994).

Cellular Localization of Galectin-3

Galectin-3 has been found in various murine tissues including embryonic bone, cartilage, liver, lung, muscle, spleen, and thymus, and the artery, lung, spleen, and thymus of the adult mice (Crittenden *et al.*, 1984). In humans, galectin-3 has been detected in capillary endothelium, colon, inflammatory cells, ovary, stomach, and other tissues (Lotan *et al.*, 1994; Lotz *et al.*, 1993; Cherayil *et al.*, 1990; Truong *et al.*, 1993). In some tissues galectin-3 expression may be regulated during development or differentiation: 1) Galectin-3 is abundant in the embryonic mouse liver but is not detectable in the liver of the adult mouse (Crittenden *et al.*, 1984); 2) In lesioned peripheral nerves, granulocytes or some cytokines induce the up-regulation of galectin-3 on the cell surface (Saada *et al.*, 1996); 3) Galectin-3 is observed predominantly in the nucleus of human normal colon epithelial cells but its pattern of expression is altered in colon cancer cells (Lotz *et al.*, 1993). Galectin-3 has been reported to be localized in the intracellular compartments like the cytoplasm or the nucleus, and on the cell surface and the extracellular matrix (Wang *et al.*, 1992). It was once believed that galectin-3 could not function in extracellular milieu because reducing reagents were thought to be essential to the carbohydrate- binding of protein. In addition, it was assumed that it was confined to the cytoplasm because of the lack of transmembrane sequences and classical signal peptides for secretion (Kasai and Hirayabashi, 1996). It has been shown, however, that reducing reagents including β-mercaptoethanol or dithiothreitol are not required for the binding to complimentary glycoconjugates and that galectin-3 may be secreted from the cytoplasm by atypical pathways: translocation across the plasma membrane or by binding to some carriers containing transmembrane sequences or signal peptides for secretion (Frigeri *et al.*, 1990; Lindstedt *et al.*, 1993). Mehul and Hughes (1997) have recently reported that the amino-terminal half of galectin-3 is sufficient to direct export of a chimeric CAT protein indicating that part of the signal for plasma membrane translocation lies in the N-terminal domain of the lectin.

GALECTIN-3 IN TUMOR METASTASIS

We have previously described the presence of galectin-3 in various human and murine tumor cells and proposed that interaction of cell surface galectin-3 with a complimentary serum glycoprotein(s) promotes homotypic aggregation of tumor cells in the circulation, thereby playing an important role in pathogenesis of metastasis. In support of this hypothesis we reported a series of experimental observations: 1) Highly metastatic cells display more galectin-3 molecules on the cell surface than their weakly metastatic counterparts (Raz *et al.*,1986); 2) Exogenous asialofetuin, a glycoprotein rich in terminal β-galactosidase residues induces homotypic aggregation, which is impaired by anti-galectin-3 antibody (Meromsky *et al.*,1986); 3) Anti-galectin-3 antibody inhibits experimental lung metastasis of i.v. injected murine melanoma and fibrosarcoma cells (Meromsky *et al.*,1986); 4) Overexpression of recombinant galectin-3 in weakly metastatic cells resulted in increased metastatic potential (Raz *et al.*, 1990); 5) Co-suspension of recombinant virus infected Sf9 insect cells with uninfected cells in the presence of asialofetuin resulted in a preferential cell-cell adhesion of the galectin-3 expressing cells (Inohara and Raz, 1995). These results strongly suggest a functional role for galectin-3 in cell adhesion and tumor cell metastasis.

The emphasis of this chapter is on the role of galectin-3 in tumor progression and metastasis. Several studies have been performed by our laboratory and various other groups which indicate a role for galectin-3 in tumor progression.

Colon Carcinomas

Irimura *et al.* (1991), studied the expression of galectin-3 in extracts prepared from colon carcinomas classified according to the Astler-Collar-modified Duke's classification, and demonstated that expression of galectin-3 increased with advaning tumor stage.They also showed a correlation of galectin-3 expression with the serum levels of carcinoma embryonic antigen (CEA). Immunohistochemical studies identified galectin-3 in the cytoplasm, but the cell surface expression could not be demonstrated by light microscopy. Lee *et al.* (1991) also reported that poorly differentiated colon carcinoma cell lines marked by increased invasiveness expressed more cell surface galectin-3 than the well differentiated cell lines. Work by Castronovo *et al.*,(1992), however, suggested a decreased expression of galectin-3 mRNA in tumors compared to the normal mucosa. Lotz *et al.* (1993), also observed a 5–10 fold decrease in galectin-3 mRNA levels in cancer compared to normal mucosa. They also reported that neoplastic progression from normal mucosa to adenoma to carcinoma is associated with a loss of galectin-3 nuclear localization.

It was necessary to resolve this conflicting data by using a very large number of samples from the known stages of colon carcinoma. Therefore, Schoppner *et al.* (1995) examined the expression of galectin-3 in 153 tissue samples including 29 adenomas containing early cancer, 66 colon carcinomas of known Duke's stage with available long term survival data, and 23 additional primary carcinomas with 35 associated metastasis. Detection of galectin-3 was performed using monoclonal antibody TIB-166, raised against N-terminal domain of the galectin-3. This antibody is specific for galectin-3 and does not recognize any other galectins with carbohydrate-terminal homology. This study showed that normal mucosa distant from areas of neoplasia was often stained only weekly for galectin-3, and staining was limited to areas near the surface of the crypt. Transitional mucosa adjacent to carcinomas became progressively more positive, with a predominancc of strong nuclear staining throughout the crypt.In most cases the transition from adenoma to moderate and high grade dysplasia could be distinguished by galectin-3 staining. Their results also indicate that a significant increase in galectin-3 staining across the five Duke's stages. Patient survival was best for patients with a score of 1 and was progressively worse for patients with a score of 2 and 3 (Schoppner *et al.*, 1995).

Additional evidence suggesting a role for galectin-3 in colon cancer progression comes from the analysis of a separate group of primary adenocarcinomas and paired metastasis derived from these tumors. Ratio estimation analysis, demonstrated a significantly higher degree of galectin-3 expression in metastases compared with primary tumors from which they were derived (Schoppner *et al.*, 1995). Analysis of galectin-3 in this large group of specimens provides strong evidence that expression of this endogenous lectin is related to neoplastic transformation and progression towards metastasis in the colon.

Breast Carcinomas

We examined the expression of galectin-3 in relation to the malignant phenotypes of five established and well characterized human breast carcinoma cell lines. MDA-MB-231 and MDA-MB-435 are metastatic cell lines, while T47D is poorly metastatic and BT-549 and SK-Br-3 are non-tumorigenic in nude mice. We found an increased expression of galectin-3 RNA, total protein and cell surface protein in the metastatic cell lines compared to the non-tumorigenic cell lines. Anchorage-independent growth of galectin-3 expressing cell lines was significantly higher than that of the galectin-3 null cell lines (Nangia-Makker *et al.*, 1995).

In order to more directly establish the role of galectin-3 in tumorigenicity of breast carcinoma cells, galectin-3 cDNA was introduced into the non-tumorigenic cell line BT-549. The galectin-3 expressing clones were injected in the mammary fat pads of nude mice. Table 1 shows the tumorigenic potential and anchorage independent growth efficiency according to galectin-3 expression of various clones. Clone 11-9-1-4 was further analyzed for its metastatic potential. When 1×10^5 cells were injected in the mammary fat pads of nude mice, the tumors micro metastasized to lungs in 4-5 weeks. Digestion of lungs and their subsequent cultures in selection medium showed the presence of lung colonies (Figure 2). Tail vein injections with

Table 1 Phenotypic Charaterization of Galectin-3 Transfected Cell Clones

Cell Clone	Gal-3 Expression	Soft Agar Growth % Efficiency	Tumorigenicity
BT-549	—	3.5 ± 0.25	0/15
11-9-1-4	+++	17.3 ± 1.2	14/14
11-8-1-1	+++	17.8 ± 7.6	10/18
11-YX-1	++++	18.3 ± 4.9	0/16
11-9-1-3	+	6.6 ± 2.0	1/10
4-1-4-2-1	—	3.3 ± 0.47	0/11

3×10^6 cells were injected into the mammary fat pad region of 6-8 weeks old nude mice. Animals were followed up for up to 120 days after injection.

1×10^6 cells resulted in macro metastasis in the lungs in 12 weeks (Figure 3). However, failure of one of the galectin-3 expressing cell lines to grow in nude mice prompted us to investigate the presence of other protein(s), whose co-expression might induce tumorigenesis in breast cancer cells.

Differential display analysis of the total RNA from non-tumorigenic and metastatic BT-549 clones showed the presence of L1 retrotransposon in the tumorigenic clones and cell lines. To establish the significance of their co-expression

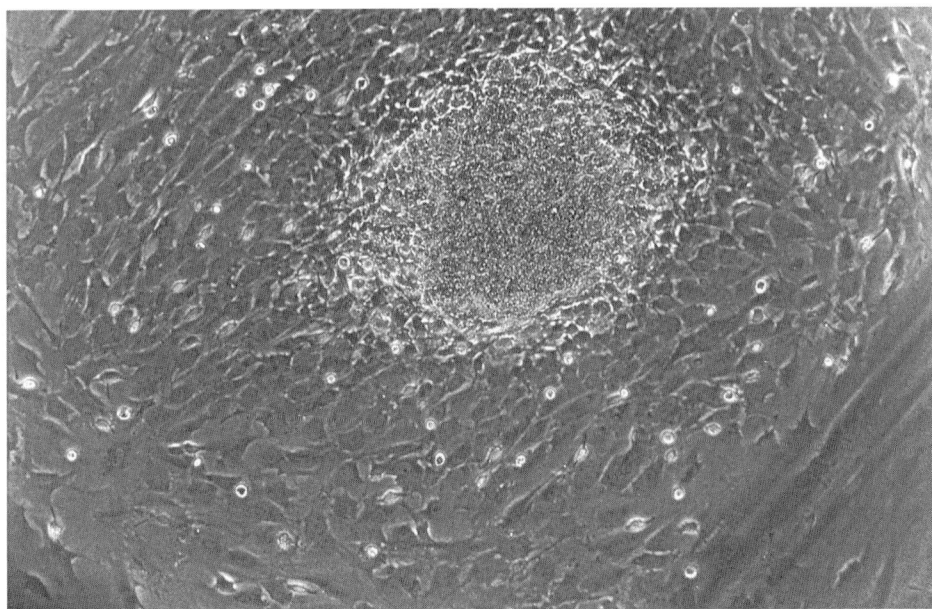

Figure 2 In vitro growth of colonies isolated from the lungs of nude mice: Galectin-3 transfected breast carcinoma cells BT-549 were injected in the mammary fat pad regions of nude mice. After 5 weeks, the mice were sacrificed and the lungs were digested with 5 mg collagenase type IV and 36 units elastase/ ml of DMEM. Tumor cells from the digested lungs were grown in selection medium used for transfection.

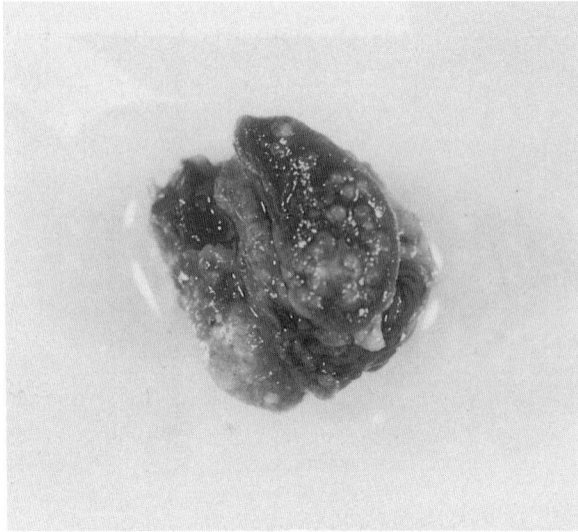

Figure 3 Lung metastasis in nude mice: Galectin-3 transfected breast carcinoma cells BT-549 were injected in the tail vein of nude mice. After 12 weeks the mice were sacrificed and lungs were fixed with Buin's fixative.

immunohistochemical studies of 9 benign and 34 malignant cell lines were performed and the following observations were made: 1) There was a weak, focal immunoreactivity for galectin-3 and L1 in histologically unremarkable benign breast tissue; 2) Fibrocystic change with hyperplasia appeared to be associated with increased immunoreactivity for L1 and galectin-3; 3) Breast carcinomas demonstrated significant expression of both proteins in a majority of tumors within the tumor cell population as well as within the host stromal cell populations (Nangia-Makker *et al.*, 1998). The colocalization and accentuation of galectin-3 and L1 immunoreactivity at the invasive front suggests a functional role in stromal cell remodeling and/or tumor cell invasion, and that there might be an association between immunoreactivity for galectin-3 and L1 and aggressive behavior in breast carcinoma.

Marer and Hughes (1996) have also shown poor expression of galectin-3 in normal breast epithelial (MTSV1-7) cells and cell lines derived from benign (242A) and infiltrating (341 and 531E) primary breast carcinomas. Cloned cell lines derived from malignant effusions in patients with metastatic breast cancers T47D, ZR75-1 and CAMA-1, however, overexpressed galectin-3.

These authors also found a stimulatory effect of galectin-3 on the invasiveness of the low galectin-3 expressing breast cancer cell line (341). They hypothesize that a localized threshold concentration of lectin at the invasive sites is more directly correlated with invasiveness of the cell line. Increased invasiveness of galectin-3 expressing breast cancer cells was also reported by Warfield *et al.*, (1997).

Thyroid Carcinomas

Histologically thyroid carcinoma is classified into four principal subtypes: papillary, follicular, medullary, and undifferentiated. Papillary and follicular carcinomas are derived from the follicular epithelium, produce thyroglobulin, and are considered to be low grade malignancies. Medullary carcinoma is derived from the parafollicular C cells, which are of neural crest origin and produce calcitonin. Undifferetiated, or anaplastic carcinomas also originate from follicular epithelium; they are highly aggressive tumors and typically present in an advanced stage. Xu *et al.*, (1995) examined the expression of galectin-1 and galectin-3 in a spectrum of surgically excised thyroid lesions to determine whether these galectins might be useful immunohistochemical markers for the differential diagnosis of benign and malignant thyroid cancers. They found that all thyroid malignancies of epithelial origin (i.e. papillary and follicular carcinomas) and a metastatic lymph node from papillary carcinoma expressed high levels of both galectin-1 and galectin-3. The medullary thyroid carcinomas, which are of parafollicular C cell origin, showed a weaker and variable expression of these proteins. In contrast neither benign thyroid tissue nor adjacent normal thyroid tissue expressed galectin-1 or galectin-3. Both the galectins were detected in tumor cells at the luminal cell surface and at cell-cell contacts. Therefore they may serve as markers for thyroid malignancy.

Similar results were reported by Fernandez *et al.* (1996), who showed that expression of galectin-3 is limited to inflammatory foci in normal and benign thyroid tissue and is a phenotypic feature of malignant thyroid neoplasms, especially papillary carcinomas. They also demonstrated an overexpression of galectin-3 mRNA in papillary carcinomas, no expression was repoted in matched normal tissue, hyperplastic nodule and follicular adenoma.

Head and Neck Squamous Cell Carcinoma

Recently, Gillenwater *et al.* (1996) evaluated patient tumor specimens and 14 cell lines from aerodigestive tract squamous cell carcinomas (SCCs) for galectin-1 and galectin-3 expression to determine whether galectins play a role in progression of HNSCC or could be used as biological markers of malignant potential. They found that cell lines expressed both the galectins on the cell surface, and galectin-3 was present in the cytoplasm as well. In the tumor specimens, galectin-1 was detected in the basal layer of normal adjacent mucosa, in connective tissue stroma, and at the periphery of invasive tumor islands. Galectin-3 localized to superficial mucosal layers, and adjacent to keratin pearls in invasive carcinoma. Honjo *et al.* (unpublished results) examined galectin-3 expression in a spectrum of surgically excised tongue lesions including squamous cell carcinoma, dysplasia, and adjacent normal mucosa. They found a progressive increase in galectin-3 expression from normal epithelium, through dysplasia, to cancer. A subgroup of patients with tumors demonstrating high galectin-3 staining score tended to have a poorer survival compared to those whose tumors exhibited a low galectin-3 staining score.

Other Carcinomas

Lotan *et al.* (1994) analyzed the specific subtypes of gastric cancer and found that there was no overall increase in galectin-3 levels during gastric cancer progression. However, levels of galectin-3 were higher in lymph node metastases than in the primary, poorly differentiated gastric carcinomas. They were also higher in liver metastases than in primary, well differentiated tubular carcinomas. These results were based on analysis of 39 patients with different types of primary gastric carcinomas and adjacent normal mucosa, as well as specimens from primary gastric carcinomas and their metastasis to lymph nodes, liver, lungs, kidney, or ovaries from 74 cases.

Bresalier *et al.* (1997) have recently demonstrated that expression of galectin-3 correlates with the malignant potential of tumors in the central nervous system. They analyzed 42 primary brain tumors and 29 metastasis and found that all glioblastomas (grade 4 astrocytomas) stained strongly for galectin-3, whereas low grade astrocytomas (grade 2) did not express the endogenous galectin-3. Anaplastic astrocytomas (grade 3) exhibited intermediate expression. Normal brain tissue and benign tumors did not express galectin-3, whereas metastasis to the brain were all positive for galectin-3 expression. Metastases expressed significantly more galectin-3 than the primary tumors from which they were derived.

FUNCTIONAL RELEVANCE OF GALECTIN-3 TO METASTASIS

The above reports strongly suggest that galectin-3 plays a significant role in the metastasis of a variety of tumors. The role of galectin-3 in individual events of the metastatic cascade, however, remains to be determined. We have recently provided evidence indicating the importance of galectin-3 in homotypic cell aggregation which may lead to tumor emboli formation. There is also evidence (unpublished) that galectin-3 may be involved in heterotypic cell aggregation. Galectin-3 expressing breast cancer cells showed a higher binding to human endothelial cells than the cells not expressing this lectin. These results were further confirmed by performing the binding assay with biotinylated rgalectin-3. An increased colony-forming efficiency by the galectin-3 expressing breast cancer cell lines compared with the non-expressing cell lines suggests its involvement in anchorage independent growth (Nangia-Makker *et al.*, 1995). Marer and Hughes (1996) have recently shown that addition of exogenous galectin-3 may stimulate the invasiveness of some breast cancer cell lines. Warfield *et al.* (1997) have also reported that endogenous galectin-3 is involved with the invasiveness of breast cancer cells.

Galectins, are a family of regulatory molecules which perform a variety of biological functions depending on their location and their specific receptors. Galectin-3 is present intracellularly as well as extracellularly. Only a few intracellular ligands like LAMP-1 and LAMP-2 (Inohara and Raz, 1994) and cytokeratins (Goletz *et al.*, 1997) have been identified since all glycoconjugates in the intracellular space lack β-galactoside residues except for proteins with O-linked N-acetylglucosamine. Non-classical secretion described above or its translocation into the nucleus or into the vesicles may indicate its possible interaction through intracellular ligands. It has been shown, for example, that galectin-3 may bind to RNA in the nuclear matrix (Dagher *et al.*, 1995; Wang *et al.*, 1995) and that through CRD galectin-3 may interact with the bcl-2 protein (Yang *et al.*, 1996) which is localized to mitochondria, endoplasmic reticulum and nuclear membranes. The former finding may imply its involvement in RNA splicing or transcription and the latter may imply a role in apoptosis. The possible localization and function of galectin-3 in the intracellular space remains to be examined.

Extracellular galectin-3 appears to be responsible for cell-cell interactions since it acts by cross-linking glycoconjugates containing β-galactoside containing sugars.

β-galactoside residues are commonly observed in almost all extracellular glycoproteins, but some structural modifications of saccharides alter the affinity for galectin-3. The disaccharide lactose has 100-fold stronger galectin-3-binding activity than the monosaccharide galactose, and the affinity of galectin-3 for GalNAcα1-3(Fucα1-2)Galβ1-4Glc is approximately 2500-fold higher than for β-galactoside (Crittenden *et al.*, 1984; Lefler and Barondes, 1986). These differences in affinity allow galectin-3 to bind to only a subset of oligosaccharides in target cells. We have previously shown that the metastatic potential of mouse melanoma and fibrosarcoma cells correlates with the expression of galectin-3 on the cell surface (Raz and Lotan, 1987). Activated macrophages, on the other hand, have been shown to secrete galectin-3 (Cherayil *et al.*, 1989). Extracellular ligands for galectin-3 include carcinoembryonic antigen (Ohannesian *et al.*, 1995), IgE (Truong *et al.*, 1993), laminin (Woo *et al.*, 1990; Ochieng *et al.*, 1992; van den Brule *et al.*, 1995; Sato and Hughes, 1992), Mac-2-binding protein (Koths *et al.*, 1993; Inohara *et al.*, 1996), and mucin (Bresalier *et al.*, 1996). This suggests the possible extracellular functions of galectin-3 in cell growth, differentiation, adhesion, immuno-reaction, signal transduction, and metastasis through interaction with a number of ligands. Laminin, for example, is a major component of the extracellular matrix, which may regulate cell growth, differentiation, migration and adhesion. Laminin has poly-*N*-acetyllactosamine carbohydrate chains which may have affinity with galectin-3. Laminin binding is not altered, however, by the exogenously-added galectin-3 nor the anti-galectin-3 antibody (Woo *et al.*, 1990; Ochieng *et al.*, 1992). Warfield *et al.* (1997) have shown that galectin-3 is essential for adhesion to laminin and collagen type 1V by breast cancer cells.

A large body of evidence, therefore, suggests a role for galectin-3 in biological processes asociated with tumor transformation and metastasis. The precise function of galectin-3 in these processes remains, however, to be determined.

REFERENCES

Barondes, S.H., Castronovo, V., Cooper, D.N.W., Cummings, R.D., Drickmer, K., Feizi, T., Gitt, M.A., Hirabayashi, J., Hughes, C., Kasai, K., Leffler, H., Liu, F.T., Lotan, R., Mercurio, A.M., Monsigny, M., Pillai, S., Poirier, F., Raz, A., Rigby, P.W.J., Rini, J.M. and Wang, J.L. (1994) Galectins : a family of animal b-galactoside-binding lectins. *Cell*, **76**, 597–598.

Bresalier, R.S., Byrd, J.C., Wang, L. and Raz, A. (1996) Colon cancer mucin : a new ligand for beta-galactoside-binding protein galectin-3. *Cancer Res.*, **56**, 4354–4357.

Bresalier, R.S., Yan, P.-S., Byrd, J.C., Lotan, R. and Raz, A. (1997) Expression of the endogenous galactoside-binding protein galectin-3 correlates with the malignant potential of tumors in the central nervous system. *Cancer*, **80**, 776–787.

Castronovo, V., Campo, E., van den Brule, F.A., ClaySmith, A.P., Cioce, V., Liu F-T., Fernandez, P.L. and Sobel, M.E. (1992) Inverse modulation of steady state messenger RNA levels of two non-integrin laminin-binding proteins in human colon carcinoma. *J. Natl.Cancer Inst.* **84**, 1161–1169.

Cherayil, B.J., Chaitovitz, S., Wong, C. and Pillai, S. (1990) Molecular cloning of a human macrophage lectin specific for galactose. *Proc. Natl. Acad. Sci. USA* **87**, 7324–7329.

Cherayil, B.J., Weiner, S.J. and Pillai, S. (1989) The Mac-2 antigen is a galactoside-specific lectin that binds IgE. *J.Exp.Med.*, **170**, 1959–1972.

Crittenden, S.L., Roff, C.F., Wang, J.L. (1984) Carbohydrate-binding protein 35: identification of the galactose-specific lectin in various tissues of mice. *Mol. Cell. Biol.*, **4**, 1252–1259.

Dagher, S.F., Wang, J.L. and Patterson, R.J. (1995) Identification of galectin-3 as a factor in pre-mRNA splicing. *Proc. Natl. Acad. Sci.USA* **92**, 1213–1217.

Drickamer, K. and Taylor, M.E. (1993) Biology of animal lectins. *Annu. Rev. Cell. Biol.*, **9**, 237–264.

Fernandez, P.L., Merino, M.J., Gomez, M., Campo, E., Medina, T., Castrononvo, V., Sanjuan, X., Cardesa, A., Liu, F.-T. and Sobel, M. (1997) Galectin-3 and laminin expression in neopastic and non-neoplastic thyroid tissue. *J. Pathol.*, **181**, 80–86.

Frigeri, L.G., Robertson, M.W. and Liu, F.T. (1990) Expression of biologically active recombinant rat IgE-binding protein in Escherichia coli. *J. Biol. Chem.*, **265**, 20763–20769.

Gillenwater, A., Xu, X.-C., El-Nagger A.K., Clayman, G.L. and Lotan, R. (1996) Expression of galectins in head and neck squamous cell carcinoma. *Head and Neck*, **18**, 422–432.

Goletz, S., Hanisch, F-G. and Karsten, U. (1997) Novel _GalNAc containing glycan on cytokeratins are recognized in vitro by galectins with carbohydrate type 11 domains. *J. Cell Sci.*, **110**, 1585–1596.

Herrmann, J., Turck, C.W., Atchison, R.E., Huflejt, M.E., Poulter, L., Gitt, M.A., Burlingame, A.I., Barondes, S.H. (1993) Primary structure of the soluble lactose binding lectin L-29 from rat and dog and interaction of its non-collgenous Proline-, Glycine-, Tyrosine-rich sequence with bacterial and tissue collagenase. *J. Biol. Chem.* **268**, 26704–26711.

Ho, M.K. and Springer, T.A. (1982) Mac-2, a novel 32, 000 Mr mouse macrophage subpopulation-specific antigen defined by monoclonal antibodies. *J. Immunol.*, **128**, 1221–1228.

Honjo, Y., Kubo, T., Yoshida, J., Hattori, K., Akahani, S., Swada, T., Raz, A., Inohara, H. Expression of galectin-3 correlates with neoplastic transformation of human squamous epithelial cell of the tongue. (submitted).

Hsu, D.K., Zuberi, R.I. and Liu, F.T. (1992) Biochemical and biophysical characterization of human recombinant IgE-binding protein, an S-type animal lectin. *J. Biol. Chem.*, **267**, 14167–14174.

Huflejt, M.E., Turck, C.W., Lindstedt, R., Barondes, S.H. and Leffler, H. (1993) L-29, a soluble lactose-binding lectin, is phosphorylated on Serine 6 and Serine 12 *in vivo* and by casein kinase I. *J. Biol. Chem.*, **268**, 26712–26718.

Inohara, H., Akahani, S., Koths, K. and Raz.A. (1996) Interactions between galectin-3 and Mac-2-binding protein mediate cell-cell adhesion. *Cancer Res.*, **56**, 4530–4534.

Inohara, H. and Raz, A. (1994) Identification of human melanoma cellular and secreted ligands for galectin-3. *Biochem. Biophys. Res. Commun.*, **201**, 1366–1375.

Inohara, H. and Raz, A. (1995) Functional evidence that cell surface galectin-3 mediates homotypic cell adhesion. *Cancer Res.* **55**: 3267–3271.

Irimura, T., Matsushita, Y., Sutton, R.C., Carralero, D., Ohannesian, D.W., Cleary, K.R., Ota, D.M., Nicolson, G.L. and Lotan, R. (1991) Increased content of an endogenous lactose-binding lectin in human colorectal carcinoma prograssed to metastatic stages. *Cancer Res.*, **51**, 387–393.

Kasai, K., Hirabayashi, J. (1996) Galectins : a family of animal lectins that decipher glycocodes. *J. Biochem.*, **119**, 1–8.

Koths, K., Taylor, E., Halenbeck, R., Casipit, C. and Wang, A. (1993) Cloning and characterization of a human Mac-2-binding protein, a new member of the super family defined by the macrophage scavenger receptor cysteine-rich domain. *J. Biol. Chem.* **268**, 14245–14249.

Lee E.C., Woo, H-J., Korzelius, C.A., Steele, G.E., Mercurio, A.M., (1991) Carbohydrate-binding protein 35 is the major cell-surface laminin-binding protein in colon carcinoma. *Arch. Surg.*, **126**, 1496–1502.

Leffler, H. and Barondes, S.H. (1986) Specificity of binding of three soluble rat lung lectins to substituted and unsubstituted mammalian beta-galactosides. *J. Biol. Chem.*, **261**, 10119–10126.

Lindstedt, R., Apodaca, G., Barondes, S.H., Mostov, K.E. and Leffler, H. (1993) Atypical secretion of a cytosolic protein by Madin-Darby canine kidney cells : evidence for polarized release of an endogenous lectin by a nonclassical secretory pathway. *J. Biol. Chem.*, **268**, 11750–11757.

Lotan, R., Belloni, P.N., Tressler, R.J., Lotan, D., Xu, X.C. and Nicolson, G.L. (1994) Expression of galectins on microvessel endothelial cells and their involvement in tumor cell adhesion. *Glycoconj. J.*, **11**, 462–468.

Lotan, R., Ito, H., Yasui, W., Yokozaki, H., Lotan, D. and Tahara, E. (1994) Expression of 31-kDa lactoside-binding lectin in normal human gastric mucosa and in primary and metastatic gastric carcinomas. *Int. J. Cancer*, **56**, 474–480.

Lotz, M., Andrews, C.W., korzelius, C.A., Lee, E.C., Steele, G.D., Clark, A. and Mercurio, A.M. (1993) Decreased expression of Mac-2 (carbohydrate-binding protein 35) and loss of its nuclear localization are associated with the neoplastic progression of colon carcinoma. *Proc. Natl. Acad. Sci. USA*, **90**, 3466–3470.

Marer, N.L. and Hughes, R.C. (1996) Effects of the carbohydrate-binding protein galectin-3 on the invasiveness of human breast carcinoma cells. *J. Cell. Physiol.*, **168**, 51–58.

Mehul, B. and Hughes, R.C. (1997) Plasma membrane targetting, vesicular budding and release of galectin-3 from the cytoplasm of mammalian cells during secretion. *J. Cell Sci.*, **110**, 1169–1178.

Meromsky L., Lotan, R. and Raz, A. (1986) Implication of endogenous tumor cell surface lectins as mediatoors of cellular interactions and lung colonization. *Cancer Res.*, **46**, 5270–5376.

Nangia-Makker P., Thompson, E., Hogan C., Ochieng, J. and Raz, A. (1995) Induction of tumorigenicity in a non-tumorigenic human breast carcinoma cell line. *Int. J. Oncol.*, **7**, 1079–1087.

Nangia-Makker P., Sarvis, R., Visscher, D.W., Bailey-Penrod, J., Raz, A. and Sarkar, F.H. Galectin-3 and L1 retrotransposons in human breast carcinomas. *Breat Cancer Res. Treat.*, **49**(2), 171–183.

Ochieng, J., Gerold, M. and Raz.A. (1992) Dichotomy in the laminin-binding properties of soluble and membrane-bound human galactoside-binding protein. *Biochem. Biophys. Res. Commun.*, **186**, 1674–1680.

Ohannesian, D.W., Lotan, D., Thomas, P., Jessup, J.M., Fukuda, M., Gabius, H.J. and Lotan.R. (1995) Carcinoembryonic antigen and other glycoconjugates act as ligands for galectin-3 in human colon carcinoma cells. *Cancer Res.*, **55**, 2191–2199.

Raz, A. and Lotan, R. (1987) Endogenous galactoside-binding lectins : a new class of functional tumor cell surface molecules related to metastasis. *Cancer Metastasis Rev.*, **6**, 433–452.

Raz, A., Meromsky, L. and Lotan, R. (1986) Differential expression of endogenous lectins on the surface of nontumorigenic, tumorigenic and metastatic cells. *Cancer Res.*, **46**, 3667–3672.

Raz, A., Zhu, D., Hogan, V., Shah, N., Raz, T., Karkash, R., Pizerini, R. and Carmi, P. (1990) Evidence for the role of 34–kDa galactoside-binding lectin in transformation and metastasis. *Int. J. Cancer*, **46**, 871–877.

Saada, A., Reichert, F. and Rotshenker, S. (1996) Granulocyte macrophage colony stimulating factor produced in lesioned peripheral nerves induces the up-regulation of cell surface expression of Mac-2 by macrophages and schwann cells. *J. Cell Biol.*, **133**, 159–167.

Sato, S. and Hughes, R.C. (1992) Binding specificity of a baby hamster kidney lectin for H type I and II chains, polylactosamine glycans and appropriately glycosylated forms of laminin and fibronectin. *J. Biol. Chem.*, **267**, 6983–6990.

Schoppner H.L., Raz, A., Ho, Samuel B. and Bresalier, R.S. (1995) Expression of an endogenous galactose-binding lectin correlates with neoplastic progression in the colon. *Cancer*, **75**, 2813–2826.

Truong, M.J., Gruart, V., Kusnierz, J.P., Papin, J.P., Louiseau, S., Capron, A. and Capron, M. (1993) Human neutrophils express immunoglobulin E (IgE)-binding proteins (Mac-3/epsilon BP) of the S-type lectin family : role in IgE-dependent activation. *J. Exp. Med.*, **177**, 243–248.

van den Brule F.A., Buici C., Sobel, M.E., Liu, F.T. and Castronovo, V. (1995) Galectin-3, a laminin binding protein, fails to modulate adhesion of human melanoma cells to laminin. *Neoplasm*, **42**, 215–219.

Wang, L., Inohara, H., Pienta, K.J. and Raz, A. (1995) Galectin-3 is a nuclear matrix protein which binds RNA. *Biochem. Biophys. Res. Commun.*, **217**, 292–303.

Wang, J.L., Werner, E.A., Laing, J.G. and Patterson, R.J. (1992) Nuclear and cytoplasmic localization of a lectin-ribonucleoprotein complex. *Biochem. Soc. Trans.*, **20**, 269–272

Warfield, P.R., Nangia-Makker, P., Raz, A. and Ochieng, J. Interactions of a breast epithelial cell line and its galectin-3 transfected subclone with Extracellular matrix proteins. *Invasion and Metastasis*, **17**, 101–112.

Woo, H.J., Shaw, L.M., Messier, J.M. and Mercurio, A.M. (1990) The major non- integrin laminin binding protein of macrophages is identical to carbohydrate binding protein 35 (Mac-2). *J. Biol. Chem.*, **265**, 7097–7099.

Xu, X.-C., El-Nagger, A. and Lotan, R. (1995) Differential expression of Galectin-1 and galectin-3 in thyroid tumors : potential diagnostic implications. *Am. J. Pathol.*, **147**, 815–822.

Yang, R.Y., Hsu, D.K. and Liu, F.T. (1996) Expression of galectin-3 modulates T-cell growth and apoptosis. *Proc. Natl. Acad. Sci. USA*, **93**, 6737–6742.

6. LAMININ-BINDING LECTINS DURING CANCER INVASION AND METASTASIS

FRÉDÉRIC A. VAN DEN BRÛLE AND VINCENT CASTRONOVO

Metastasis Research Laboratory, Pathology B23, Sart Tilman, B-4000 Liège, Belgium

INTRODUCTION

Interactions of cancer cells with the glycoprotein laminin constitute key events during basement membrane crossing, cancer invasion and metastasis formation. A number of cell surface binding proteins including several members of the integrin family have been characterized as laminin-binding proteins. Recently, it appeared that the abundant poly-N-acetyllactosamine repeats on laminin can be recognized by lectins such as members of the galectin family. Some of these soluble lactose-specific lectins exhibit modulated expression in cancer cells, and several biological functions, related or not to their ability to bind laminin, have been described. In this chapter, we will review the current knowledge about galectins and their possible implication in cancer invasion and metastasis.

BASEMENT MEMBRANE INVASION: A KEY EVENT DURING THE METASTATIC CASCADE

Metastasis formation from the primary tumor constitutes the primary cause for cancer mortality (Liotta, 1986). The complex events underlying metastasis formation, known as the metastatic cascade (Figure 1), result from multiple interactions between cancer cells and host tissues. Indeed, to form a metastatic deposit, cancer cells must actively cross several times specialized structures, the basement membranes (BMs), when leaving the primary tumor, during intra- and extravasation, as well as during muscle and nerve penetration. The BMs are specialized extracellular matrices that separate different tissues such as epithelia and connective tissues. They contain, among others, laminin, type IV collagen, heparan sulfate, entactin and fibronectin, and do not contain pores large enough to allow passive crossing of normal or cancer cells (Vracko, 1974). Basement membrane invasion can be viewed as repeating cycles of three main steps (Liotta, 1986): the first step implies adhesion of the tumor cell to the BM components such as laminin, type IV collagen and fibronectin; then, the cell secretes, or induces secretion by host cells of degradative enzymes such as type IV collagenases; finally, the cell migrates actively through the zone of matrix degradation (Figure 2).

Correspondence: F.A. van den Brûle, Metastasis Research Laboratory Pathology B23, Sart Tilman, B-4000 Liège, Belgium. E-mail: f.vandenbrule@chu.ulg.ac.be

F.A. VAN DEN BRÛLE and V. CASTRONOVO

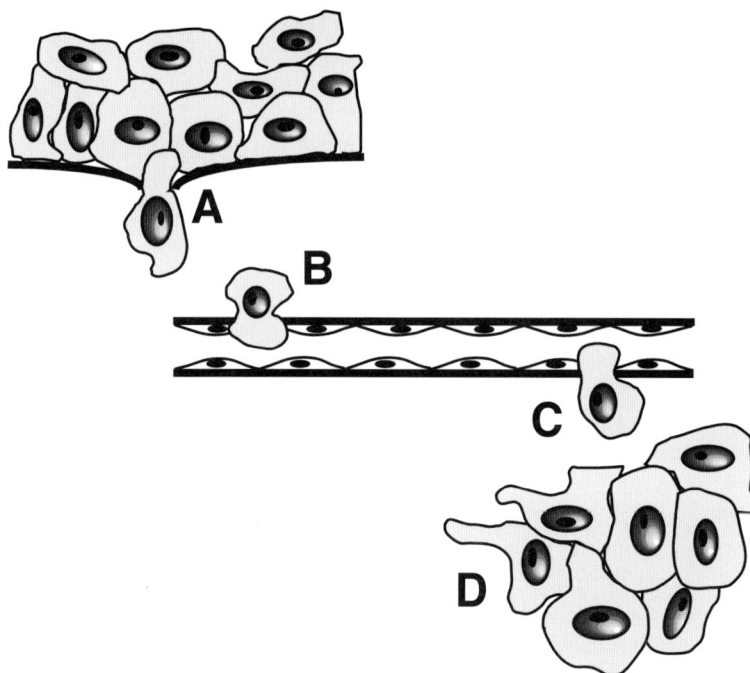

Figure 1 Schematic description of the metastatic cascade. In order to create a metastatic colony, cancer cells must leave the primary tumor and invade the surrounding connective tissues (A), cross the vascular wall (intravasation, B), enter the blood stream, attach to the vascular endothelium and cross the vascular wall again (extravasation, C), migrate in the parenchyma of the target organ and form the metastatic colony (D).

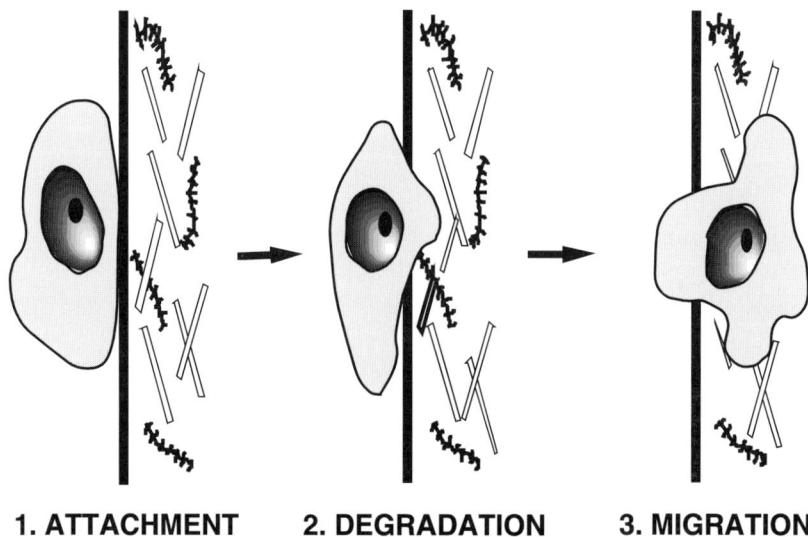

1. ATTACHMENT   2. DEGRADATION   3. MIGRATION

Figure 2 Three-step model of basement membrane invasion. In order to cross the basement membrane, cancer cells must attach to the basement membrane components including laminin (1), induce local proteolysis (2) and migrate through the newly created defect (3).

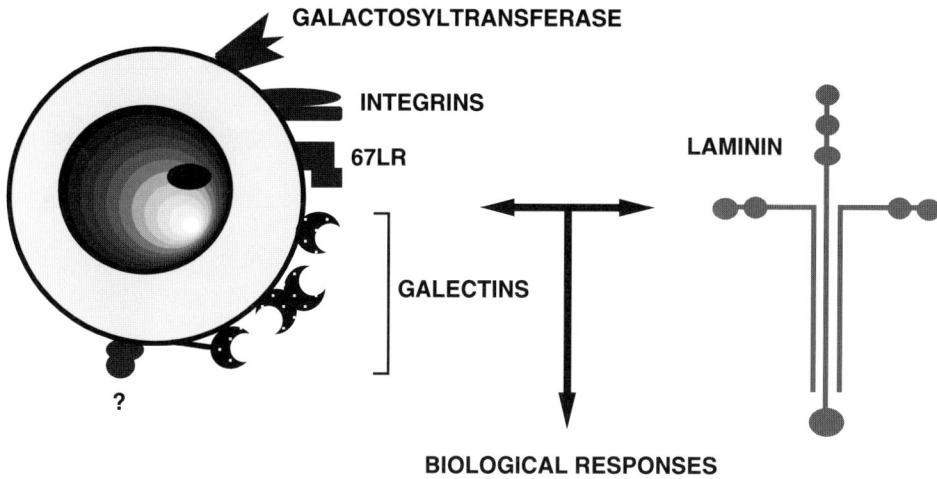

Figure 3 Interactions of cancer cells with laminin. The multiple biological activities of laminin are mediated through numerous cell surface binding proteins including the galactosyltransferase, members of the integrin family, the 67 kDa laminin binding protein and galectins.

INTERACTIONS BETWEEN CANCER CELLS AND LAMININ

Laminin is a major BM glycoprotein that is involved in several biological processes such as attachment, spreading, migration, proliferation and differentiation of normal as well as cancer cells (Kleinman *et al.*, 1985). It has been purified and characterized in 1979 from a mouse fibrosarcoma tumor, the Engelbreth-Holm-Swarm (EHS) tumor (Timpl *et al.*, 1979). This cross-shaped molecule is composed of three subunit polypeptides, the chains A (440 kDa) (Sasaki *et al.*, 1988), B1 (210 kDa) (Sasaki *et al.*, 1987) and B2 (200 kDa) (Sasaki and Yamada, 1987). The molecule contains approximately 30% of complex N-linked carbohydrate chains containing abundant repeats of [3Galß1,4GlcNAcß1]n, or poly-N-acetyllactosamine sequence (Dennis *et al.*, 1984; Arumugham *et al.*, 1986; Fujiwara *et al.*, 1988). The existence of several laminin isoforms has defined a laminin family (Engvall, 1993, Timpl and Brown, 1994). Most of the studies have been performed on the widely available EHS laminin, also known as type 1 laminin.

It rapidly appeared that laminin is composed of several functional polypeptidic domains (Beck *et al.*, 1990) able to interact with multiple laminin-binding proteins (Figure 3) (Castronovo, 1993). The first molecule defined as a laminin-binding protein is the 67-kDa laminin receptor (67LR) (Castronovo, 1993), whose structure is still under characterization (Landowski *et al.*, 1995). Several members of the integrin family of cell surface receptors including $\alpha_1\beta_1$ (Ignatius and Reichardt, 1988), $\alpha_2\beta_1$ (Elices and Hemler, 1989; Languino *et al.*, 1989), $\alpha_3\beta_1$ (Gehlsen *et al.*, 1988; Peltonen *et al.*, 1989; Elices *et al.*, 1991; Gehlsen *et al.*, 1992), $\alpha_6\beta_1$ (Kramer *et al.*, 1990; Cooper *et al.*, 1991), $\alpha_7\beta_1$ (Kramer *et al.*, 1991) and $\alpha_6\beta_4$ (Sonnenberg *et al.*, 1988; Sonnenberg *et al.*, 1991) have been reported to bind laminin. Other molecules such as the enzyme β_1-4-galactosyltransferase (Shur, 1982, Runyan *et al.*,

DIMERIC GALECTINS   **BI-CRD GALECTINS**

GAL-1   GAL-2   GAL-4   GAL-6   GAL-8   GAL-9

MONO-CRD GALECTINS

GAL-3   GAL-5   GAL-7

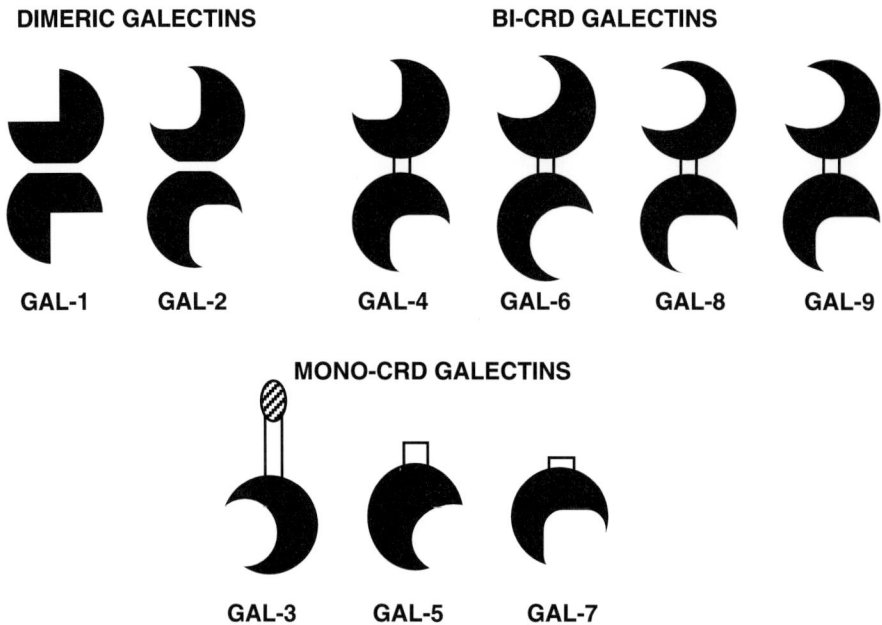

Figure 4   Schematic representation of the galectin family. The lectins are grouped according to their protein structure in dimeric, bi-CRD and mono-CRD galectins.

1988, Shur, 1989, Penno *et al.*, 1989) also bind laminin. The presence of carbohydrate chains on laminin has been related to biological functions such as cell adhesion (Fujiwara *et al.*, 1988; Dean *et al.*, 1990), suggesting the presence of carbohydrate-binding structures on the cell surface.

GALECTINS: A FAMILY OF LACTOSE-BINDING LECTINS

The galectins are a family of soluble lactose-binding lectins that share affinity for ß-galactosides moieties and significant sequence similarity in their carbohydrate-binding site (Barondes *et al.*, 1994b; Barondes *et al.*, 1994a) (Figures 4 and 5, and Table 1). These lectins were first described by a variety of names, but a consensus was adopted in 1994 by most investigators in the field (Barondes *et al.*, 1994a). To date, 9 members of this family have been characterized. In addition, a previously known protein, the Charcot-Leyden crystal (CLC) protein, has a carbohydrate recognition domain (CRD) characteristic of the galectins. The best studied members among the galectins are galectin-1 and galectin-3 (Barondes *et al.*, 1994b). The galectins have been grouped in proto type (galectin-1, 2, 5, 7), chimera type (galectin-3) and tandem-repeat type (galectin-4, 6, 8, 9); according to the overall protein structure (Hirabayashi and Kasai, 1993) (schematized in Figure 4). Another classification involves CRD specificity, and defines type I, or conserved CRD (galectin-1), and type II, or variable CRD (other galectins) (Ahmed and Vasta, 1994). A general overview of the galectins is presented in Table 1.

```
hG1      ---------- ---------- ---------- ---------- ----------
hG2      ---------- ---------- ---------- ---------- ----------
hG3      ADNFSLHDAL SGSGNPNPQG WPGAWGNQPA GAGGYPGASY PGAYPGQAPP   50
hG4 1-175  --------M ---------- ---------- ---------- ----------    1
hG4 176-end ---------- ---------- ---------- ---------- ----------
rG5      ---------- ---------- ---------- ---------- ----------
hG7      ---------- ---------- ---------- ---------- ----------
hG8 1-184  --------M L--------- ---------- ---------- ----------    2
hG8 185-end ---------- ---------- ---------- ---------- ----------
hG9 1-174  --------M ---------- ---------- ---------- ----------    1
hG9 175-end ---------- ---------- ---------- ---------- ----------


hG1      ---------- --------AC G--------- ---------- ----------    3
hG2      ---------- --------MT G--------- ---------- ----------    3
hG3      GAYPGQAPPG AYHGAPGAYP GAPAPGVYPG PPSGPGAYPS SGQPSAPGAY  100
hG4 1-175  ---------- --------AYV PAP------- ---------- -------GY    9
hG4 176-end ---------- --------LN SLP--TMEG- ---------- ----------    9
rG5      ---------- --------SS FSTQTP---- ---------- ----------    8
hG7      ---------- ---------S N--------- ---------- ----------    2
hG8 1-184  ---------- -------SLN NLQ------- ---------- --------N-    9
hG8 185-end ---------- ---------- ---------- ---------- ----------
hG9 1-174  ---------- -------AFS GSQ------- ---------- --------A-    8
hG9 175-end ---------- --------MF STPAIPPMM- ---------- ----------   11


hG1      ---------- ----LV-ASN L--NLKPGEC LRVRGEVAPD AKS--FVLNL   34
hG2      ---------- ---ELE-VKN M--DMKPGST LKTIGSIADG TDG--FVINL   35
hG3      PATGPYGAPA GPLIVPYNLP LPGGVVPRML ITILG-TVKP NAN-RIALDF  148
hG4 1-175  Q---PTYNPT ----LPYYQP IPGGLNVGMS VYIQG-VASE HMK-RFFVNF   50
hG4 176-end -------PPT FNPPVPYFGR LQGGLTARRT IIIKGVVPPT GKS--FAINF   50
rG5      -------YPN --LAVPFFTS IPNGLYPSKS IVISGVVLSD AKR--FQINL   47
hG7      ---------- ----VPHKSS LPEGIRPGTV LRIRGLVPPN ASRFHVNLLC   38
hG8 1-184  ----IIYNPV ----IPFVGT IPDQLDPGTL IVIRGHVPSD ADRFQVDLQN   51
hG8 185-end ---------- ----PFAAR LNTPMGPGRT VVVQGEVANN AKS--FNVDL   33
hG9 1-174  ----PYLSPA ----VPFSGT IQGGLQDGLQ ITVNGIVLSS SGT-RFAVNF   49
hG9 175-end -------YPH PAYPMPFITT ILGGLYPSKS ILLSGTVLPS AQR--FHINL   52


hG1      GK---DSNNL CIHFNPRFNA HGDANTIVCN SKDGGAWGTE QREAVFPFQP   81
hG2      GQ---GTDKL NLHFNPRFS- ---ESTIVCN SLDGSNWGQE QREDHLCFSP   78
hG3      Q----RGNDV AFHFNPRFNE NN-RRVIVCN TKLDNNWGRE ERQSVFPFES  193
hG4 1-175  VVGQDPGSDV AFHFNPRFDG WD-KVV--FN TLQGGKWGSE ERKRSMPFKK   97
hG4 176-end KVGSS--GDI AIHINPRMGN G--TVVRNSL LNGSWGSEEK KI-THNFFGP   95
rG5      RCG----GDI AFHLNPRFDE N--AVVRNTQ INNSWGPEER SLPGSMPFSR   91
hG7      GEE--QGSDA AIHFNPRLDT SE--VVF---N SKEQGSWGRE ERGPGVFFQR   82
hG8 1-184  GSSVKPRADV AFHFNPRFKR AG-CI--VCN TLINEKWGRE EITYDTFFKR   98
hG8 185-end LAGKS--KDI AIHLNPRLNI K--AFVRNSF LQESWGEEER NI-TSFPFSP   78
hG9 1-174  QTGF-SGNDI AFHFNPRFED GG-YVV--CN TRQNGSWGPE ERRTHMFPQK   95
hG9 175-end CSG----NHI AFHLNPRFDE N--AVVRNTQ IDNSWGSEER SLPRKMFEVR   96


hG1      GSVAEVCITF DQANLTVKLP DGYEFKFPNR L-NLEAINYM AADGDFKIKC  130
hG2      GSEVKFTVTF ESDKFKVKLP DGHELTFPNR L-GHSHLSYL SVRGGFNMSS  127
hG3      GKPFKIQVLV EPDHFKVAVN DAHLLQYNHR L---KKLNEI -------SK  232
hG4 1-175  GAAFELVFIV LAEHYKVVVN GNPFYEYGHR L-PLQMVTHL QVDGDLQLQS  146
hG4 176-end GQFFDLSIRC GLDRFKVYAN GQHLFDFAHR LSAFQRVDTL EIQGDVTLSY  145
rG5      GQRFSVWILC EGHCFKVAVD GQHICEYSHR LMNLPDINTL EVAGDIQLTH  141
hG7      GQPFEVLIIA SDDGFKAVVG DAQYHHFRHR L-PLARVRLV EVGGDVQLDS  131
hG8 1-184  EKSFEIVIMV LKDKFQVAVN GKHTLLYGHR I-GPEKIDTL GIYGKVNIHS  147
hG8 185-end GMYFEMIIYC DVREFKVAVN GVHSLEYKHR FKELSSIDTL EINGDIHLLE  128
hG9 1-174  GMPFDLCFLV QSSDFKVMVN GILFVQYFHR V-PFHRVDTI FVNGSVQLSY  144
hG9 175-end GQSFSVWILC GAHCLKVAVD GQHLFEYYHR LRNLPTTINRL EVGGDIQLTH  146


hG1      VAFD------ ---------- ---------- -------                134
hG2      FKLKE----- ---------- ---------- -------                132
hG3      LGISCDIDLT CASYT  -- --MI------ -------                249
hG4 1-175  INFIGGQPLR PQGPP----- --MMPPYPGP GHC-HQQ               175
hG4 176-end VQI------- ---------- ---------- -------                148
rG5      VET------- ---------- ---------- -------                144
hG7      VRIF------ ---------- ---------- -------                135
hG8 1-184  IGFSFSSDLQ STQASSLELT EIVRENVPKS GTPQLSL               184
hG8 185-end VRSW------ ---------- ---------- -------                132
hG9 1-174  ISFQPPGVWP ANPAP----- --ITQTVIHT VQSAPGQ               174
hG9 175-end VQT------- ---------- ---------- -------                149
```

Figure 5 Amino acid sequence homology between members of the galectin family. Alignment of the sequences was performed using the GeneWorks software version 2.5.1, from IntelliGenetics, Inc., using the following amino acid sequences: human galectin-1 (hG1), p09382; human galectin-2 (hG2), p05162; human galectin-3 (hG3), p17931; human galectin-4 (hG4), ab006781_1; rat galectin-5 (rG5), p47967; human galectin-7 (hG7), p47929; human galectin-8 (hG8), o00214; human galectin-9 (hG9), o00182.

Table 1   Overview of the Structure of Galectin Members

Name	Schema	Previous names	Structure	Subunit MW*	Type*	Initial characterization	Peptide sequence	Cloning
Galectin-1		L-14, galaptin, HLBP14	Homodimer	14 kDa	1-CRD, proto-type galectin	protein purification[a]	Many sources[a]	Many sources[b]
Galectin-2		L-14-II	Homodimer	14 kDa	1-CRD, proto-type galectin	cDNA cloning from expression library[c]	Not reported	Hepatoma[c]
Galectin-3		Mac-2, CBP-30, CBP-35, L-29, L-34, IgEBP, eBP, HLBP31	Monomer	29-35 kDa	1 CRD; specific N-terminal domain with variable number of repeats, phosphorylation sites on Ser6 and Ser12, homologies to hnRNP members, α_1 collagen (sensitivity to collagenases), SRF	protein purification[d]	Many sources[d]	Many sources[e]
Galectin-4		RI-H, L-36	Monomer	36-37 kDa	2-CRD, tandem-repeat type galectin	purification of proteolytic degradation fragment[f]	Purified mainly from intestine[f]	Intestine, colon cancer cells[g]
Galectin-5		RL-18	Monomer	16-18 kDa	1-CRD, proto-type galectin	protein purification[h]	Intestine, erythrocytes[h]	reticulocyte library[j]
Galectin-6			Monomer	33 kDa	2-CRD, tandem-repeat type galectin	nr*	nr	nr

Table 1 Continued.

Name	Schema	Previous names	Structure	Subunit MW*	Type*	Initial characterization	Peptide sequence	Cloning
Galectin-7		IEF 17	Monomer	14 kDa	1-CRD, proto-type galectin	cDNA cloning[j]	Keratinocyte database[k]	Keratinocytes[j]
Galectin-8		PTCA-1	Monomer	35 kDa	2-CRD, tandem-repeat type galectin	cDNA cloning[l]	Rat liver[m]	rat liver, prostate carcinoma cells[l]
Galectin-9			Monomer	36 kDa	2-CRD, tandem-repeat type galectin	cDNA cloning[n]	nr	embryonic kidney, Hodgkin diseased tissue[n]

*MW, molecular weight (kDa); CRD, carbohydrate recognition domain; N-terminal, amino-terminal; hnRNP, human nuclear ribonucleprotein; SRF, serum response factor; nr, not reported.

[a] Galectin-1 is an ubiquitous lectin that has been first purified from the electric organ of the eel (Teichberg et al., 1975). Amino acid sequence of galectin-1 has been obtained from several sources (Gitt and Barondes, 1986, Clerch et al., 1988, Paroutaid et al., 1987, Hirabayashi et al., 1987a, Hirabayashi et al., 1987b, Southan et al., 1987, Hirabayashi and Kasai, 1988, Sharma et al., 1990, Ohsawa et al., 1990, Bladier et al., 1991, Ozeki et al., 1991a, Wells and Malucci, 1991, Castronovo et al., 1992b, Ahmed et al., 1996b).

[b] Cloning of galectin-1 cDNA was performed from many cells and tissues (Ohyama et al., 1986, Raz et al., 1987a, Raz et al., 1988, Wilson et al., 1989, Abbott et al., 1989, Cooper and Barondes, 1990, Hynes et al., 1990, Wells and Malucci, 1991, Poirier et al., 1992, Gitt and Barondes, 1986, Abbott and Feizi, 1989, Hirabayashi et al., 1989, Couraud et al., 1989, Allen et al., 1991a).

[c] Galectin-2 was cloned from a human hepatoma expression library (Gitt and Barondes, 1986, Gitt et al., 1992).

[d] Galectin-3 was purified from chicken intestine (Beyer et al., 1980) and mouse 3T3 and human SL66 fibroblasts (Roff et al., 1983, Roff and Wang, 1983) by affinity chromatography on lactose or glycoconjugates; it was also characterized as an IgE binding protein (Liu and Orida, 1984, Liu et al., 1985, Laing et al., 1989). Numerous other sources were subsequently used for purification (Crittenden et al., 1984, Gabius et al., 1985, Cerra et al., 1985, Leffler and Barondes, 1986, Raz et al., 1987b, Sparrow et al., 1987, Lotan et al., 1989a, Foddy et al., 1990, Woo et al., 1990, Lotan et al., 1991, Oda et al., 1991, Bachar-Lustig et al., 1991, Lee et al., 1991, Herrmann et al., 1993, Huflejt et al., 1993, Lindstedt et al., 1993, Jung and Fujimoto, 1994, Ohannesian et al., 1995). Amino acid sequence of galectin-3 was obtained from many sources (Woo et al., 1990, Sato et al., 1993, Lindstedt et al., 1993, Herrmann et al., 1993).

[e] Galectin-3 complementary DNA sequence was obtained from various cells and tissues (Liu et al., 1985, Raz et al., 1987a, Albrandt et al., 1987, Jia et al., 1987, Jia and Wang, 1988, Raz et al., 1988, Raz et al., 1989, Cherayil et al., 1989, Cherayil et al., 1990, Robertson et al., 1990, Oda et al., 1991, Raz et al., 1991, Lotz et al., 1993, Herrmann et al., 1993, Mehul et al., 1994, Gaudin et al., 1995, Aubin et al., 1996).

[f] A proteolytic degradation fragment of galectin-4 was first purified from intestine by affinity chromatography on lactose (Leffler et al., 1989, Ruggiero-Lopez et al., 1992, Oda et al., 1993, Tardy et al., 1995); the intact protein was purified thereafter from desmosome-enriched pig tongue epithelium extracts (Chiu et al., 1994) and from the colon carcinoma T84 cell line (Huflejt et al., 1997).

[g] Sequence of galectin-4 cDNA was obtained from intestine (Oda et al., 1993), pig tongue epithelium (Chiu et al., 1994) and the colon carcinoma T84 cell line (Huflejt et al., 1997). It was also obtained by differential screening from colon cancer tissue, where its expression is down regulated (Rechreche et al., 1997).

[h] Galectin-5 was purified from intestine (Leffler et al., 1989) and lung, but the erythrocytes were the source of the protein (Gitt et al., 1995).

[i] cDNA sequence of galectin-5 was obtained from a reticulocyte library (Gitt et al., 1995).

[j] Galectin-7 was cloned from a human epidermal cDNA library by differential screening in search of keratinocyte-specific cDNAs (Magnaldo et al., 1995), and from a keratinocyte library using oligonucleotides derived from peptide sequences present in the keratinocyte protein database (Madsen et al., 1995).

[k] The only amino acid sequence available for galectin-7 originates from the human keratinocyte protein database (IEF 17) (Madsen et al., 1995).

[l] Galectin-8 cDNA was obtained from rat liver (Hadari et al., 1995) and from the human prostate carcinoma LNCaP cell line by the technique of surface epitope masking (Su et al., 1996).

[m] Galectin-8 has been purified from rat liver (Hadari et al., 1995).

[n] Galectin-9 has been cloned from mouse embryonic kidney using a degenerate primer strategy (Wada and Kanwar, 1997) and from a Hodgkin's diseased tissue expression library (Türeci et al., 1997).

Galectin-1 is a pleiotropic homodimeric galectin that has been purified from numerous sources by lactose affinity chromatography (Table 1). It is a 14 kDa polypeptide that assembles in homodimers and is almost exclusively composed of a single CRD. Maintenance of the lectin activity needs the presence of reducing agents (Den and Malinzak, 1977; Briles *et al.*, 1979; Powell and Whitney, 1980; Levi and Teichberg, 1981). Mutagenesis and chemical modification studies have contributed to define the role of hydrophilic amino acid residues in the binding of galectin-1 to saccharides (Abbott and Feizi, 1991; Hirabayashi and Kasai, 1991; Hirabayashi and Kasai, 1994; Ahmed *et al.*, 1996a) and specifically to poly-N-acetyllactosamine chains present on glycoconjugates (Merkle and Cummings, 1988; Zhou and Cummings, 1993). The three-dimensional ß-sheet topology of galectin-1 termed "jelly roll motif" and characteristic of numerous lectins including from legumes, has been determined by X-ray crystallographic analysis of the recombinant protein in the presence or the absence of the ligand (Bourne *et al.*, 1994; Liao *et al.*, 1994; Rini, 1995). Galectin-1 developmental modulation of expression in several organs has suggested that it is involved in cell differentiation and embryogenesis (Poirier *et al.*, 1992; Akimoto *et al.*, 1992; Marschal *et al.*, 1994; Colnot *et al.*, 1996; Maquoi *et al.*, 1997; van den Brûle *et al.*, 1997b). However, mice carrying a null mutation in the galectin-1 gene develop apparently normally (Poirier and Robertson, 1993), suggesting that the functions played by this lectin could be carried over by other molecules.

Less information is available for galectin-2 (Table 1). It was first cloned from a human hepatoma expression library using an anti-lectin antiserum (Gitt and Barondes, 1986; Gitt *et al.*, 1992). Similarly to galectin-1, it is a homodimeric galectin composed of 14 kDa monomers. Its X-ray structure, alone or in combination with lactose, and the topology of its CRD, were the first to be defined in the galectin family and are similar to that found in leguminous plant lectins (Lobsanov *et al.*, 1993).

Galectin-3 has been purified and characterized as a lactose-binding lectin and as an IgE binding protein (Liu and Orida, 1984; Liu *et al.*, 1985; Laing *et al.*, 1989) from numerous sources (Table 1). Galectin-3 is identical to the previously described macrophage Mac-2 antigen detected by the monoclonal antibody M3/38 (Ho and Springer, 1982; Cherayil *et al.*, 1989). Galectin-3 is a 30 kDa lectin composed of two structural domains of approximative equal size (Figures 4 and 5, Table 1). The carboxy-terminal half is the carbohydrate-binding site, homologous to the typical galectin CRD. The amino-terminal domain consists of multiple repetitive sequences of 7–10 amino acids with a consensus sequence $PGAYPG(X)_{1-4}$ (where X is any amino acid); the number of repeats varies according to the species considered and accounts for the observed differences in molecular weight. This domain bears homologies to the human nuclear ribonucleoprotein complex members (Jia and Wang, 1988), with a part of the amino-terminal region of the serum response factor (SRF), a transcription factor (Oda *et al.*, 1991), and with the amino-terminal part of the α_1 collagen (II) chain (Raz *et al.*, 1989). This latter domain is sensitive to bacterial collagenases (Raz *et al.*, 1989; Herrmann *et al.*, 1993; Mehul *et al.*, 1994) and human metalloproteinases (Mehul *et al.*, 1994; Ochieng *et al.*, 1994); digestion with these enzymes provides a 16 to 22 kDa galectin-3 fragment that still binds carbohydrates but does not oligomerize (see below). Galectin-3 can be phosphorylated on Ser6 and Ser12 in vivo (Huflejt *et al.*, 1993), but to date, no biological significance has been determined for these events. Previous data revealed that galectin-3 behaves as a monomeric lectin, a finding that is surprising as galectin-3 possesses hemagglutination properties (Feizi *et al.*, 1994). In addition to a study reporting dimerization using Cys186 (Woo *et al.*, 1990), several reports have demonstrated that galectin-3 can oligomerize through its amino-terminal domain (Hsu *et al.*, 1992; Massa *et al.*, 1993; Ochieng *et al.*, 1993). Moreover, the fact that transglutaminase can covalently cross-link galectin-3 to itself using Gln and Lys residues (Mehul *et al.*, 1995) shed new light on the possible functions of galectin-3. It was recently reported that galectin-3 bears a NWGR motif that is homologous to that motif on Bcl-2 and allows these two molecules to interact together (Yang *et al.*, 1996). Like galectin-1, galectin-3 is developmentally modulated in several organs (Foddy *et al.*, 1990; Weitlauf and Knisley, 1992; Fowlis *et al.*, 1995; Aubin *et al.*, 1996, Colnot *et al.*, 1996) and during human placentation (van den Brûle *et al.*, 1994b; Maquoi *et al.*, 1997).

Galectin-4 was purified from intestine by affinity chromatography on lactose (Table 1). This 36 kDa polypeptide is composed by two CRDs joined by a link peptide that is similar to that of galectin-3 (Oda *et al.*, 1993; Chiu *et al.*, 1994), and is easily subject to proteolytic degradation, giving rise to smaller fragments. Galectin-4 has been cloned by differential screening from colon cancer tissue, where its expression is down regulated (Rechreche *et al.*, 1997).

Galectin-5 has been purified from rat erythrocytes and intestine (Leffler *et al.*, 1989, Gitt *et al.*, 1995). It is a 16 kDa monomeric protein composed of one CRD (Gitt *et al.*, 1995). Galectin-6 is a tandem-type galectin reportedly found in the intestine (Colnot *et al.*, 1996), but to date, no published data are available. Galectin-7 has been cloned from a human epidermal cDNA library by differential screening in search of keratinocyte-specific cDNAs (Magnaldo *et al.*, 1995), and from a keratinocyte library using oligonucleotides derived from peptide sequences present in the keratinocyte protein database (Madsen *et al.*, 1995). It encodes a 14 kDa polypeptide that is expressed in keratinocytes. Galectin-8 was cloned from rat liver (Hadari *et al.*, 1995) and from the human prostate carcinoma LNCaP cell line by the technique of surface epitope masking (Su *et al.*, 1996). The protein was purified from rat liver (Hadari *et al.*, 1995) and is a 35 kDa polypeptide composed of two CRDs joined by a link peptide (Hadari *et al.*, 1995). Galectin-9 has been cloned from mouse embryonic kidney using a degenerate primer strategy (Wada and Kanwar, 1997) and from a Hodgkin's diseased tissue expression library (Türeci *et al.*, 1997). This 36 kDa protein is composed of two CRDs joined by a link peptide. The Charcot-Leyden crystal (CLC) protein, a 16.5 kDa major autocrystallizing constituent of the eosinophils and basophils, is constituted of one CRD, has lysophospholipase activity (Ackerman *et al.*, 1993) and binds lactose (Dyer and Rosenberg, 1996).

FIRST DESCRIPTION OF GALECTINS IN CANCER

The first description of galectins in the field of experimental oncology was related to the seminal observation that cancer cells aggregate in the presence of glycoproteins such as asialofetuin, and that protein extracts from these cells induce hemagglutination (Raz and Lotan, 1981). A monoclonal antibody, 5D7, raised against B16 melanoma cell extracts enriched in lectin activity, was developed and recognized 68, 34 and 14 kDa bands on Western blots (Raz *et al.*, 1984; Raz *et al.*, 1986). This antibody displays the functional ability to inhibit asialofetuin-induced homotypic melanoma and fibrosarcoma cell aggregation and adhesion of cancer cells, but not untransformed 3T3 fibroblasts, to substratum (Meromsky *et al.*, 1986). Moreover, preincubation of B16 melanoma and UV-2237 fibrosarcoma cells with this monoclonal antibody decreased (–90%) their ability to form lung colonies in syngeneic mice (Meromsky *et al.*, 1986). FACS analyses performed with the antibody suggested increased lectin expression on the surface of neoplastic cells (Raz *et al.*, 1986). The monoclonal antibody allowed to clone complementary DNAs corresponding to galectin-1 and galectin-3 (Raz *et al.*, 1987a; Raz *et al.*, 1989; Raz *et al.*, 1991). However, the molecular mechanisms of galectin implication in cancer biology were still to be unveiled.

GALECTINS BIND LAMININ

To date, two members of the galectin family, galectin-1 and galectin-3, were shown to bind laminin (Table 2). Several groups, including ours (Castronovo *et al.*, 1992b), have demonstrated that previously described lactose-binding lectins were able to bind EHS laminin (Figures 6 and 7). A first report from 1990 demonstrates that galectin-1, previously known as L-14, purified from calf heart, binds laminin as determined by ligand blotting and solid-phase radioligand binding assays (Zhou and Cummings, 1990). Competition studies demonstrated that lactose and N-acetyllactosamine best inhibited binding to laminin. The carbohydrate specificity of this interaction was further defined by digestion experiments. Whereas exoglycosidase treatments demonstrated that binding of galectin-1 to laminin is not dependent on terminal sialyl-, fucosyl-, β- or α-linked galactosyl residues, treatment of immobilized laminin with endo-ß-galactosidase, an enzyme that cuts between non sulfated galactosides and glucosamine or N-acetylglucosamine (GlcNAc) particularly in poly-N-acetyllactosamine repeats (Maley *et al.*, 1989), significantly decreased galectin-1 binding (Zhou and Cummings, 1990; Castronovo *et al.*, 1992b). Galectin-1 binding to laminin was further confirmed by several studies using solid phase radioligand assays (Massa *et al.*, 1993; Mahanthappa *et al.*, 1994), affinity chromatography (Cooper *et al.*, 1991; Castronovo *et al.*, 1992b; Zhou and Cummings, 1993; Gu *et al.*, 1994; Ozeki *et al.*, 1995) and quantitative precipitation (Zhou and Cummings, 1993). Interestingly, it was reported that galectin-1 binds to the $\alpha_7\beta_1$ integrin and can modulate binding of the latter molecule to laminin (Gu *et al.*, 1994), suggesting that galectin-1 activity could be dependent on the molecular microenvironment of the cell.

At the same time, a report related that galectin-3 (previously known as carbohydrate protein 35, CBP35) is the major cell surface laminin-binding protein on murine macrophages as defined by laminin-Sepharose affinity chromatography and high salt elution (Woo *et al.*, 1990). Other experiments including solid phase (Ochieng *et al.*, 1992) and ligand blotting (Ochieng *et al.*, 1992; Sato and Hughes, 1992; Massa *et al.*, 1993; Ochieng and Warfield, 1995; Ohannesian, 1995), quantitative precipitation (Knibbs *et al.*, 1993), coimmunoprecipitation and affinity chromatography (Castronovo *et al.*, 1992b; Castronovo *et al.*, 1992a; Ohannesian *et al.*, 1995) confirmed the galectin-3-laminin interaction. As mentioned above, galectin-3 binding to laminin

Table 2 Overview of Galectins Ligands

Ligands	Gal-1*	Gal-2	Gal-3	Gal-4	Gal-5	Gal-6	Gal-7	Gal-8	Gal-9
Laminin	Yes	nr	Yes	nr	nr	nr	nr	nr	nr
Lamps	Yes	nr	Yes	nr	nr	nr	nr	nr	nr
Mac-2 BP	nr	nr	Yes	nr	nr	nr	nr	nr	nr
Fibronectin	Yes	nr	Yes	nr	nr	nr	nr	nr	nr
Mucin	Yes	nr	Yes	nr	nr	nr	nr	nr	nr
CEA	Yes	nr	Yes	nr	nr	nr	nr	nr	nr
a7ß1 integrin	Yes	nr	nr	nr	nr	nr	nr	nr	nr
glycolipids	Yes	nr	nr	nr	nr	nr	nr	nr	nr
CD43	Yes	nr	nr	nr	nr	nr	nr	nr	nr
CD45	Yes	nr	nr	nr	nr	nr	nr	nr	nr
hsp90-homologous protein	Yes	nr	nr	nr	nr	nr	nr	nr	nr
milk glycoproteins	Yes	nr	nr	nr	nr	nr	nr	nr	nr
actin	Yes	nr	nr	nr	nr	nr	nr	nr	nr
IgE	nr	nr	Yes	nr	nr	nr	nr	nr	nr
neural recognition proteins	nr	nr	Yes	nr	nr	nr	nr	nr	nr
AGE products	nr	nr	Yes	nr	nr	nr	nr	nr	nr
Mac-1 integrin	nr	nr	Yes	nr	nr	nr	nr	nr	nr
CD98 heavy chain	nr	nr	Yes	nr	nr	nr	nr	nr	nr
LPS	nr	nr	Yes	nr	nr	nr	nr	nr	nr
bcl-2	nr	nr	Yes	nr	nr	nr	nr	nr	nr
CBP70	nr	nr	Yes	nr	nr	nr	nr	nr	nr
DNA, RNA	nr	nr	Yes	nr	nr	nr	nr	nr	nr
cytokeratins	nr	nr	Yes	nr	nr	nr	nr	nr	nr
adherens junction complex component	nr	nr	nr	Yes	nr	nr	nr	nr	nr

*Gal, Galectin; nr, not reported.

was characterized by positive cooperativity and reversible oligomerization through its amino-terminal domain (Massa *et al.*, 1993). Recent affinity chromatography experiments indicate that galectin-3 oligomers obtained by transglutaminase treatment bind immobilized type 1 laminin (van den Brûle *et al.*, 1998). Some reports indicate that galectin-3 binds the laminin A chain with better efficiency that the B chains on ligand blotting, and galectin-3 binds all laminin proteolytic fragments on ligand blots (Ochieng *et al.*, 1992). Due to differences in glycosylation, galectin-3 binds blotted EHS laminin better that human laminin (Ochieng and Warfield, 1995).

OTHER GALECTIN LIGANDS

Many other galectin ligands have been characterized so far (Table 2). Most of them are glycoproteins and are involved in cell surface recognition. Other described ligands are rather thought to be involved in intracellular interactions.

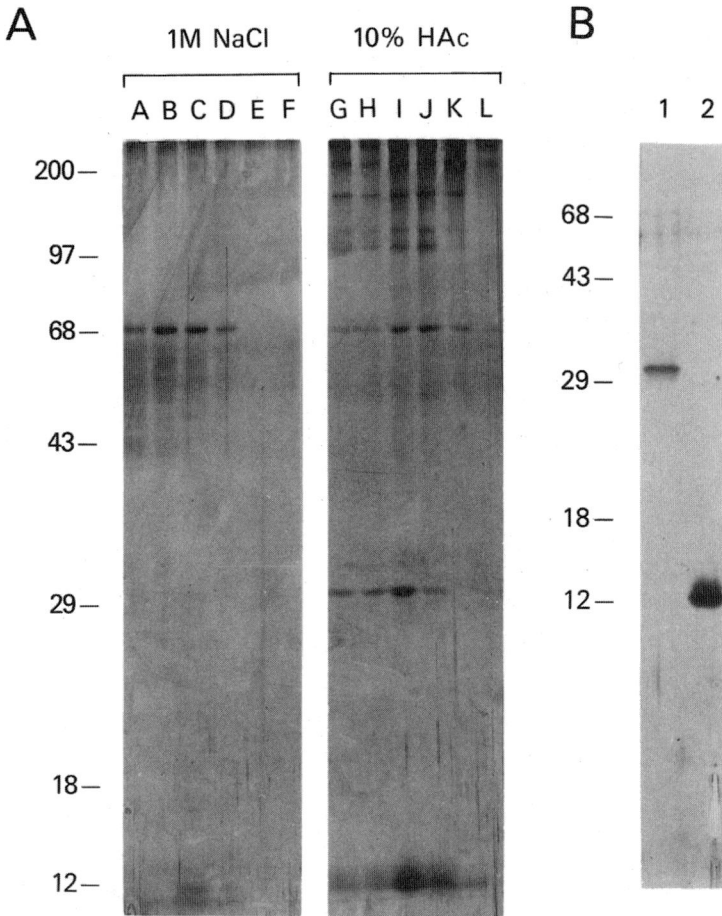

Figure 6 Purification of laminin-binding proteins from A2058 human melanoma cells by affinity chromatography on type 1 laminin, and electroelution of galectin-1 and galectin-3. (A) Silver staining of the SDS-polyacrylamide gel performed after affinity chromatography. The EHS laminin-Affigel 10 (BioRad) affinity matrix was incubated with a human A2058 melanoma cell lysate, packed, washed with the incubation buffer, and laminin-binding proteins were eluted successively with 1M NaCl and 10% acetic acid (HAc). Concentrated elution fractions were electrophoresed on a 12.5% SDS-polyacrylamide gel. A to F, high salt elution fractions; G to L, 10% acetic acid elution fractions. (B) Purity assessment of electroeluted galectin-3 and galectin-1 by 15% SDS-polyacrylamide gel electrophoresis. Lane 1, galectin-3; lane 2, galectin-1. Protein markers are indicated in the left margins and expressed in kDa.

Lamps

Lamps (lysosomal associated membrane proteins), including the most studied lamp-1 and lamp-2, are glycoproteins located on lysosomal membranes and sometimes on the cell surface. They are major carriers of poly-N-acetyllactosamine sequences, and cell surface expression of these glycoproteins is increased in highly metastatic tumor cells (Fukuda, 1991). Affinity chromatography and ligand blotting experiments

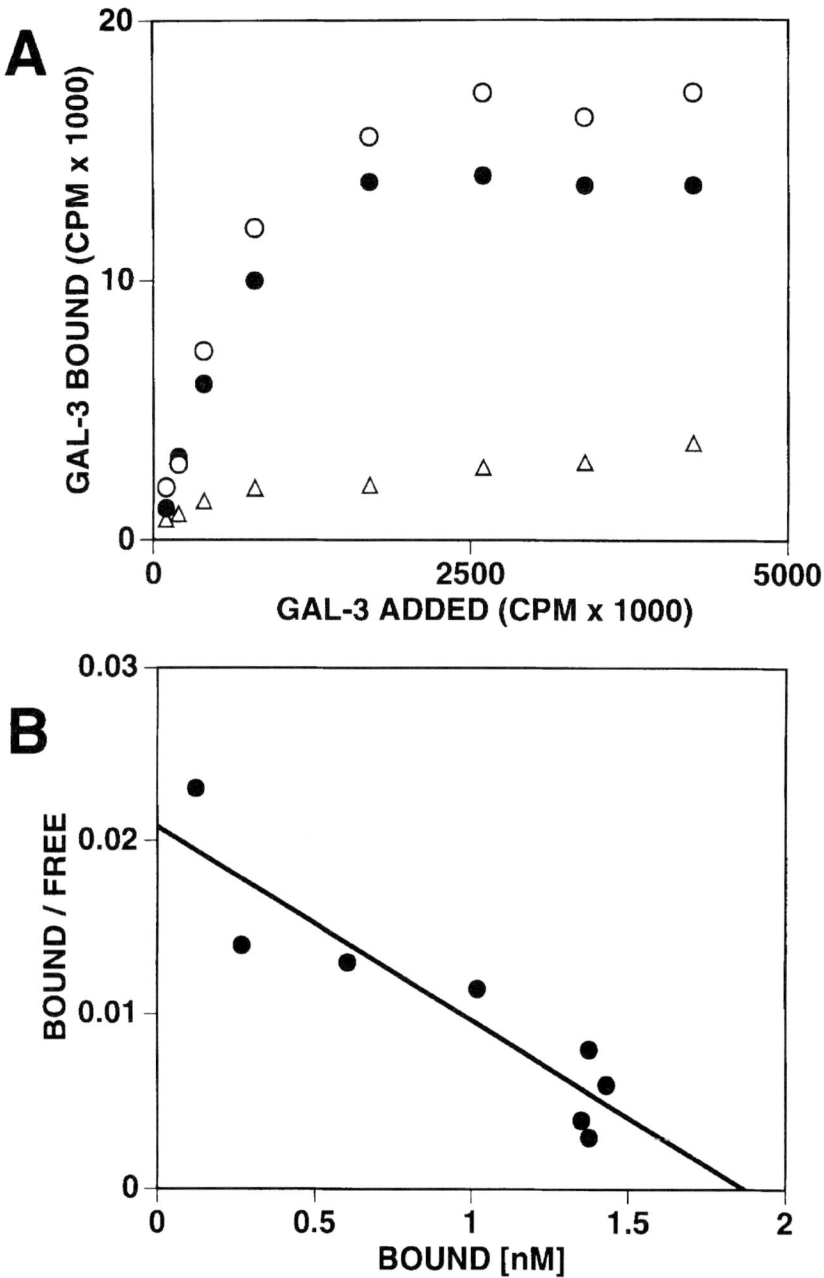

Figure 7 Recombinant galectin-3 (GAL-3) binds to laminin in a solid-phase radioligand binding assay. (A) Radioiodinated recombinant galectin-3 was added to laminin-coated microwells for 4 hours at 4°C, and the bound radioactivity was determined by scintillation counting. Specific binding (closed circles) represents the difference between the values obtained in the absence (open circles) or the presence (open triangles) of unlabelled galectin-3. (B) Scatchard plot of galectin-3 binding to laminin.

demonstrated that lamps are ligands for galectin-1 in various cell lines including the A121 ovary carcinoma and the KM12 colon carcinoma cell lines (Do *et al.*, 1990; Skrincosky *et al.*, 1993; Ohannesian *et al.*, 1994; Ohannesian *et al.*, 1995). Similar data from affinity chromatography experiments were obtained for galectin-3 in A375 melanoma and KM12 cells (Inohara and Raz, 1994b; Ohannesian *et al.*, 1995; Dong and Hughes, 1997).

Mac-2 Binding Protein

Coimmunoprecipitation and immunoaffinity chromatography experiments revealed the existence of two related glycoproteins present in colon cancer cell lines, Mac-2 binding protein-1 and -2 (M2BPs), able to bind galectin-3 (Rosenberg *et al.*, 1991). The same molecule, previously designated tumor L3 antigen (Linsley *et al.*, 1986) and melanoma-associated antigen (MAA) (Natali *et al.*, 1982) was also characterized from the human breast carcinoma SK-BR-3 cell line (Koths *et al.*, 1993), and found by affinity chromatography to be a ligand for galectin-3 in human A375 melanoma cells (Inohara and Raz, 1994b).

Fibronectin

Affinity chromatography experiments demonstrated that placenta and amniotic, but not plasma fibronectin, bind galectin-1 (Ozeki *et al.*, 1995). Galectin-3 similarly binds fetal fibronectin but not the adult form, a fact due to differences in glycosylation of the two forms of fibronectin (Sato and Hughes, 1992).

Mucin

Affinity chromatography and histochemical experiments demonstrated that galectin-1 binds mucin from the quail and rat intestine (Fang *et al.*, 1993; Wasano and Hirakawa, 1997). Early reports suggested interactions of galectin-3 with intestinal goblet mucin (Beyer and Barondes, 1982). Subsequent solidphase assay experiments demonstrated that galectin-3 binds mucin, with greater affinity for that purified from the highly metastatic LS-Lim6 human colon cancer cells that from the parental, low metastatic cell line (Bresalier *et al.*, 1996).

CEA

Galectin-1 and galectin-3 bind the carcinoembryonic antigen (CEA) in human KM12 colon carcinoma cells as determined by coimmunoprecipitation, affinity chromatography and ligand blotting experiments (Ohannesian *et al.*, 1994; Ohannesian *et al.*, 1995). A cell surface CEA-related glycoproteins from neutrophils, NCA-160, also designated CD66, also binds galectin-3 (Yamaoka *et al.*, 1995).

Other ligands

Other galectin-1 ligands include the $\alpha_7\beta_1$ integrin (Gu *et al.*, 1994); glycolipids (Caron *et al.*, 1993; Solomon *et al.*, 1991); the T lymphocyte cell surface glycoproteins CD43 and CD45 (Baum *et al.*, 1995); a brain protein homologous to hsp90 (Chadli *et al.*, 1997); and milk glycoproteins (Takeuchi *et al.*, 1982). In addition, galectin-1 binds actin by non-carbohydrate interactions (Joubert *et al.*, 1992).

As mentioned previously, galectin-3 binds IgE (Liu and Orida, 1984; Liu *et al.*, 1985; Laing *et al.*, 1989; Hsu *et al.*, 1992). It also binds neural recognition molecules (Probstmeier *et al.*, 1995); advanced glycation end (AGE) products, the reactive derivatives of nonenzymatic glucose-protein condensation reactions (Vlassara *et al.*, 1995); the alpha subunit (CD11b) of the CD11b-CD18 integrin (Mac-1, a receptor for C3b-opsonized particles, fibrinogen and ICAM-1); and the heavy chain of CD98 present on macrophages and activated T cells (Dong and Hughes, 1997). In addition, galectin-3 binds bacterial lipopolysaccharides through both its CRD and amino-terminal domain (Mey *et al.*, 1996). It forms heterodimers with Bcl-2 by a NWGR motif (Yang *et al.*, 1996). Galectin-3 also binds the carbohydrate-binding protein 70 (CBP70), a C-type lectin located in the cell nucleus, in a lactose-dependent manner (Seve *et al.*, 1993; Seve *et al.*, 1994). Galectin-3 is associated with the nuclear matrix, binds single-stranded DNA and RNA (Wang *et al.*, 1995); RNA seems to be the preferred nucleic acid ligand for galectin-3 (Hubert *et al.*, 1995). Recent work describes αGalNAc glycans-containing cytokeratins as the first described cytoplasmic glycoconjugate ligands for galectin-3 (Goletz *et al.*, 1997).

Galectin-4 binds a 120-kDa glycoprotein component of the adherens junction complex on ligand blotting in a lactose-dependent manner (Chiu *et al.*, 1994).

SUBCELLULAR LOCALIZATION OF GALECTINS

Galectins do not have a signal peptide; they are ubiquitous soluble proteins found in various cellular compartments and can be secreted in the extracellular medium.

Galectin-1 is synthesized on free ribosomes in the cytoplasm (Wilson *et al.*, 1989). Its subcellular localization has been precisely determined in CHO cells (Cho and Cummings, 1995): half of total galectin-1 is intracellular and the other half is on the cell surface, associated with carbohydrate structures. Its externalization half life has been evaluated to 20 hours by pulse-chase labelling. Galectin-1 accumulates in the culture medium in an oxidized inactive form (Whitney *et al.*, 1986). Secretion has been examined in myoblasts, where it is concentrated under the plasma membrane; evagination of the latter structure forms vesicles that are released in a process called ectocytosis (Cooper and Barondes, 1990). Galectin-1 localization has also been examined in cancer cells such as melanoma cells, and found both in the cytoplasm and associated to the cell surface (Figure 8) (van den Brûle *et al.*, 1995a).

Galectin-3 has also been located in various cell compartments. In COS-7 cells, plasma membrane targeting, vesicular budding and galectin-3 release is dependent on the amino-terminal domain of the molecule (Mehul and Hughes, 1997). Secretion has been examined in the MDCK and BHK models and found to use a non classical pathway (Lindstedt *et al.*, 1993; Sato *et al.*, 1993). Secretion is usually proportional to cell surface expression in macrophages, and cell surface galectin-3 is associated to carbohydrate structures (Sato and Hughes, 1994). Unexpectedly for a lectin, galectin-3 has been detected in the nucleus of several cell types, including 3T3 and SL66 fibroblasts (Moutsatsos *et al.*, 1986; Moutsatsos *et al.*, 1987; Hubert *et al.*, 1995)

GAL-1　　　　　　　　GAL-3

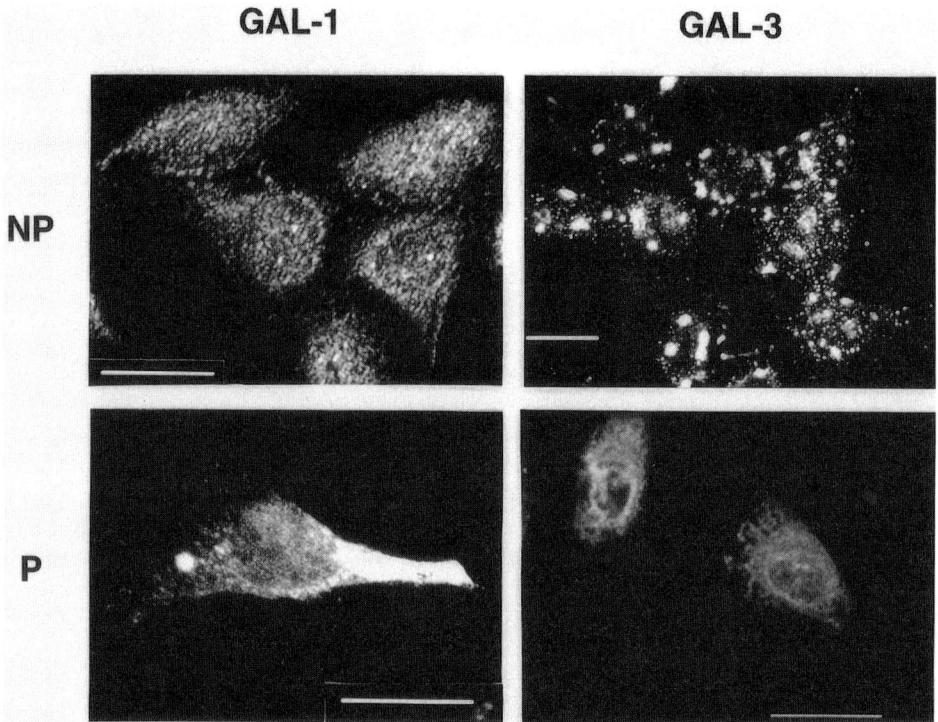

Figure 8 Subcellular localization of galectin-1 and galectin-3 in A2058 human melanoma cells. Both galectins were found intracellularly (cytoplasm and/or nucleus) as well as on the cell surface. Adherent cells were fixed and permeabilized (P) or not (NP), incubated with the anti-galectin-1 (GAL-1) or anti-galectin-3 (GAL-3) antiserum, and then with a fluorescein-conjugated secondary antibody. Bar = 25 microns.

and colon epithelial cells (Lotz *et al.*, 1993). Immunolocalization and 2D-electrophoresis studies on fibroblasts demonstrated that the cytosol contains the phosphorylated form of galectin-3; the nuclear compartment contains both phosphorylated and non phosphorylated variants (Cowles *et al.*, 1990). All fractions increase to variable extents in proliferative fibroblasts (Cowles *et al.*, 1990), but the biological significance of this event is unknown. Galectin-3 is localized intracellularly and on the cell surface of human melanoma cells, where it shows a typical appearance (Figure 8) (van den Brûle *et al.*, 1995b).

Galectin-4 has been localized in regions between desmosomes in pig tongue epithelium (Chiu *et al.*, 1992). Immunolocalization of galectin-3 and galectin-4 in the T-84 human colon carcinoma cell line demonstrated that in confluent cells, galectin-4 is mostly cytosolic and concentrated at the basement membrane pole, whereas galectin-3 is in apical granular inclusions. In subconfluent cells, galectin-4 is located at the leading edge of the cell, and galectin-3 more proximally on lamellipodia (Huflejt *et al.*, 1997), suggesting specific functions for these two proteins. Galectin-8 is present in prostate carcinoma cells and their conditioned medium (Su *et al.*, 1996).

EXPRESSION OF GALECTINS IN CANCER

Expression of galectins has been evaluated in in vitro and animal models, and in human tumors. Early reports using a multispecific monoclonal antibody in FACS demonstrated increased expression of galectins on the cell surface of metastatic neoplastic cells such as B16 and K1735 melanoma, and UV-2237 fibrosarcoma cells, compared to untransformed cells like 3T3 fibroblasts (Raz et al., 1986). It was also demonstrated that tumor extracts induced more hemagglutination than extracts from normal tissues (Allen et al., 1987). Subsequent studies examined expression of selected specific galectins.

Pleiotropic expression of galectin-1 was demonstrated in several cancer cell lines and in ovary cancer cells from effusions and cell lines (Table 3). Additional information came from studies comparing expression in cell lines. Several studies demonstrated increased expression of galectin-1 in cancer or transformed cells compared to parental cells, and in higher compared to lower metastatic cells, in several cell line models from various sources (Table 3). However, this was not the case in all models. Galectin-1 expression was examined in human carcinomas by several groups. The most abundant reports concerned colon and thyroid carcinomas (Table 3). Most of these studies reported increased expression of galectin-1 in carcinomas from colon, endometrium and thyroid, compared to the corresponding normal tissues, at the mRNA or protein level.

Modulation of galectin-1 expression was examined in in vitro models. Cell differentiation by agents such as cyclic AMP, retinoid acid, dimethylsulfoxide and butyrate was accompanied by increased galectin-1 expression in KM12P colon carcinoma cells; decreased expression in S20 neuroblastoma, MDA-MB175 breast carcinoma, and HL-60 and THP-1 leukemia cells; and by no modulation in K1735P melanoma cells (Lotan et al., 1989a, Lotan et al., 1989b). Glucocorticoid treatment of the human CEM C7 leukemia cell line increased galectin-1 mRNA levels (Goldstone and Lavin, 1991). Fusion of cells not expressing galectin-1 with undifferentiated or tumorigenic cells induced increased galectin-1 expression (Chiariotti et al., 1994). The galectin-1 gene has been cloned and characterized as a 4-exon gene (Ohyama and Kasai, 1988; Gitt and Barondes, 1991; Chiariotti et al., 1991). It is located on chromosome 22q12-13 (Mehrabian et al., 1993). Its promoter activity has been examined and found dependent on methylation status (Salvatore et al., 1995; Benvenuto et al., 1996). However, its promoter activity has not been examined yet in cancer cells.

Expression of galectin-3 was the subject of numerous studies that showed apparently conflicting results. Galectin-3 protein and/or mRNA has been detected in many cancer cell lines including melanoma, fibrosarcoma, leukemia, colon, breast and squamous carcinoma cell lines (Table 4). Comparison of galectin-3 expression in in vitro models using cell lines usually demonstrated increased expression in cancer cell lines from colon, kidney, and melanoma and sarcoma cell lines, compared to parental or other "normal" cell lines, as determined at the protein and/or mRNA level (Table 4). Galectin-3 expression is increased in metastatic UV-2237 compared to normal embryonal fibroblasts (Raz et al., 1987b), and in highly metastatic UV-2237 fibrosarcoma cells compared to low metastatic clones (Raz et al., 1987a). A similar increase in galectin-3 protein and mRNA levels was found in Kirsten murine sarcoma virus-transformed 3T3 cells compared to parental cells (Moutsatsos et al.,

Table 3   Galectin-1 expression and modulations in cancer

SOURCE	METHOD	RESULTS	REFERENCES
CELL LINES			
BREAST CARCINOMA			
MDA-MB 175 cells	WB*	detection of galectin-1; decreased by RA	(Lotan et al., 1989b)
MDA-MB435 cells	WB	detection of galectin-1; not modulated by RA	(Lotan et al., 1989b)
T47D, MDA-MB 231,	FACS,	detection of galectin-1 in all but MDA-MB 435 and SK-Br-3 cell lines	(Nangia-Makker et al., 1995)
BT-549, MDA-MB 435, SK-Br-3 cell lines	WB, NB		
CHORIOCARCINOMA			
JEG-3 choriocarcinoma cells	NB	detection of galectin-1	(Couraud et al., 1989)
EMBRYONAL CARCINOMA			
F9 cells	WB	detection of galectin-1; increased by RA	(Lotan et al., 1989b)
SARCOMA			
UV-2237 fibrosarcoma cells, rat embryonic fibroblasts	IPPT	identical expression	(Raz et al., 1987b)
3T3 and subclones	IPPT	identical expression	(Raz et al., 1987b)
Highly metastatic UV-2237 cells, low metastatic subclones	RNA SB	increased expression in highly metastatic cells	(Raz et al., 1987a)
SV-40-transformed 3T3, parental 3T3 cells	NB	increased in SV-40-transformed 3T3 cells	(Raz et al., 1988)
ras-transformed 3T3, parental 3T3 cells	NB	increased in ras-transformed 3T3 cells	(Hebert and Monsigny, 1994)
GASTROINTESTINAL			
SK hepatoma cell line	NB	detection of galectin-1	(Couraud et al., 1989)
KM12P cell line	WB	absence of galectin-1; increased by butyrate	(Lotan et al., 1989b)
21 colon cancer cell lines	WB	detection in 7/21 cell lines	(Ohannesian et al., 1995)
HEAD AND NECK			
14 HNSCC cell lines	WB	detection in 12/14 cell lines; variable modulation with butyrate, RA, TGF-ß, hydrocortisone	(Gillenwater et al., 1996)

Table 3 Continued.

SOURCE	METHOD	RESULTS	REFERENCES
KIDNEY			
NRK-RSV, parental NRK cells	NB	identical expression	(Raz et al., 1988)
LEUKEMIA			
HL-60 leukemia cells	WB, NB	detection of galectin-1	(Couraud et al., 1989, Allen et al., 1991a)
HL-60 leukemia cells	WB	detection of galectin-1; decreased by RA, DMSO	(Lotan et al., 1989b)
THP-1 leukemia cells	WB	detection of galectin-1; decreased by RA	(Lotan et al., 1989b)
KG-1a erythroleukemia cell line	NB	detection of galectin-1	(Couraud et al., 1989)
CEM C7 leukemia cell line	NB	no basal expression; induction by DXM	(Goldstone and Lavin, 1991)
MELANOMA			
B16-F1 melanoma, MDF cells	NB	identical expression	(Raz et al., 1988)
B16-F1 melanoma cells	WB	detection of galectin-1; not modulated by RA or cAMP	(Lotan et al., 1989b)
K-1735P melanoma cells	WB, IF	detection of galectin-1; not modulated by RA or cAMP	(Lotan et al., 1989b; Lotan et al., 1989a)
A2058 melanoma cells	NB, WB, IF	detection of galectin-1	(van den Brûle et al., 1995a; van den Brûle et al., 1996)
A375 melanoma cells	NB, WB, IF	detection of galectin-1	(van den Brûle et al., 1995a)
NEUROBLASTOMA			
S20 neuroblastoma cells	WB	detection of galectin-1, decreased by cAMP	(Lotan et al., 1989b)
OVARY			
A121 ovary carcinoma cells	ELISA, IPPT, WB	detection of galectin-1	(Allen et al., 1990)
OVCAR-3 ovary carcinoma cells	NB, WB	absence of galectin-1	(van den Brûle et al., 1996)

Table 3 Continued.

SOURCE	METHOD	RESULTS	REFERENCES
THYMUS			
primary cultures from thymomas	WB	detection of galectin-1	(Hafer-Macko et al., 1996)
THYROID			
17 thyroid cancer cell lines	NB	increased expression in malignant cells	(Chiariotti et al., 1992)
TISSUES			
BREAST			
24 breast carcinoma samples	IHC	occasional staining in carcinomas	(Gabius et al., 1986)
COLON			
48 colorectal carcinomas	WB	detection of galectin-1	(Irimura et al., 1991)
20 colorectal carcinomas, 6 adenomas	IHC	positivity in 5/20 carcinomas, 5/6 adenomas	(Irimura et al., 1991)
46 colon carcinomas	WB	no modulation	(Lotan et al., 1991)
46 colon carcinomas	IHC	associated with secreted material	(Lotan et al., 1991)
15 adenomas, 25 colorectal carcinomas, 11 metastases	IHC	progressive overexpression from normal to carcinomas	(Sanjuan et al., 1997)
HEAD AND NECK			
35 primary HNSCC samples	IHC	detection in the periphery of invasive cells	(Gillenwater et al., 1996)
GYN TUMORS			
GYN carcinoma cells from effusions	IHC	detection of galectin-1	(Allen et al., 1991b)
20 advanced endometrial carcinomas	IHC	increased expression vs. normal	(van den Brûle et al., 1996)
SKIN			
basal cell carcinomas	IHC	absence of galectin-1	(Allen et al., 1991b)

Table 3 Continued.

SOURCE	METHOD	RESULTS	REFERENCES
THYMUS			
27 hyperplastic including 8 thymoma from myasthenia gravis patients	IHC, WB	detection of galectin-1 in the samples	(Hafer-Macko et al., 1996)
THYROID			
10 thyroid papillary carcinoma, 3 normal, 4 benign samples	NB	increased expression in malignant samples	(Chiariotti et al., 1992)
58 thyroid carcinomas, 11 benign, 6 normal samples	NB, IHC	increased expression in 28/40 papillary carcinomas, in 6/7 anaplastic carcinomas; identical in folllicular carcinomas and adenomas, vs. normal	(Chiariotti et al., 1995)
32 thyroid carcinomas, 10 adenomas, 33 normal tissue samples	IHC, WB	increased expression in papillary and follicular carcinomas, variable in medullary samples, vs. normal	(Xu et al., 1995)

*Abbreviations used: ELISA, Enzyme-linked immunosorbent assay; IF, immunofluorescence; IHC, immunohistochemistry; IPPT, immunoprecipitation; NB, Northern blotting; SB, slot blotting; WB, Western blotting; cAMP, cyclic AMP; DMSO, dimethylsulfoxide; DXM, dexamethasone; RA; retinoic acid.

Table 4 Galectin-3 expression and modulations in cancer

SOURCE	METHOD	RESULTS	REFERENCES
CELL LINES			
BREAST CARCINOMA			
MDA-MB435 cell line	WB*	absence of galectin-3; no effect of RA	(Lotan *et al.*, 1989b)
MDA-MB 175 cell line	WB	detection of galectin-3; no effect of RA	(Lotan *et al.*, 1989b)
MCF-7 cell line	NB, WB	detection of galectin-3; increased by R5020 progestin	(van den Brûle *et al.*, 1992)
T47D cell line	NB, WB	detection of galectin-3; increased by R5020 progestin	(van den Brûle *et al.*, 1992)
T47D, MDA-MB 231, BT-549,	FACS,	detection of galectin-3 in all but BT-549 and SK-Br-3 cell lines;	(Nangia-Makker *et al.*, 1995)
MDA-MB 435, SK-Br-3 cell lines	WB, NB	expression correlates with reported tumorigenicity	
T47D, ZR75, CAMA, 531E, 242A,	WB, FACS	detection of galectin-3; increased expression in malignant cell lines	(Le Marer and Hughes, 1996)
341, MTSV1-7 cell lines			
EMBRYONAL CARCINOMA			
F9 cells	WB	absence of galectin-3; no effect of RA	(Lotan *et al.*, 1989b)
F9 cells	WB	absence of galectin-3	(Foddy *et al.*, 1990)
GASTROINTESTINAL			
KM12P cell line	WB	absence of galectin-3; no effect of butyrate	(Lotan *et al.*, 1989b)
HT-29 colon carcinoma cell line	NB	detection of galectin-3	(Cherayil *et al.*, 1990, Foddy *et al.*, 1990)
CaCo-2 colon carcinoma cell line	NB	detection of galectin-3	(Cherayil *et al.*, 1990)
8 colon carcinoma cell lines	FACS	increased surface expression in poorly differentiated cell lines	(Lee *et al.*, 1991)
21 colon cancer cell lines	WB	detection in 20/21 cell lines	(Ohannesian *et al.*, 1995)
GYN TUMORS			
HeLa cervix carcinoma cell line	NB	detection of galectin-3	(Cherayil *et al.*, 1990, Raz *et al.*, 1991)
A431 vulvar squamous carcinoma	IPPT	absence of galectin-3	(Foddy *et al.*, 1990)
cell line			

Table 4 Continued.

SOURCE	METHOD	RESULTS	REFERENCES
HEAD AND NECK			
SCC cell line	NB	detection of galectin-3	(Cherayil et al., 1990)
14 HNSCC cell lines	WB	detection in all cell lines; variable modulation with butyrate, RA, TGF-ß, hydrocortisone	(Gillenwater et al., 1996)
KIDNEY			
NRK-RSV, parental NRK cells	NB	increased expression in transformed cells	(Raz et al., 1988)
LEUKEMIA			
HL-60 leukemia cells	WB	absence of galectin-3; no effect of RA, DMSO	(Lotan et al., 1989b)
HL-60 leukemia cells	NB	detection of galectin-3; increased by DMSO or PMA	(Cherayil et al., 1990)
HL-60 leukemia cells	NB, WB	detection of low amounts of galectin-3; no effect of RA; increased by TPA	(Nangia-Makker et al., 1993)
THP-1 leukemia cells	WB	absence of galectin-3; no effect of RA	(Lotan et al., 1989b)
THP-1 leukemia cells	NB	detection of galectin-3	(Cherayil et al., 1990)
MELANOMA			
B16-F1 melanoma, MDF cells	NB	increased expression in B16 cell line	(Raz et al., 1988)
B16-F1 melanoma cells	WB	detection of galectin-3; decreased by RA or cAMP	(Lotan et al., 1989b)
K-1735P melanoma cells	WB, IF	detection of galectin-3; decrease by RA or cAMP	(Lotan et al., 1989b, Lotan et al., 1989a)
A375 melanoma cells	NB	detection of galectin-3	(Raz et al., 1991)
A375 melanoma cells	NB, WB, IF	detection of galectin-3	(van den Brûle et al., 1995a)
A2058 melanoma cells	NB, WB, IF	detection of galectin-3	(van den Brûle et al., 1995a; van den Brûle et al., 1996)
3 melanoma cell lines and 5 subclones	FACS, CM, WB	detection of galectin-3; extracellular galectin-3 containing vesicles in metastatic cell lines	(Mey et al., 1994)

Table 4   Continued.

SOURCE	METHOD	RESULTS	REFERENCES
NEUROBLASTOMA			
S20 neuroblastoma cells	WB	absence of galectin-3, no effect of cAMP	(Lotan et al., 1989b)
SARCOMA			
Highly metastatic UV-2237 cells, low metastatic subclones	RNA SB	increased expression in highly metastatic cells	(Raz et al., 1987a)
UV-2237 fibrosarcoma cells, rat embryonic fibroblasts	IPPT	increased expression in UV-2237 cells	(Raz et al., 1987b)
3T3 and subclones	IPPT	increased expression in metastatic cells	(Raz et al., 1987b)
SV-40-transformed 3T3, 3T3 cells	NB	increased in SV-40-transformed 3T3 cells	(Raz et al., 1988)
HT-1080 fibrosarcoma cells	NB	detection of galectin-3	(Raz et al., 1991)
transformed 3T3, parental 3T3 cell lines	WB, IF, NB	increased expression in transformed cells	(Crittenden et al., 1984; Moutsatsos et al., 1987; Agrwal et al., 1989)
ras-transformed 3T3, parental 3T3 cells	NB	increased in ras-transformed 3T3 cells	(Hebert and Monsigny, 1994)
ras-transformed 3T3, parental 3T3 cell lines	WB	increased expression in transformed cells	(Chammas et al., 1996)
ras-transformed HOS, parental HOS cell lines	NB, CM	increased expression in transformed cells	(Hebert et al., 1996)
TISSUES			
BREAST			
24 breast carcinoma samples	IHC	detection in almost all carcinomas, in luminal cells and secreted products, in benign lesions	(Gabius et al., 1986)
12 fibroadenomas, 15 fibrocystic lesions, 22 in situ carcinomas, 49 infiltrating ductal carcinomas	IHC	decreased expression in carcinomas, particularly if node infiltration	(Castronovo et al., 1996)

Table 4 Continued.

SOURCE	METHOD	RESULTS	REFERENCES
COLON			
48 colorectal carcinomas	WB	detection of galectin-3; increased in Dukes' D stage tumors	(Irimura et al., 1991)
20 colorectal carcinomas, 6 adenomas	IHC	positivity in 9/20 carcinomas, 0/6 adenomas	(Irimura et al., 1991)
46 colon carcinomas	WB	increased in Dukes'D stage tumors	(Lotan et al., 1991)
46 colon carcinomas	IHC	positivity in cytoplasm of tumor cells	(Lotan et al., 1991)
21 pairs of matched colon carcinomas and normal mucosas	NB, WB	decreased expression in carcinomas, particularly in Dukes' D stage tumors	(Castronovo et al., 1992a)
13 colon carcinomas, 9 adenomas and normal mucosa	NB, WB, IHC	decreased expression, loss of nuclear localization in carcinomas	(Lotz et al., 1993)
29 adenomas, 66 carcinomas, 23 carcinomas with 35 associated metastases	IHC	increased expression in high-grade dysplasia and early invasive cancer, compared to adenomas, and in metastases compared to primary tumors; increased expression in the primary tumors correlates with decreased survival	(Schoeppner et al., 1995)
183 colon carcinomas	IHC	expression not associated with survival	(Lise et al., 1997)
25 adenomas, 87 colorectal carcinomas, 39 node metastases	IHC	down-regulation at initial stages, and increased cytoplasmic expression at later stages	(Sanjuan et al., 1997)
HEAD AND NECK			
35 primary HNSCC samples	IHC	detection in the periphery of invasive cells	(Gillenwater et al., 1996)
GASTRIC			
26 carcinomas and adjacent mucosa	WB	similar expression in 14, increased expression in 9 carcinomas, vs. normal mucosa	(Lotan et al., 1994b)
39 carcinomas and adjacent mucosa	IHC	higher expression in 55% of the well differentiated carcinomas, and in 50% of the stage III and IV carcinomas, vs. normal tissue	(Lotan et al., 1994b)

Table 4 Continued.

SOURCE	METHOD	RESULTS	REFERENCES
GYN TUMORS			
20 advanced endometrial carcinomas	IHC	decreased expression vs. normal; loss of nuclear localization in myometrium-invading tumors	(van den Brûle et al., 1996)
30 ovary carcinomas	WB, NB, IHC	decreased expression in carcinomas vs. normal tissue	(van den Brûle et al., 1994a)
LYMPHOMA			
43 leukemia samples	IHC	positivity in all large cell lymphomas, and only in 2/35 other types	(Konstantinov et al., 1996)
THYROID			
32 thyroid carcinomas, 10 adenomas, 33 normal tissue samples	IHC, WB	increased expression in carcinomas, vs. benign and normal tissues	(Xu et al., 1995)
18 papillary, 8 follicular, 3 poorly differentiated, 5 anaplastic, 6 medullary and 1 Hürtle cell carcinomas	IHC, NB	high positivity in 18 papillary carcinomas, less intense staining in anaplastic, poorly differentiated, medullary and follicular carcinomas	(Fernandez et al., 1997)

*Abbreviations used: PMA, phorbol-12myristate 13-acetate; TPA, 12-O-tetradecanoylphorbol-13-acetate; CM, confocal microscop; FACS, fluorescence-activated cell sorting; IF, immunofluorescence; IHC, immunohistochemistry; IPPT, immunoprecipitation; NB, Northern blotting; SB, slot blotting; WB, Western blotting; cAMP, cyclic AMP; DMSO, dimethylsulfoxide; RA; retinoic acid.

1987; Agrwal *et al.*, 1989), in ras-transformed 3T3 fibroblasts (Hebert and Monsigny, 1994; Chammas *et al.*, 1996), and in transformed HOS cells (Hebert *et al.*, 1996). An other report described increased galectin-3 mRNA levels in 3 pairs of normal and transformed cells (Raz *et al.*, 1988). In a panel of 8 colon carcinoma cell lines, a higher cell surface galectin-3 expression has been correlated with undifferentiation (Lee *et al.*, 1991). Increased galectin-3 expression correlates with tumorigenicity in a panel of breast carcinoma cell lines (Nangia-Makker *et al.*, 1995; Le Marer and Hughes, 1996).

Galectin-3 expression was examined in human carcinoma samples. Many studies have been performed in colon cancer, and their various results constitute a basis for a general discussion about the studies of galectin-3 expression in carcinomas. To date, three studies have detected increased expression of galectin-3. The first one demonstrated increased expression in Dukes' D carcinomas, using Western blotting and histochemical staining, and an polyclonal antiserum raised against a mixture of bovine lung lectins (Irimura *et al.*, 1991). A second study found increased expression in Dukes' D tumors as determined by Western blotting with polyclonal antibodies (Lotan *et al.*, 1991). A third, immunohistochemical study of 153 samples using the anti-Mac-2 monoclonal antibody M3/38 demonstrated increased galectin-3 expression in high grade dysplasia and early invasive cancer, compared to adenomatous tissue, and in metastases compared to primary tumors; increased expression in primary tumors was correlated with decreased survival of the patients (Schoeppner *et al.*, 1995). On the other hand, several independent studies demonstrated decreased expression in carcinomas. In a population of 21 pairs of colon carcinoma and corresponding matched normal colonic mucosa, decreased expression of galectin-3 was found by Northern blotting, particularly in metastatic tumors (Dukes' D) (Castronovo *et al.*, 1992a). Decreased galectin-3 expression was further demonstrated in 13 colon carcinomas at the protein and mRNA levels, with loss of nuclear localization of the lectin on immunohistochemical slides using the M3/38 monoclonal antibody (Lotz *et al.*, 1993). In another study, galectin-3 expression was not associated with survival in a series of 183 colon samples (Lise *et al.*, 1997). A recent, extensive survey about 39 normal colonic mucosas, 25 adenomas, 87 carcinomas and 39 lymph node metastases demonstrated down regulation of galectin-3 at the initial stages of colon cancer progression, with subsequent increased cytoplasmic expression in advanced phases (Sanjuan *et al.*, 1997), a finding that draws an interesting picture of galectin-3 expression during colon cancer progression using a large population of samples. Expression of galectin-3 in breast carcinomas has been examined in two studies. The first reported staining in normal luminal cells and their secretory products, benign lesions and almost all 24 mammary carcinomas examined (Gabius *et al.*, 1986). A second analysis in a population of 98 breast lesions demonstrated clearly decreased galectin-3 expression in carcinomas compared to normal and benign tissues as determined by immunohistochemistry (Figure 9) (Castronovo *et al.*, 1996). Galectin-3 expression in gastric cancer samples has been examined in two studies performed from the same group. Western blotting experiments demonstrated similar expression in 14, and increased expression in 9/26 samples, compared to normal mucosa; higher galectin-3 expression was determined by imunohistochemical staining in 55% of the well differentiated carcinomas, and in 50% of the advanced (stages III and IV) carcinomas, compared to normal tissues, in 39 paired primary cancers and normal mucosa. Examination

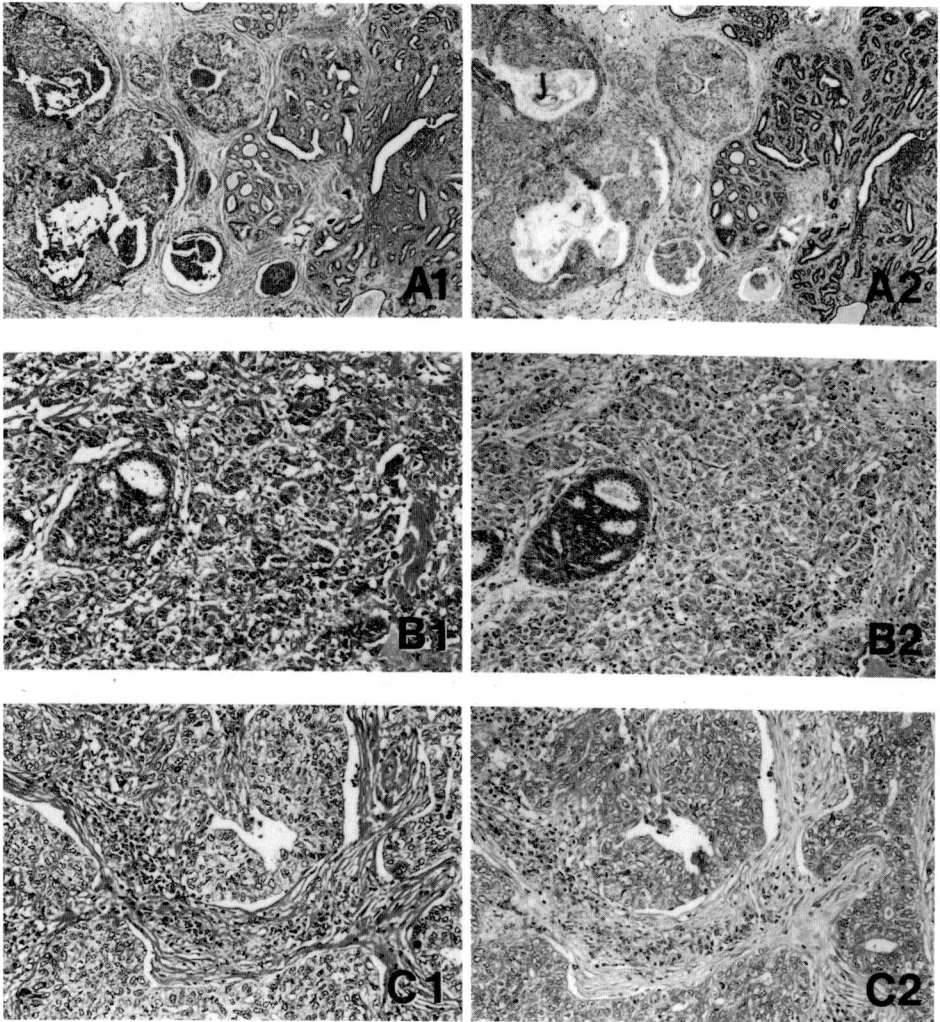

Figure 9 Decreased expression of galectin-3 in human breast carcinoma. Paraffin slides of breast lesions were stained by hematoxylin and eosin (A1, B1, C1) and immunostained by an anti-recombinant galectin-3 polyclonal antiserum (A2, B2, C2). Strong immunostaining for galectin-3 is found in normal breast epithelium (A2, right) and in benign lesions. Intraductal carcinoma samples were characterized by intermediate positivity for galectin-3 (DCIS, B2). Decreased expression is found in invasive ductal carcinoma (A2, left; B2; C2). (See Colour Plate IV)

of 74 paired primary carcinomas and metastases revealed higher expression in liver metastases compaired to well-differentiated primaries in 31% of the cases, and in lymph node metastases compaired to poorly differentiated carcinomas in 38% of the cases (Lotan *et al.*, 1994b). Other results concern carcinomas of various origins. Decreased galectin-3 expression was demonstrated immunohistochemically in a population of 20 advanced uterine adenocarcinomas, and tumors that invaded the myometrium were characterized by loss of nuclear galectin-3 (van den Brûle *et al.*, 1996), a finding also reported in colon carcinoma cells (Lotz *et al.*, 1993). Increased galectin-3 protein expression was found in 32 thyroid cancer samples, compaired to benign and normal tissue (Xu *et al.*, 1995). High expression was also found at the protein and mRNA levels in thyroid papillary carcinoma, but less prominent positivity was found in less differentiated, medullary and follicular carcinomas (Fernandez *et al.*, 1997). Galectin-3 is a marker of anaplastic large-cell lymphoma (Konstantinov *et al.*, 1996), and is expressed in head and neck squamous cell carcinomas (Gillenwater *et al.*, 1996).

The apparently discordant data of the numerous reports available (increased or decreased galectin-3 expression in carcinoma cells, compared to normal tissue) could be related to the fact that some studies examined different stages of colon carcinoma progression, while other discrepancies could be related to the use of different evaluation techniques: Western and Northern blotting have a poor ability to specifically determine specific galectin expression in cancer cells originating from heterogenous tissues containing stromal cells and stromal matrix; immunohistochemical staining can efficiently discriminate these cancer cells from the surrounding tissue, but needs adequate antibodies and adsorbtion and specificity controls, that have been performed only in several studies (Castronovo *et al.*, 1996; Sanjuan *et al.*, 1997). It seems however that the data, such as from colon carcinoma, obtained from various studies using different sample populations and tools, could finally fit in a global study (Sanjuan *et al.*, 1997). In this case, galectin-3 expression would decrease during the initial phases of colon cancer progression, and could increase as the tumor later spreads. Of course, this pattern of expression remains to be demonstrated for other cancer types.

Modulation of galectin-3 expression has been reported in various systems (Table 4). Differentiation of melanoma cell by cyclic AMP, retinoic acid, dimethylsulfoxide or butyrate decreased galectin-3 protein levels (Lotan *et al.*, 1989a; Lotan *et al.*, 1989b). Galectin-3 expression is increased in HL60 cells during differentiation by dimethylsulfoxide or phorbol esters (Cherayil *et al.*, 1990, Nangia-Makker *et al.*, 1993). Estrogen and progestin able to affect the metastatic phenotype increase galectin-3 mRNA and protein levels in T47D and MCF-7 cell lines (Figure 10) (van den Brûle *et al.*, 1992). The galectin-3 gene has been localized on chromosome 14q21-22 (Raimond *et al.*, 1997), and has been cloned and characterized a a 6-exon gene (Gritzmacher *et al.*, 1992, Rosenberg *et al.*, 1993, Voss *et al.*, 1994). Strong promoter activity has been characterized has been characterized in the second intron; interestingly, this activity is down-regulated by wild type p53, as opposed to mutated p53 (Raimond *et al.*, 1995), a finding that could be related to increased galectin-3 expression in mutated p53 tumors.

Few data is available for the other described galectins. Galectin-4 mRNA expression is down regulated in 18/19 human colon cancer samples compared to the corresponding normal mucosa; only two cell lines (HT29 and LS174T) out of five express galectin-4 (Tardy *et al.*, 1995, Rechreche *et al.*, 1997). Analysis of proteome profiles of squamous cell bladder carcinomas determined that galectin-7 expression is decreased in poorly differentiated tumors (Ostergaard *et al.*, 1997). Galectin-8 is expressed at the surface of various human prostate carcinoma cell lines; RT-PCR analysis and immunohistochemistry demonstrated that galectin-8 is clearly expressed in human prostate carcinoma, but not in the corresponding normal tissue; prostatic intraepithelial neoplasia (PIN) is weakly positive (Su *et al.*, 1996). Galectin-9 mRNA has been detected in Hodgkin diseased tissues (Türeci *et al.*, 1997).

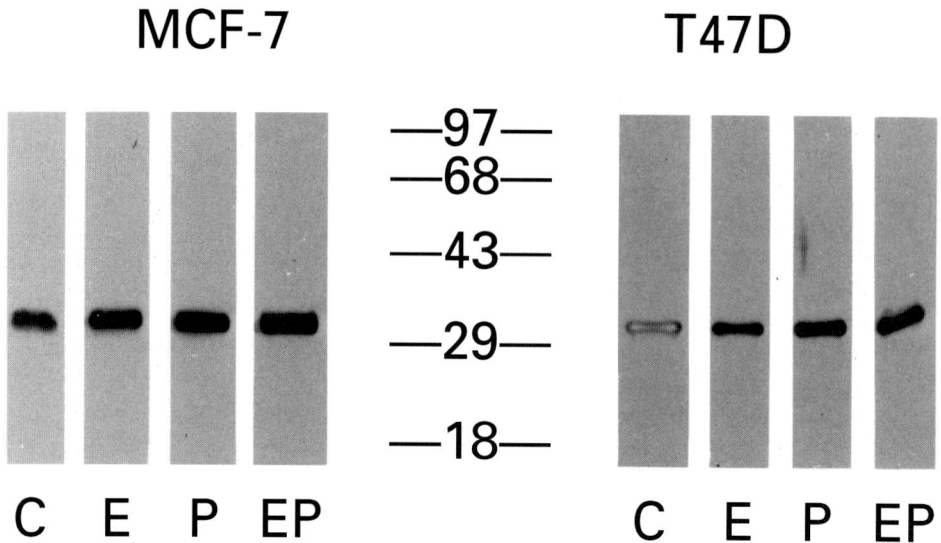

Figure 10 Estrogen and progestin modulate galectin-3 expression in steroid-sensitive breast carcinoma cells. MCF-7 and T47D human breast carcinoma cells were treated with 17ß-estradiol (10 nM, E), the progestin R5020 (10 nM, P) or both steroids (10 nM each, EP) for 48 hours, and expression of galectin-3 was determined by Western blotting.

POSSIBLE FUNCTIONS OF GALECTINS DURING CANCER INVASION AND METASTASIS

Galectins plays various roles in the eukaryotic cell, during cell-matrix and cell-cell interactions (Figure 11), but also during other, apparently unrelated biological events (Figure 12). We will describe here the important functions that could be important during cancer progression (Figure 13).

As initially suggested from early hemagglutination studies, galectin-1 is involved in homotypic lactose-inhibitable aggregation of dissociated embryonic brain cells (Joubert et al., 1985; Joubert et al., 1986; Joubert et al., 1987; Caron et al., 1987), erythroblasts (Harrison and Chesterton, 1980), thymocytes (Levi and Teichberg, 1983; Levi and Teichberg, 1985), and baby hamster kidney cells (Stojanovic et al., 1983). Galectin-1 is also implicated in aggregation of thymoma cells to T lymphocytic cells (Hafer-Macko et al., 1996). This property of galectin-1 could modulate homotypic aggregation of cancer cells in the primary tumor and in the metastatic deposit, and disaggregation when leaving the primary tumor (Figure 13).

Moreover, a large body of evidence implicates galectin-1 in the promotion of cell adhesion. For instance, immobilized galectin-1 mediates attachment of BHK cells (Stojanovic et al., 1983) and lymphoblastoid cells (Ahmed et al., 1992). Immobilized galectin-1 promotes adhesion of A121 ovary carcinoma cells (Skrincosky et al., 1993), rhabdomyosarcoma cells (Ozeki et al., 1991b) and HT29 colon carcinoma cells (Woynarowska et al., 1996). Pretreatment of A121 ovary carcinoma cells with galectin-1 inhibits adhesion to endothelial cell-derived ectracellular matrix (Allen

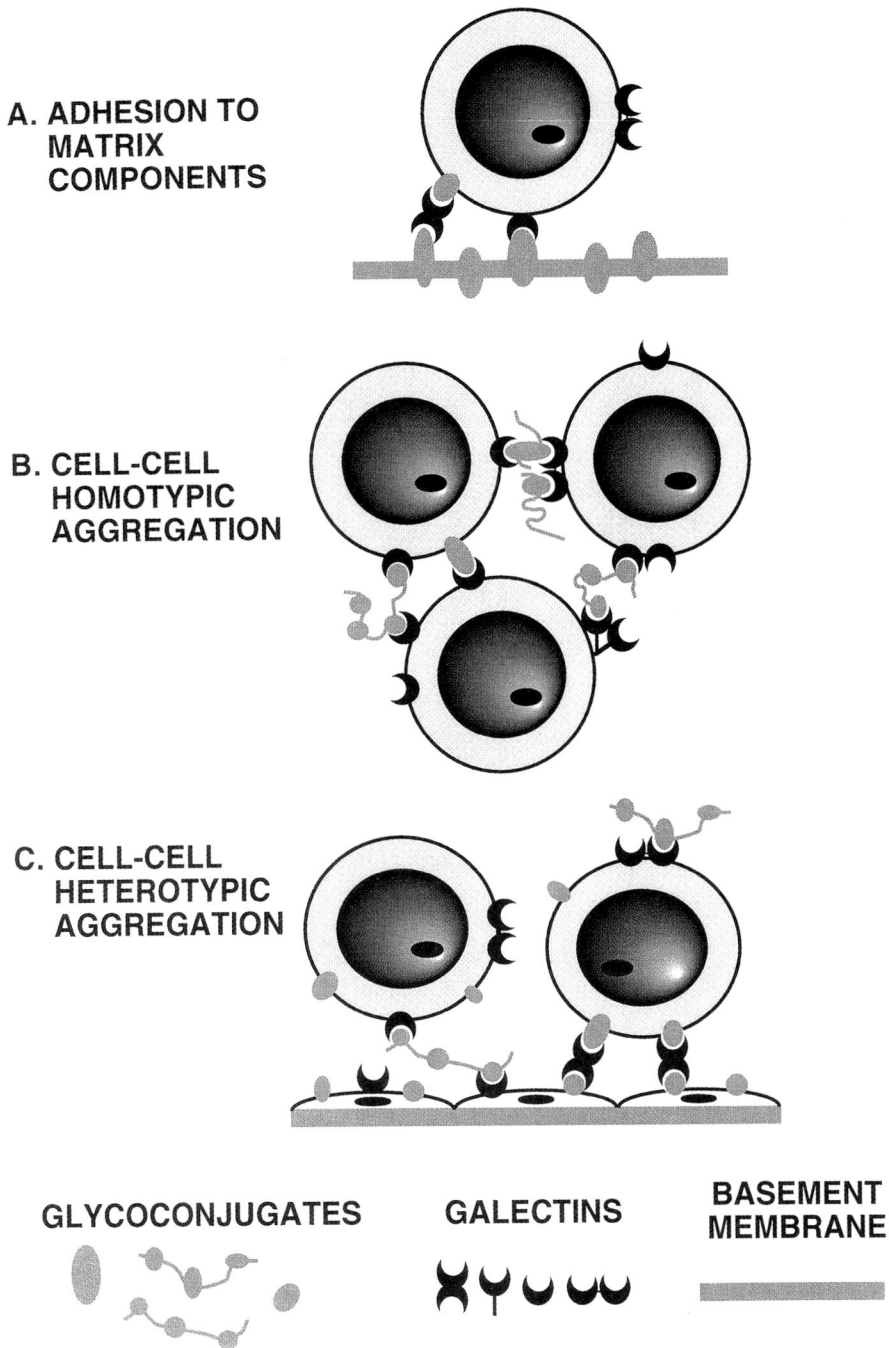

A. ADHESION TO
   MATRIX
   COMPONENTS

B. CELL-CELL
   HOMOTYPIC
   AGGREGATION

C. CELL-CELL
   HETEROTYPIC
   AGGREGATION

GLYCOCONJUGATES   GALECTINS   BASEMENT
                                             MEMBRANE

Figure 11 Schematic representation of galectin functions during cell-matrix and cell-cell interactions. Galectins are involved in cell adhesion to matrix glycoconjugates (A), for instance during cancer invasion; in homotypic (B) or heterotypic cell-cell interactions (C), such as in cancer cell aggregation, and during attachment of cancer cells to endothelial cells, prior to extravasation (see text).

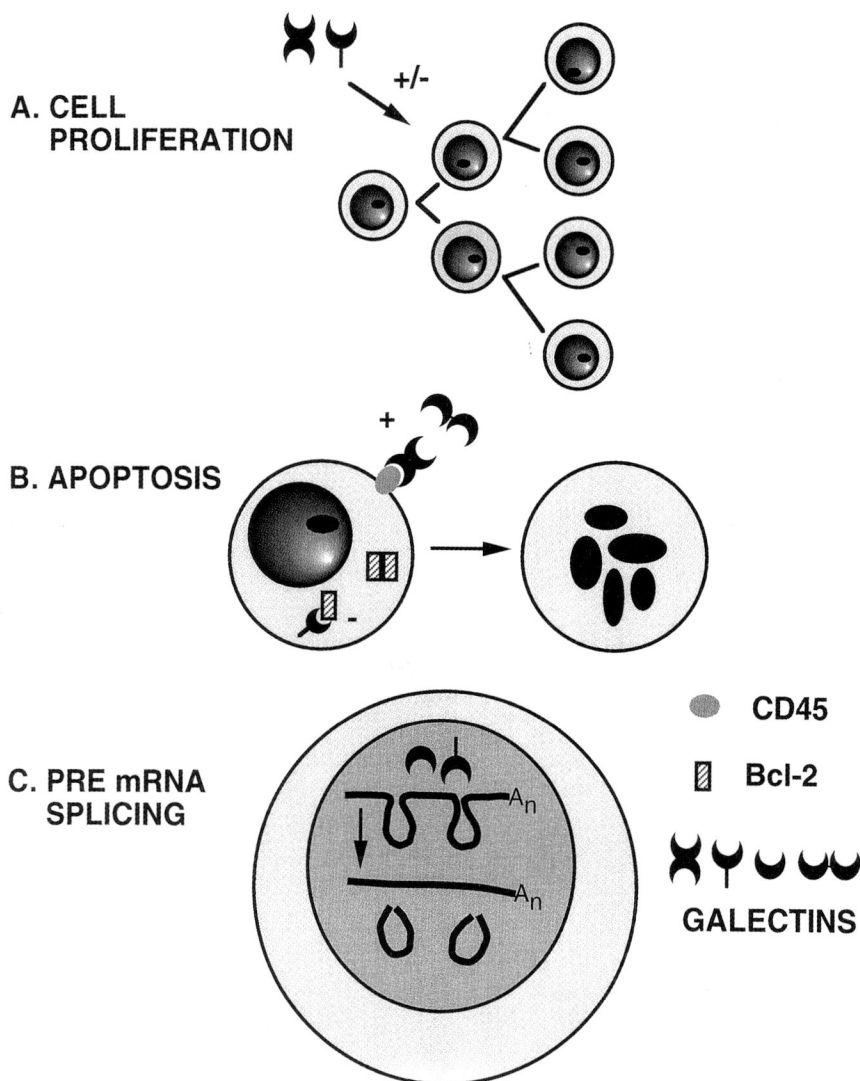

Figure 12 Galectins play roles in the regulation of cell proliferation (A), apoptosis (B) and pre-mRNA splicing (C) (see text).

et al., 1990). Numerous studies have demonstrated implication of galectin-1 in cell attachement to laminin. Galectin-1 promotes attachment of F9 teratocarcinoma cells, CHO cells and olfactory neurons to immobilized type 1 laminin in a lactose-dependent manner (Zhou and Cummings, 1993; Mahanthappa *et al.*, 1994). Similar data were reported for galectin-1-mediated promotion of A2058 and A375 human melanoma cell adhesion to laminin; in addition, this effect is lactose-dependent and baseline adhesion to laminin is decreased by addition of a polyclonal anti-galectin-

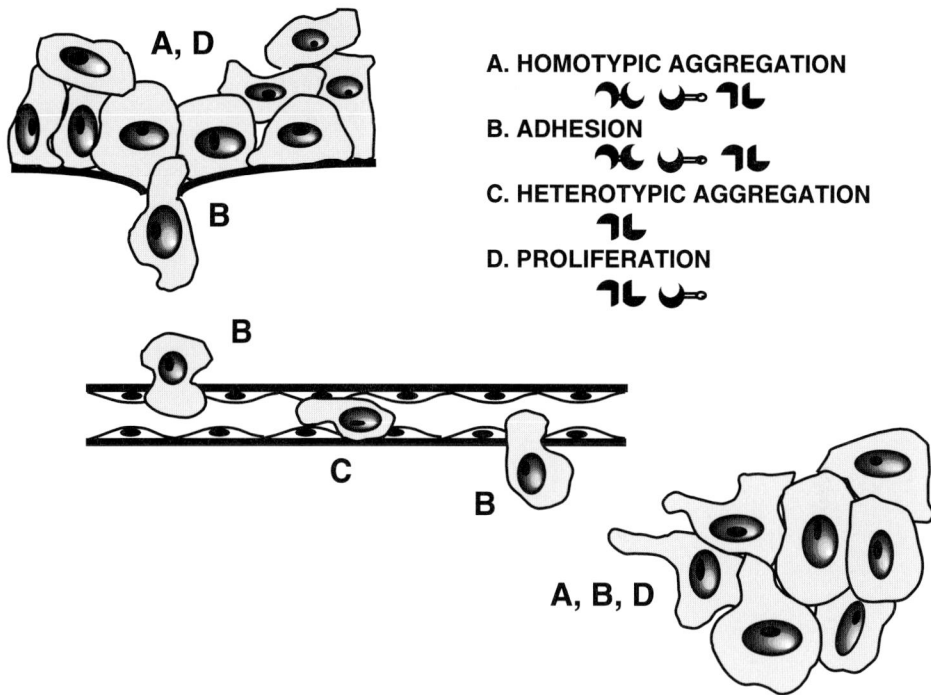

A. HOMOTYPIC AGGREGATION

B. ADHESION

C. HETEROTYPIC AGGREGATION

D. PROLIFERATION

Figure 13 Schematic representation of the functions played by galectins during the metastatic cascade. Galectins modulate homotypic aggregation (A), adhesion to matrix components (B), heterotypic aggregation (C) and cell proliferation (D). Capital letters are placed on the Figure to designate the steps of the metastatic cascade where galectins are thought to be involved.

1 antiserum (van den Brûle *et al.*, 1995a). Interestingly, galectin-1 can inhibit myoblast adhesion and spreading to laminin, as determined by addition of exogenous galectin-1 or galectin-1 cDNA transfection (Cooper *et al.*, 1991), a finding that suggests that the molecular microenvironment such as the presence of the $\alpha_7\beta_1$ integrin (Gu *et al.*, 1994) on myoblasts could modulate galectin-1 functions or that, alternatively, galectin-1 could modulate integrin functions. Galectin-1 is also implicated in the promotion of rhabdomyosarcoma cell adhesion to fibronectin, in a lactose-independent, RGD-dependent manner (Ozeki *et al.*, 1995). Galectin-1 also promotes adhesion of large lymphoma cells to endothelial cells, as this event is inhibited by an anti-galectin-1 antiserum (Lotan *et al.*, 1994a). These adhesive properties of galectin-1 could be of importance during the first step of basement membrane crossing (Figure 2) as well as during host tissue invasion (Figure 13).

Numerous reports implicate galectin-1 in cell growth regulation. Discordant results were first reported. Galectin-1 has been classified as a transforming growth factor (TGF) as it induces DNA synthesis and promotes growth of 3T3 cells, and was purified from conditioned medium of the 77N1 avian sarcoma virus-transformed rat cell line; however, the lectin did not compete with EGF for EGF receptors (Hirai *et al.*, 1983; Yamaoka *et al.*, 1984). Transfection of galectin-1 in 3T3 fibroblasts induced acquisition of the transformed phenotype, with loss of anchorage dependence, reduced contact inhibition, colony formation in soft agar and tumor formation in nude mice; protein extracts of transfected cells induce DNA synthesis in 3T3 cells, but this effect is lactose-dependent (Yamaoka *et al.*, 1991). Other data reported galectin-1-mediated cell cycle inhibition of mouse embryo fibroblasts as demonstrated by addition of exogenous galectin-1 (Wells and Mallucci, 1983; Wells and Malucci, 1991; Wells and Mallucci, 1992). A recent report highlights the relationship between the presence of two intramolecular disulfide bridges and the TGF activity (Yamaoka *et al.*, 1996). Another study indicated that high concentrations of galectin-1 exerts lactose-independent growth inhibition; however, low concentrations promote proliferation in a lactose dependent manner (Adams *et al.*, 1996). It could thus be possible that galectin-1 regulates cancer cell proliferation at various stages of cancer dissemination (Figure 13).

Galectin-1 is also implicated in the process of apoptosis. A first report indicated overexpression during apoptosis induction by dexamethasone in a human leukemia cell line (Goldstone and Lavin, 1991). Subsequently, galectin-1 was demonstrated to induce apoptosis in activated T cells and T leukemia cell lines in a lactose-dependent manner; this interaction between either soluble or cell surface galectin-1 is dependent on appropriately glycosylated CD45 on lymphocytes (Perillo *et al.*, 1995). Galectin-1 is also implicated in apoptosis of thymocytes (Perillo *et al.*, 1997). Galectin-1 seems to be also implicated in pre-mRNA splicing (Vyakarnam *et al.*, 1997).

The role played by galectin-1 in vivo is still to be completely defined. Indeed, initial examination of galectin-1 knock-out mice did not reveal a specific phenotype, and fertility was apparently normal (Poirier and Robertson, 1993). However, closer examination of the knock-out animals revealed aberrant topography of olfactory axons (Puche *et al.*, 1996).

Galectin-3 is also involved in several cell-cell aggregation and cell-matrix adhesion processes that could be of importance for cell cohesion in the primary tumor and metastases, as well as during cell detachment prior to tissue invasion, and for attachment to matrix components. Molecular interactions between galectin-3 and Mac-2BP are involved in homotypic cell aggregation (Inohara *et al.*, 1996). Various data are available concerning the implication of galectin-3 in attachment to laminin, where it seems to be variable from one to another cell line. For instance, low doses (10 µg/ml) of galectin-3 inhibit attachement of baby hamster kidney cells to laminin, and higher doses (45 µg/ml) inhibit spreading but not attachment to that glycoprotein; attachment and spreading are not affected by an anti-galectin-3 polyclonal antiserum (Sato and Hughes, 1992). Galectin-3 does not seem to be involved in SCM153 breast epithelial cell adhesion to laminin, since this event is not affected by addition of an anti-galectin-3 antiserum or by lactose (Ochieng *et al.*, 1992). Similarly, attachment of A2058 or A375 melanoma cells to laminin is not affected by addition of galectin-3 or an anti-galectin-3 antiserum (van den Brûle *et al.*, 1995b). However, galectin-3 promotes adhesion of human neutrophils to laminin in a calcium-dependent manner, and this is dependent on the integrity of the amino-terminal region of galectin-3 (Kuwabara and Liu, 1996). Galectin-3 also promotes attachment of neutrophils to fibronectin, and this event is dependent on the ß2 integrin chain (Kuwabara and Liu, 1996). Interestingly, addition of transglutaminase-oligomerized galectin-3 preparations do not alter melanoma cell adhesion or migration to laminin, but markedly promote cell spreading on that glycoprotein (van den Brûle *et al.*, 1998). The importance of this phenomenon for cancer invasion remains to be defined.

In vitro studies demonstrated that Matrigel invasion by breast carcinoma cells is promoted by 10 μg/ml soluble galectin-3; interestingly, lower doses (1-2 μg/ml) are required when galectin-3 is bound to Matrigel (Le Marer and Hughes, 1996). Transfection of galectin-3 cDNA in fibroblasts results in acquisition of anchorage-independent growth, morphological transformation in vitro but not in tumorigenicity in vivo; interestingly, weakly metastatic fibrosarcoma cells transfected with galectin-3 resulted in an increase of experimental lung metastases in mice (Raz *et al.*, 1990).

Other functions have been demonstrated for galectin-3. In vitro experiments demonstrated that galectin-3 is involved in mRNA splicing: indeed, lactose, and galectin depletion in the nuclear extracts, by lactose affinity chromatography inhibits splicing; the activity is restored after addition of recombinant galectin-1 or galectin-3 (Dagher *et al.*, 1995; Vyakarnam *et al.*, 1997). Galectin-3 subcellular localization and expression have been correlated to proliferation in fibroblasts. Galectin-3 protein and mRNA levels are increased in proliferating versus serum-starved fibroblasts (Moutsatsos *et al.*, 1987; Agrwal *et al.*, 1989; Cowles *et al.*, 1989; Cowles *et al.*, 1990; Hamann *et al.*, 1991; Hubert *et al.*, 1995), mainly by increase of a mRNA variant that differs by its initiation site and splicing (Voss *et al.*, 1994). This regulation is lost in late passage cells (Cowles *et al.*, 1989). Galectin-3 is also involved in T cell growth and apoptosis regulation: indeed, antisense-transfected Jurkat T cells exhibit increased growth rates and resistance to apoptosis, as galectin-3 is able to interact with Bcl-2 in a lactose-dependent manner (Yang *et al.*, 1996). Interestingly, antisense-mediated decreased galectin-3 expression in MDA-MB435 breast cancer cells results in decreased thymidine incorporation (van den Brûle *et al.*, 1997a). Several functions related to the immune system have been described for galectin-3 (Lipsick *et al.*, 1980; Jeng *et al.*, 1994; Liu *et al.*, 1995; Mathews *et al.*, 1995; Yamaoka *et al.*, 1995; Dong and Hughes, 1996, Kuwabara and Liu, 1996, Liu *et al.*, 1996).

In comparison to the large body of data available for galectin-1 and galectin-3, relatively few reports examined the functions of the other galectins. Immobilized galectin-4 has been implicated in promotion of T84 colon cancer cell attachment (Huflejt *et al.*, 1997). Recent work demonstrates that low doses of galectin-9 promote lactose-inhibitable apoptosis of mouse thymocytes, and higher doses induce homotypic aggregation (Wada *et al.*, 1997). The relevance of these findings to carcinoma cells needs further evaluation.

INSIGHTS INTO THERAPEUTIC APPLICATIONS

Several results suggest possible therapeutic approaches that could modulate galectin-based interactions and impair cancer progression. Different approaches have been explored, using either galectins or galectin fragments, anti-galectin antibodies, glycoconjugate ligands, or saccharide analogs resulting in altered glycosylation of glycoconjugates. An early report indicates that a galectin-3-derived peptide that contains part of the CRD (amino acid 172-209) of the lectin decreases lung colonization of B16-F1 melanoma cells by 80% when injected with the cells (Raz *et al.*, 1989). This activity is similar to that of an anti-lectin monoclonal antibody

(Meromsky *et al.*, 1986). Citrus pectin (CP) and its pH-modified derivative (MCP) are complex polysaccharides rich in galactoside residues, that bind cell surface and galectin-3 in a lactose-dependent manner. Experiments demonstrated that CP increases in vitro homotypic cell-cell aggregation and in vivo lung colonization in mice; interestingly, MCP significantly decreases B16-F1 experimental metastasis (Platt and Raz, 1992). In addition, MCP, but not CP, inhibits melanoma cell adhesion to laminin and asialofetuin-mediated homotypic aggregation, and both pectins inhibit anchorage-independent growth (Inohara and Raz, 1994a). MCP decreases adhesion of rat prostate carcinoma cells to endothelium in vitro, and inhibit colony formation in agarose (Pienta *et al.*, 1995). Oral administration of MCP to rats decreases lung colony formation by prostate carcinoma cells, but does not affect primary tumor growth (Pienta *et al.*, 1995). Adhesion of colon carcinoma cells to galectin-1 and homotypic aggregation can be decreased by pretreatment of the cells with a 4-fluoro-glucosamine analog, that impairs glycosylation of molecules such as lamp-1, lamp-2 and CEA, and reduces surface exppression of terminal lactosamine residues (Woynarowska *et al.*, 1996).

CONCLUSIONS

Great progress has been performed in the field of galectins. Indeed, numerous members of the family have been characterized, their ligands have been defined, and several biological functions have been demonstrated, allowing to draw tentative sketches of galectin implications in eukaryotic cell biology. However, besides the definition of putative unknown functions and the characterization of new members of that family, many questions remain to be answered concerning galectin implication in cancer progression. For instance, the expression profiles of galectins, and their regulation of expression in human cancer, the effectively relevant and important functions of galectins in cancer cells, and the possible redundance between galectins for each considered functions, remain to be precisely determined. The quest to these answers will effectively help understanding the role of galectins in cancer progression and will constitute future challenges in basic cell biology and oncology.

AKNOWLEDGEMENTS

Frédéric A. van den Brûle is a Research Associate, and Vincent Castronovo is a Senior Research Associate of the National Fund for Scientific Research (NFSR), Belgium. Our work has been supported by the NFSR, the European BIOMED contract # BMH1-CT92-0520, the Association Sportive contre le Cancer, and the Association Contre le Cancer, Belgium. We want to aknowledge fruitful collaboration with Douglas N.W. Cooper, PhD (University of California at San Francisco, CA, USA), Fu-Tong Liu, PhD, MD (La Jolla Institute for Allergy and Immunology, La Jolla, CA, USA), Pedro L. Fernandez, MD, PhD, and E. Campo, MD, PhD (Department of Pathology, Hospital Clinic, Barcelona, Spain). Figure 4 was performed in collaboration with Dr. Hakon Leffler (University of Lund, Lund, Sweden).

REFERENCES

Abbott, W.M. and Feizi, T. (1989) *Biochem J.,* **259,** 291–294.

Abbott, W.M. and Feizi, T. (1991) *J. Biol. Chem.,* **266,** 5552–5557.

Abbott, W.M., Mellor, A., Edwards, Y. and Feizi, T. (1989) *Biochem. J.,* **259,** 283–290.

Ackerman, S.J., Corrette, S.E., Rosenberg, H.F., Bennett, J.C., Mastrianni, D.M., Nicholson-Weller, A., Weller, P. F., Chin, D.T. and Tenen, D.G. (1993) *J. Immunol.,* **150,** 456–468.

Adams, L., Scott, G.K. and Weinberg, C. S. (1996) *Biochim. Biophys Acta,* **1312,** 137–144.

Agrwal, N., Wang, J.L. and Voss, P. G. (1989) *J. Biol. Chem.,* **264,** 17236–17242.

Ahmed, H., Fink, N.E., Pohl, J. and Vasta, G. R. (1996a) *J. Biochem. (Tokyo),* **120,** 1007–1019.

Ahmed, H., Pohl, J., Fink, N.E., Strobel, F. and Vasta, G.R. (1996b) *J. Biol. Chem.,* **271,** 33083–33094.

Ahmed, H., Sharma, A., DiCioccio, R.A. and Allen, H.J. (1992) *J. Mol. Recognition,* **5,** 1–8.

Ahmed, H. and Vasta, G. R. (1994) *Glycobiology,* **4,** 545–548.

Akimoto, Y., Kawakami, H., Oda, Y., Obinata, A., Endo, H., Kasai, K. and Hirano, H. (1992) *Exp Cell Res,* **199,** 297–304.

Albrandt, K., Orida, N. K. and Liu, F. T. (1987) *Proc Natl Acad Sci USA,* **84,** 6859–6863.

Allen, H. J., Gottstine, S., Sharma, A., DiCioccio, R. A., Swank, R. T. and Li, H. (1991a) *Biochemistry,* **30,** 8904–8910.

Allen, H. J., Karakousis, C., Piver, M. S., Gamarra, M., Nava, H., Forsyth, B., Matecki, B., Jazayeri, A., Sucato, D., Kisailus, E. and DiCioccio, R. (1987) *Tumor Biol,* **8,** 218–229.

Allen, H. J., Sucato, D., Gottstine, S., Kisailus, E., Nava, H., Petrelli, N., Castillo, N. and Wilson, D. (1991b) *Tumor Biol,* **12,** 52–60.

Allen, H. J., Sucato, D., Woynarowska, B., Gottstine, S., Sharma, A. and Bernacki, R. J. (1990) *J Cell Biochem,* **43,** 43–57.

Arumugham, R. G., Hsieh, T. C. Y., Tanzer, M. L. and Laine, R. (1986) *Biochem Biophys Acta,* **883,** 112–126.

Aubin, J. E., Gupta, A. K., Bhargava, U. and Turksen, K. (1996) *J Cell Physiol,* **169,** 468–480.

Bachar-Lustig, E., Gan, Y. and Reisner, Y. (1991) *Carbohydrate Res,* **213,** 345–352.

Barondes, S. H., Castronovo, V., Cooper, D. N. W., Cummings, R. D., Drickamer, D., Feizi, T., Gitt, M. A., Hirabayashi, J., Hughes, C., Kasai, K. I., Leffler, H., Liu, F. T., Lotan, R., Mercurio, A. M., Monsigny, M., Pillai, S., Poirier, F., Raz, A., Rigby, P. W. J., Rini, J. M. and Wang, J. L. (1994a) *Cell,* **78,** 597–598.

Barondes, S. H., Cooper, D. N. W., Gitt, M. A. and Leffler, H. (1994b) *J. Biol. Chem.,* **269,** 28807–28810.

Baum, L. G., Pang, M., Perillo, N. L., Wu, T., Delegeane, A., Uittenbogaart, C. H., Fukuda, M. and Seilhamer, J. J. (1995) *J Exp Med,* **181,** 877–887.

Beck, K., Hunter, I. and Engel, J. (1990) *FASEB J,* **4,** 148–160.

Benvenuto, G., Carpentieri, M. L., Salvatore, P., Cindolo, L., Bruni, C. B. and Chiariotti, L. (1996) *Mol Cell Biol,* **16,** 2736–2743.

Beyer, E. C. and Barondes, S. H. (1982) *J Cell Biol,* **92,** 28–33.

Beyer, E. C., Zweig, S. E. and Barondes, S. H. (1980) *J. Biol. Chem.,* **255,** 4236–4239.

Bladier, D., Le Caer, J. P., Joubert, R., Caron, M. and Rossier, J. (1991) *Neurochem Int,* **18,** 275–281.

Bourne, Y., Bolgiano, B., Liao, D. I., Strecker, G., Cantau, P., Herzberg, O., Feizi, T. and Cambillau, C. (1994) *Nat Struct Biol,* **1,** 863–870.

Bresalier, R. S., Byrd, J. C., Wang, L. and Raz, A. (1996) *Cancer Res,* **56,** 1354–1357.

Briles, E. B., Gregory, W., Fletcher, P. and Kornfeld, S. (1979) *J Cell Biol,* **81,** 528–537.

Caron, M., Joubert, R. and Bladier, D. (1987) *Biochim Biophys Acta,* **925,** 290–296.

Caron, M., Joubert-Caron, R., Cartier, J. R., Chadli, A. and Bladier, D. (1993) *J Chromatogr,* **646,** 327–333.

Castronovo, V. (1993) *Inv Metastasis,* **13,** 1–30.

Castronovo, V., Campo, E., van den Brûle, F. A., Claysmith, A. P., Cioce, V., Liu, F. T., Fernandez, P. L. and Sobel, M. E. (1992a) *J Natl Cancer Inst,* **84,** 1161–1169.

Castronovo, V., Luyten, F., van den Brûle, F. and Sobel, M. E. (1992b) *Arch. Biochem. Biophys.,* **297,** 132–138.

Castronovo, V., van den Brûle, F. A., Jackers, P., Clausse, N., Liu, F. T., Gillet, C. and Sobel, M. E. (1996) *J Pathol,* **179,** 43–48.

Cerra, R. F., Gitt, M. A. and Barondes, S. H. (1985) *J. Biol. Chem.,* **260,** 10474–10477.

Chadli, A., LeCaer, J. P., Bladier, D., Joubert-Caron, R. and Caron, M. (1997) *J Neurochem,* **68,** 1640–1647.

Chammas, R., Jasiulionis, M.G., Ventura, A.M., Travassos, L.R. and Brentani, R.R. (1996) *Braz. J. Med. Biol. Res.*, **29**, 1141–1149.

Cherayil, B.J., Chaitovitz, S., Wong, C. and Pillai, S. (1990) *Proc Natl Acad Sci USA*, **87**, 7324–7328.

Cherayil, B.J., Weiner, S. J. and Pillai, S. (1989) *J. Exp. Med.*, **170**, 1959–1972.

Chiariotti, L., Benvenuto, G., Zarrilli, R., Rossi, E., Salvatore, P., Colantuoni, V. and Bruni, C.B. (1994) *Cell Growth Differ,* **5**, 769–775.

Chiariotti, L., Berlingieri, M.T., Battaglia, C., Benvenuto, G., Martelli, M.L., Salvatore, P., Chiappetta, G., Bruni, C.B. and Fusco, A. (1995) *Int. J. Cancer,* **64**, 171–175.

Chiariotti, L., Berlingieri, M.T., De Rosa, P., Battaglia, C., Berger, N., Bruni, C.B. and Fusco, A. (1992) *Oncogene,* **7**, 2507–2511.

Chiariotti, L., Wells, V., Bruni, C.B. and Mallucci, L. (1991) *Biochim Biophys. Acta,* **1089**, 54–60.

Chiu, M.L., Jones, J.C. and O'Keefe, E.J. (1992) *J. Cell Biol.,* **119**, 1189–1700.

Chiu, M.L., Parry, D.A., Feldman, S.R., Klapper, D.G. and O'Keefe, E.J. (1994) *J. Biol. Chem.,* **269**, 31770–31776.

Cho, M. and Cummings, R.D. (1995) *J. Biol. Chem.,* **270**, 5198–5206.

Clerch, L. B., Whitney, P., Hass, M., Brew, K., Miller, T., Werner, R. and Massaro, D. (1988) *Biochemistry,* **27**, 692–699.

Colnot, C., Ripoche, M.-A., Scaerou, F., Fowlis, D. and Poirier, F. (1996) *Biochem. Soc. Trans.,* **24**, 141–146.

Cooper, D.N. and Barondes, S.H. (1990) *J. Cell Biol.,* **110**, 1681–1691.

Cooper, H.M., Massa, S.M., Barondes, S.H. (1991) *J. Cell Biol.,* **115**, 1437–1448.

Couraud, P.O., Casentini-Borocz, D., Bringman, T.S., Griffith, J., McGrogan, M. and Nedwin, G.E. (1989) *J. Biol. Chem.,* **264**, 1310–1316.

Cowles, E.A., Agrwal, N., Anderson, R.L. and Wang, J.L. (1990) *J. Biol. Chem.,* **265**, 17706–17712.

Cowles, E.A., Moutsatsos, I.K., Wang, J.L. and Anderson, R.L. (1989) *Exp. Gerontol.,* **24**, 577–585.

Crittenden, S.L., Roff, C.F. and Wang, J.L. (1984) *Mol. Cell Biol.,* **4**, 1252–1259.

Dagher, S.F., Wang, J.L. and Patterson, R.J. (1995) *Proc. Natl. Acad. Sci. USA,* **92**, 1213–1217.

Dean, J.W., III, Chandrasekaran, S. and Tanzer, M.L. (1990) *J. Biol. Chem.,* **265**, 12553–12562.

Den, H. and Malinzak, D.A. (1977) *J. Biol. Chem.,* **252**, 5444–5448.

Dennis, J.W., Waller, C.A. and Schirrmacher, V. (1984) *J. Cell Biol.,* **99**, 1416–1423.

Do, K.Y., Smith, D.F. and Cummings, R.D. (1990) *Biochem. Biophys. Res. Commun.,* **173**, 1123–1128.

Dong, S. and Hughes, R.C. (1996) *FEBS Lett.,* **395**, 165–169.

Dong, S. and Hughes, R.C. (1997) *Glycoconj. J.,* **14**, 267–274.

Dyer, K.D. and Rosenberg, H.F. (1996) *Life Sci.,* **58**, 2073–2082.

Elices, M.J. and Hemler, M.E. (1989) *Proc. Natl. Acad. Sci. USA,* **86**, 9906–9910.

Elices, M.J., Urry, L.A. and Hemler, M.E. (1991) *J. Cell Biol.,* **112**, 169–181.

Engvall, E. (1993) *Kidney Int,* **43**, 2–6.

Fang, R., Mantle, M. and Ceri, H. (1993) *Biochem. J.,* **293**, 867–872.

Feizi, T., Solomon, J.C., Yuen, C.-T., Jeng, K.C.G., Frigeri, L.G., Hsu, D.K. and Liu, F.T. (1994) *Biochemistry,* **33**, 6342–6349.

Fernandez, P.L., Merino, M. J., Gomez, M., Campo, E., Medina, T., Castronovo, V., Sanjuan, X., Cardesa, A., Liu, F.T. and Sobel, M.E. (1997) *J. Pathol.,* **181**, 80–86.

Foddy, L., Stamatoglou, S.C. and Hughes, R.C. (1990) *J. Cell Sci.,* **97**, 139–148.

Fowlis, D., Colnot, C., Ripoche, M.-A. and Poirier, F. (1995) *Dev. Dynamics,* **203**, 241–251.

Fujiwara, S., Shinkai, H., Deutzmann, R., Paulsson, M. and Timpl, R. (1988) *Biochem. J.,* **252**, 453–461.

Fukuda, M. (1991) *J. Biol. Chem.,* **266**, 21327–21330.

Gabius, H. J., Brehler, R., Schauer, A. and Cramer, F. (1986) *Virchows Arch. (Cell Pathol),* **52**, 107–115.

Gabius, H. J., Vehmeyer, K., Engelhardt, R., Nagel, G.A. and Cramer, F. (1985) *Cell Tissue Res.,* **241**, 9–15.

Gaudin, J.C., Monsigny, M. and Legrand, A. (1995) *Gene,* **163**, 249–252.

Gehlsen, K.R., Dillner, L., Engvall, E. and Ruoslahti, E. (1988) *Science,* **241**, 1228–1229.

Gehlsen, K.R., Sriramarao, P., Furcht, L.T. and Skubitz, A.P. (1992) *J. Cell Biol.,* **117**, 449–459.

Gillenwater, A., Xu, X.C., el-Naggar, A. K., Clayman, G.L. and Lotan, R. (1996) *Head Neck,* **18**, 422–432.

Gitt, M.A. and Barondes, S.H. (1986) *Proc. Natl. Acad. Sci. USA,* **83**, 7603–7607.

Gitt, M.A. and Barondes, S.H. (1991) *Biochemistry,* **30**, 82–89.

Gitt, M.A., Massa, S. M., Leffler, H. and Barondes, S. H. (1992) *J. Biol. Chem.,* **267**, 10601–10606.

Gitt, M.A., Wiser, M.F., Leffler, H., Herrmann, J., Xia, Y.R., Massa, S.M., Cooper, D.N., Lusis, A.J. and Barondes, S.H. (1995) *J. Biol. Chem.*, **270**, 5032–5038.

Goldstone, S.D. and Lavin, M.F. (1991) *Biochem. Biophys. Res. Commun.*, **178**, 746–750.

Goletz, S., Hanisch, F.G. and Karsten, U. (1997) *J. Cell Sci.*, **110**, 1585–1596.

Gritzmacher, C.A., Mehl, V.S. and Liu, F.T. (1992) *Biochemistry*, **31**, 9533–9538.

Gu, M., Wang, W., Song, W.K., Cooper, D.N.W. and Kaufmen, S.J. (1994) *J. Cell Sci.*, **107**, 175–181.

Hadari, Y.R., Paz, K., Dekel, R., Mestrovic, T., Accili, D. and Zick, Y. (1995) *J. Biol. Chem.*, **270**, 3447–3453.

Hafer-Macko, C., Pang, M., Seilhamer, J.J. and Baum, L.G. (1996) *Glycoconj. .J*, **13**, 591–597.

Hamann, K.K., Cowles, E. A., Wang, J.L. and Anderson, R.L. (1991) *Exp. Cell Res.*, **196**, 82–91.

Harrison, F.L. and Chesterton, C.J. (1980) *Nature*, **286**, 502–504.

Hebert, E. and Monsigny, M. (1994) *Biol Cell*, **81**, 73–76.

Hebert, E., Roche, A.C., Nachtigal, M. and Monsigny, M. (1996) *C.R. Acad. Sci. III*, **319**, 871–877.

Herrmann, J., Turck, C.W., Atchison, R.E., Huflejt, M.E., Poulter, L., Gitt, M.A., Burlingame, A. L., Barondes, S. H. and Leffler, H. (1993) *J. Biol. Chem.*, **268**, 26704–26711.

Hirabayashi, J., Hitoshi, A., Soma, G.I. and Kasai, K.I. (1989) *Biochim. Biophy.s Acta*, **1008**, 85–91.

Hirabayashi, J. and Kasai, K. (1993) *Glycobiology*, **3**, 297–304.

Hirabayashi, J. and Kasai, K.I. (1988) *J. Biochem.*, **104**, 1–4.

Hirabayashi, J. and Kasai, K.I. (1991) *J. Biol. Chem.*, **266**, 23648–23653.

Hirabayashi, J. and Kasai, K.I. (1994) *Glycoconj. J.*, **11**, 437–442.

Hirabayashi, J., Kawasaki, H., Suzuki, K. and Kasai, K.I. (1987a) *J. Biochem.*, **101**, 775–787.

Hirabayashi, J., Kawasaki, H., Suzuki, K. and Kasai, K.I. (1987b) *J. Biochem.*, **101**, 987–995.

Hirai, R., Yamaoka, K. and Mitsui, H. (1983) *Cancer Res.*, **43**, 5742–5746.

Ho, M.K. and Springer, T.A. (1982) *J. Immunol.*, **128**, 1221–1228.

Hsu, D.K., Zuberi, R.I. and Liu, F.T. (1992) *J. Biol. Chem.*, **267**, 14167–14174.

Hubert, M., Wang, S.Y., Wang, J.L., Seve, A.P. and Hubert, J. (1995) *Exp. Cell Res.*, **220**, 397–406.

Huflejt, M.E., Jordan, E.T., Gitt, M.A., Barondes, S.H. and Leffler, H. (1997) *J. Biol. Chem.*, **272**, 14294–14303.

Huflejt, M.E., Turck, C.W., Linstedt, R., Barondes, S.H. and Leffler, H. (1993) *J. Biol. Chem.*, **268**, 26712–26718.

Hynes, M.A., Gitt, M., Barondes, S.H., Jessel, T.M. and Buck, L.B. (1990) *J. Neurosci.*, **10**, 1004–1013.

Ignatius, M. J. and Reichardt, L. F. (1988) *Neuron*, **1**, 713–725.

Inohara, H., Akahani, S., Koths, K. and Raz, A. (1996) *Cancer Res*, **56**, 4530–4534.

Inohara, H. and Raz, A. (1994a) *Glyconconj. J.*, **11**, 527–532.

Inohara, H. and Raz, A. (1994b) *Biochem. Biophys. Res. Commun.*, **201**, 1366–1375.

Irimura, T., Matsushita, Y., Sutton, R.C., Carralero, D., Ohannesian, D.W., Cleary, K.R., Ota, D.M., Nicolson, G.L. and Lotan, R. (1991) *Cancer Res*, **51**, 387–393.

Jeng, K.C., Frigeri, L.G. and Liu, F.T. (1994) *Immunol. Lett.*, **42**, 113–116.

Jia, S., Mee, R.P., Morford, G., Agrwal, N., Voss, P.G., Moutsatsos, I.K. and Wang, J.L. (1987) *Gene*, **60**, 197–204.

Jia, S. and Wang, J.L. (1988) *J. Biol. Chem.*, **263**, 6009–6011.

Joubert, R., Caron, M., Avellana-Adalid, V., Mornet, D. and Bladier, D. (1992) *J. Neurochem.*, **58**, 200–203.

Joubert, R., Caron, M. and Bladier, D. (1987) *Brain Res.*, **433**, 146–150.

Joubert, R., Caron, M., Deugnier, M.A. and Bisconte, J.C. (1986) In *Lectins*, Vol. 5 (Ed, de Gruyter, W.) Berlin, New York, pp. 213–219.

Joubert, R., Caron, M., Deugnier, M.A., Rioux, F., Sensenbrenner, F. and Bisconte, J.C. (1985) *Cell Mol. Biol.*, **31**, 131–138.

Jung, S.K. and Fujimoto, D. (1994) *J. Biochem.*, **116**, 547–553.

Kleinman, H.K., Cannon, F.B., Laurie, G.W., Hassel, J.R., Aumailley, M., Terranova, V.P., Martin, G.R. and Dubois-Dalcq, M. (1985) *J. Biol. Chem.*, **260**, 317–325.

Knibbs, R.N., Agrwal, N., Wang, J.L. and Goldstein, I.J. (1993) *J. Biol. Chem.*, **268**, 14940–14947.

Konstantinov, K.N., Robbins, B.A. and Liu, F.T. (1996) *Am J Pathol*, **148**, 25–30.

Koths, K., Taylor, E., Halenbeck, R., Casipit, C. and Wang, A. (1993) *J. Biol. Chem.*, **268**, 14245–14249.

Kramer, R H., Cheng, Y.F. and Clyman, R. (1990) *J. Cell Biol.*, **111**, 1233–1243.

Kramer, R.H., Vu, M.P., Cheng, Y.F., Ramos, D.M., Timpl, R. and Waleh, N. (1991) *Cell Regul.*, **2**, 805–817.

Kuwabara, I. and Liu, F.T. (1996) *J Immunol*, **156**, 3939–3944.

Laing, J.G., Robertson, M.W., Gritzmacher, C.A., Wang, J.L. and Liu, F.T. (1989) *J. Biol. Chem.*, **264**, 1097–1010.

Landowski, T.H., Dratz, E.A. and Starkey, J.R. (1995) *Biochemistry*, **34**, 11276–11287.

Languino, L.R., Gehlsen, K.R., Wayner, E., Carter, W.G., Engvall, E. and Ruoslahti, E. (1989) *J. Cell Biol*, **109**, 2455–2462.

Le Marer, N. and Hughes, R.C. (1996) *J. Cell. Physiol.*, **168**, 51–58.

Lee, E.C., Woo, H.J., Korzelius, C.A., Steele, G.D.J. and Mercurio, A.M. (1991) *Arch. Surg.*, **126**, 1498–1502.

Leffler, H. and Barondes, S.H. (1986) *J. Biol. Chem.*, **261**, 10119–10126.

Leffler, H., Masiarz, F. R. and Barondes, S. H. (1989) *Biochemistry*, **28**, 9222–9229.

Levi, G. and Teichberg, V.I. (1981) *J. Biol. Chem.*, **256**, 5735–5740.

Levi, G. and Teichberg, V.I. (1983) *Immunol. Lettr.*, **7**, 35–39.

Levi, G. and Teichberg, V.I. (1985) *Biochem. J.*, **226**, 379–384.

Liao, D. I., Kapadia, G., Ahmed, H., Vasta, G.R. and Herzberg, O. (1994) *Proc. Natl. Acad. Sci. USA*, **91**, 1428–1432.

Lindstedt, R., Apodaca, G., Barondes, S.H., Mostov, K.E. and Leffler, H. (1993) *J. Biol. Chem.*, **268**, 11750–11757.

Linsley, P.S., Horn, D., Marquardt, H., Brown, J.P., Hellstrom, I., Hellstrom, K.E., Ochs, V. and Tolentino, E. (1986) *Biochemistry*, **25**, 2978–2986.

Liotta, L.A. (1986) *Cancer Res*, **46**, 1–7.

Lipsick, J.S., Beyer, E.C., Barondes, S.H. and Kaplan, N.O. (1980) *Biochem Biophys Res Commun*, **97**, 56–61.

Lise, M., Loda, M., Fiorentino, M., Mercurio, A.M., Summerhayes, I C., Lavin, P.T. and Jessup, J.M. (1997) *Ann. Surg. Oncol.*, **4**, 176–183.

Liu, F.T., Albrandt, K., Mendel, E., Kulczycki, A.J. and Orida, N.K. (1985) *Proc. Natl. Acad. Sci. USA*, **82**, 4100–4104.

Liu, F.T., Hsu, D.K., Zuberi, R. I., Hill, P.N., Shenhav, A., Kuwabara, I. and Chen, S.S. (1996) *Biochemistry*, **35**, 6073–6079.

Liu, F.T., Hsu, D.K., Zuberi, R. I., Kuwabara, I., Chi, E.Y. and Henderson, W.R.J. (1995) *Am. J. Pathol.*, **147**, 1016–1028.

Liu, F.T. and Orida, N. (1984) *J. Biol. Chem.*, **259**, 10649–10652.

Lobsanov, Y.D., Gitt, M.A., Leffler, H., Barondes, S.H. and Rini, J.M. (1993) *J. Biol. Chem.*, **268**, 27034–27038.

Lotan, R., Belloni, P.N., Tressler, R.J., Lotan, D., Xu, X.C. and Nicolson, G.L. (1994a) *Glycoconj. J.*, **11**, 462–468.

Lotan, R., Carralero, D., Lotan, D. and Raz, A. (1989a) *Cancer Res.*, **49**, 1261–1268.

Lotan, R., Ito, H., Yasui, W., Yokozaki, H., Lotan, D. and Tahara, E. (1994b) *Int J Cancer*, **56**, 474–480.

Lotan, R., Lotan, D. and Carralero, D.M. (1989b) *Cancer Lett.*, **48**, 115–122.

Lotan, R., Matsushita, Y., Ohannesian, D., Carralero, D., Ota, D.M., Cleary, K.R., Nicolson, G.L. and Irimura, T. (1991) *Carbohydr. Res.*, **213**, 47–57.

Lotz, M.M., Andrews, C.W.J., Korzelius, C.A., Lee, E.C., Steele, G.D.J., Clarcke, A. and Mercurio, A.M. (1993) *Proc. Natl. Acad. Sci. USA*, **90**, 3466–3470.

Madsen, P., Rasmussen, H.H., Flint, T., Gromov, P., Kruse, T.A., Honore, B., Vorum, H. and Celis, J.E. (1995) *J. Biol. Chem.*, **270**, 5823–5829.

Magnaldo, T., Bernerd, F. and Darmon, M. (1995) *Dev. Biol.*, **168**, 259–271.

Mahanthappa, N.K., Cooper, D.N.W., Barondes, S. H. and Schwarting, G.A. (1994) *Development*, **120**, 1373–1384.

Maley, F., Trimble, R.B., Tarentino, A.L. and Plummer, T.H.J. (1989) *Anal Biochem.*, **180**, 195–204.

Maquoi, E., van den Brûle, F.A., Castronovo, V. and Foidart, J.M. (1997) *Placenta*, **18**, 433–439.

Marschal, P., Cannon, V., Barondes, S.H. and Cooper, D.N.W. (1994) *Glycobiology*, **4**, 297–305.

Massa, S. M., Cooper, D.N.W., Leffler, H. and Barondes, S.H. (1993) *Biochemistry*, **32**, 260–267.

Mathews, K.P., Konstantinov, K.N., Kuwabara, I., Hill, P.N., Hsu, D.K., Zuraw, B.L. and Liu, F.T. (1995) *J. Clin. Immunol.*, **15**, 329–337.

Mehrabian, M., Gitt, M.A., Sparkes, R.S., Leffler, H., Barondes, S.H. and Lusis, A.J. (1993) *Genomics*, **15**, 418–420.

Mehul, B., Bawumia, S. and Hughes, R.C. (1995) *FEBS Lett*, **360**, 160–164.

Mehul, B., Bawumia, S., Martin, S.R. and Hughes, R.C. (1994) *J. Biol. Chem.*, **269**, 18250–18258.

Mehul, B. and Hughes, R. (1997) *J. Cell Sci.*, **110**, 1169–1178.

Merkle, R. K. and Cummings, R. D. (1988) *J. Biol. Chem.*, **263**, 16143–16149.

Meromsky, L., Lotan, R. and Raz, A. (1986) *Cancer Res.*, **46**, 5270–5275.

Mey, A., Leffler, H., Hmama, Z., Normier, G. and Revillard, J.P. (1996) *J. Immunol.*, **156**, 1572–1577.

Mey, O., Berthier-Vergnes, O., Apoil, P. A., Dore, J.F. and Revillard, J.-P. (1994) *Cancer Lett.*, **81**, 155–163.

Moutsatsos, I.K., Davis, J. M. and Wang, J.L. (1986) *J Cell Biol*, **102**, 477–483.

Moutsatsos, I.K., Wade, M., Schindler, M. and Wang, J.L. (1987) *Proc. Natl. Acad. Sci. USA*, **4**, 6452–6456.

Nangia-Makker, P., Ochieng, J., Christman, J.K. and Raz, A. (1993) *Cancer Res.*, **53**, 5033–5037.

Nangia-Makker, P., Thompson, E., Hogan, C., Ochieng, J. and Raz, A. (1995) *Int. J. Oncol.*, **7**, 1079–1087.

Natali, P.G., Wilson, B.S., Imai, K., Bigotti, A. and Ferrone, S. (1982) *Cancer Res.*, **42**, 583–589.

Ochieng, J., Fridman, R., Nangia-Makker, P., Kleiner, D.E., Liotta, L.A., Stetler-Stevenson, W.G. and Raz, A. (1994) *Biochemistry*, **33**, 14109–14114.

Ochieng, J., Gerold, M. and Raz, A. (1992) *Biochem. Biophys. Res. Comm.*, **186**, 1674–1680.

Ochieng, J., Platt, D., Tait, L., Hogan, V., Raz, T., Carmi, P. and Raz, A. (1993) *Biochemistry*, **32**, 4455–4460.

Ochieng, J. and Warfield, P. (1995) *Biochem. Biophys. Res. Commun.*, **217**, 402–406.

Oda, Y., Herrmann, J., Gitt, M.A., Turck, C.W., Burlingame, A.L., Barondes, S.H. and Leffler, H. (1993) *J. Biol. Chem.*, **268**, 5929–5939.

Oda, Y., Leffler, H., Sakakura, Y., Kasai, K.I. and Barondes, S.H. (1991) *Gene*, **99**, 279–283.

Ohannesian, D.W. (1995) *Diss Abstr Int*, **56**, 33B.

Ohannesian, D.W., Lotan, D. and Lotan, R. (1994) *Cancer Res.*, **54**, 5992–6000.

Ohannesian, D.W., Lotan, D., Thomas, P., Jessup, J. M., Fukuda, M., Gabius, H.J. and Lotan, R. (1995) *Cancer Res.*, **55**, 2191–2199.

Ohsawa, F., Hirano, F. and Natori, S. (1990) *J. Biochem.*, **107**, 431–434.

Ohyama, Y., Hirabayashi, J., Oda, Y., Ohno, S., Kawasaki, H., Suzuki, K. and Kasai, K.I. (1986) *Biochem. Biophys. Res. Comm.*, **134**, 51–66.

Ohyama, Y. and Kasai, K. I. (1988) *J. Biochem.*, **104**, 173–177.

Ostergaard, M., Rasmussen, H.H., Nielsen, H.V., Vorum, H., Orntoft, T.F., Wolf, H. and Celis, J.E. (1997) *Cancer Res*, **57**, 4111–4117.

Ozeki, Y., Matsui, T., Nitta, K., Kawauchi, H., Takayanagi, Y. and Titani, K. (1991a) *Biochem. Biophys. Res. Comm.*, **178**, 407–413.

Ozeki, Y., Matsui, T. and Titani, K. (1991b) *FEBS Lett.*, **289**, 145–147.

Ozeki, Y., Matsui, T., Yamamoto, Y., Funahashi, M., Hamako, J. and Titani, K. (1995) *Glycobiology*, **5**, 255–261.

Paroutaud, P., Levi, G., Teichberg, V. I. and Strosberg, A.D. (1987) *Proc. Natl. Acad. Sci. USA*, **84**, 6345–6348.

Peltonen, J., Larajava, H., Jakkola, S., Gralnik, H., Akiyama, S.H., Yamada, S.S., Yamada, K.M. and Uitto, J. (1989) *Am. Soc. Clin. Invest.*, **84**, 1916–1923.

Penno, M.B., Passaniti, A., Fridman, R., Hart, G.W., Jordan, C., Kumar, S. and Scott, A.F. (1989) *Proc. Natl. Acad. Sci. USA*, **86**, 6057–6061.

Perillo, N.L., Pace, K.E., Seilhamer, J.J. and Baum, L.G. (1995) *Nature*, **378**, 736–739.

Perillo, N.L., Uittenbogaart, C. H., Nguyen, J.T. and Baum, L.G. (1997) *J. Exp. Med.*, **185**, 1851–1858.

Pienta, K.J., Naik, H., Akhtar, A., Yamazaki, K., Replogle, T.S., Lehr, J., Donat, T.L., Tait, L., Hogan, V. and Raz, A. (1995) *J. Natl. Cancer Inst.*, **87**, 348–353.

Platt, D. and Raz, A. (1992) *J. Natl. Cancer Inst.*, **84**, 438–442.

Poirier, F. and Robertson, E.J. (1993) *Development*, **119**, 1229–1236.

Poirier, F., Timmons, P.M., Chan, C.T.J., Guénet, J.L. and Rigby, P.W.J. (1992) *Development*, **115**, 143–155.

Powell, J.T. and Whitney, P.L. (1980) *Biochem. J.*, **188**, 1–8.

Probstmeier, R., Montag, D. and Schachner, M. (1995) *J. Neurochem.*, **64**, 2465–2472.

Puche, A.C., Poirier, F., Hair, M., Bartlett, P.F. and Key, B. (1996) *Dev. Biol.*, **179**, 274–287.

Raimond, J., Rouleux, F., Monsigny, M. and Legrand, A. (1995) *FEBS Lett.*, **363**, 165–169.

Raimond, J., Zimonjic, D. B., Mignon, C., Mattei, M., Popescu, N.C., Monsigny, M. and Legrand, A. (1997) *Mamm Genome*, **8**, 706–707.

Raz, A., Avivi, A., Pazerini, G. and Carmi, P. (1987a) *Exp. Cell Res.*, **173**, 109–116.

Raz, A., Carmi, P. and Pazerini, G. (1988) *Cancer Res.*, **48**, 645–649.

Raz, A., Carmi, P., Raz, T., Hogan, V., Mohamed, A. and Wolman, S. R. (1991) *Cancer Res.*, **51**, 2173–2178.

Raz, A. and Lotan, R. (1981) *Cancer Res.*, **41**, 3642–3647.

Raz, A., Meromsky, L., Carmi, P., Karakash, R., Lotan, D. and Lotan, R. (1984) *Embo J.*, **3**, 2979–2983.

Raz, A., Meromsky, L. and Lotan, R. (1986) *Cancer Res.*, **46**, 3667–3672.

Raz, A., Meromsky, L., Zvibel, I. and Lotan, R. (1987b) *Int. J. Cancer*, **39**, 353–360.

Raz, A., Pazerini, G. and Carmi, P. (1989) *Cancer Res.*, **49**, 3489–3493.

Raz, A., Zhu, D. G., Hogan, V., Shah, N., Raz, T., Karkash, R., Pazerini, G. and Carmi, P. (1990) *Int. J. Cancer*, **46**, 871–877.

Rechreche, H., Mallo, G.V., Montalto, G., Dagorn, J.C. and Iovanna, J. L. (1997) *Eur. J. Biochem.*, **248**, 225–230.

Rini, J.M. (1995) *Curr. Opin. Struct. Biol.*, **5**, 617–621.

Robertson, M.W., Albrandt, K., Keller, D. and Liu, F. T. (1990) *Biochemistry*, **29**, 8093–8100.

Roff, C.F., Rosevear, P. R., Wang, J.L. and Barker, R. (1983) *Biochem. J.*, **211**, 625–629.

Roff, C.F. and Wang, J.L. (1983) *J. Biol. Chem.*, **258**, 10657–10663.

Rosenberg, I., Cherayil, B.J., Isselbacher, K.J. and Pillai, S. (1991) *J. Biol. Chem.*, **266**, 18731–18736.

Rosenberg, I. M., Iyer, R., Cherayil, B., Chiodino, C. and Pillai, S. (1993) *J. Biol. Chem.*, **268**, 12393–12400.

Ruggiero-Lopez, D., Louisot, P. and Martin, A. (1992) *Biochem. Biophys. Res. Commun.*, **185**, 617–623.

Runyan, R.B., Versalovic, J. and Shur, B.D. (1988) *J. Cell Biol.*, **107**, 1863–1871.

Salvatore, P., Contursi, C., Benvenuto, G., Bruni, C.B. and Chiariotti, L. (1995) *FEBS Lett.*, **373**, 159–163.

Sanjuan, X., Fernandez, P.L., Castells, A., Castronovo, V., van den Brûle, F.A., Liu, F.T., Cardesa, A. and Campo, E. (1997) *Gastroenterology*, **113**, 1906–1915.

Sasaki, M., Kato, S., Kohno, K., Martin, G.R. and Yamada, Y. (1987) *Proc. Natl. Acad. Sci. USA*, **84**, 935–939.

Sasaki, M., Kleinman, H.K., Huber, H., Deutamann, R. and Yamada, Y. (1988) *J. Biol. Chem.*, **263**, 6936–6944.

Sasaki, M. and Yamada, Y. (1987) *J. Biol. Chem.*, **262**, 17111–17117.

Sato, S., Burdett, I. and Hughes, R.C. (1993) *Exp Cell Res*, **207**, 8–18.

Sato, S. and Hughes, R.C. (1992) *J. Biol. Chem.*, **267**, 6983–6990.

Sato, S. and Hughes, R.C. (1994) *J. Biol. Chem.*, **269**, 4424–4430.

Schoeppner, H.L., Raz, A., Ho, S.B. and Bresalier, R.S. (1995) *Cancer*, **75**, 2818–2826.

Seve, A.P., Felin, M., Doyennette-Moyne, M.A., Sahraoui, T., Aubery, M. and Hubert, J. (1993) *Glycobiology*, **3**, 23–30.

Seve, A. P., Hadj-Sahraoui, Y., Felin, M., Doyennette-Moyne, M.-A., Aubery, M. and Hubert, J. (1994) *Exp. Cell Res.*, **213**, 191–197.

Sharma, A., Chemelli, R. and Allen, H.J. (1990) *Biochemistry*, **29**, 5309–5314.

Shur, B.D. (1982) *J. Biol. Chem.*, **257**, 6871–6878.

Shur, B.D. (1989) *Curr. Opin. Cell Biol.*, **1**, 905–912.

Skrincosky, D.M., Allen, H.J. and Bernacki, R.J. (1993) *Cancer Res.*, **53**, 2667–2675.

Solomon, J.C., Stoll, M. S., Penfold, P., Abbott, W.M., Childs, R.A., Hanfland, P. and Feizi, T. (1991) *Carbohydr Res*, **213**, 293–307.

Sonnenberg, A., Gehlsen, K.R., Aumailley, M. and Timpl, R. (1991) *Exp. Cell Res.*, **197**, 234–244.

Sonnenberg, A., Modderman, P.W. and Hogervorst, F. (1988) *Nature*, **336**, 487–489.

Southan, C., Aitken, A., Childs, R.A., Abbott, W.M. and Feizi, T. (1987) *FEBS Letters*, **214**, 301–304.

Sparrow, C. P., Leffler, H. and Barondes, S.H. (1987) *J. Biol. Chem.*, **262**, 7383–7390.

Stojanovic, D., Hughes, R.C., Feizi, T. and Childs, R.A. (1983) *J. Cell. Biochem.*, **21**, 119–127.

Su, Z.Z., Lin, J., Shen, R., Fisher, P.E., Goldstein, N.I. and Fisher, P.B. (1996) *Proc. Natl. Acad. Sci. USA*, **93**, 7252–7257.

Takeuchi, M., Yoshikawa, M., Sasaki, R. and Chiba, H. (1982) *Agric. Biol. Chem.*, **46**, 2741–2747.

Tardy, F., Deviller, P., Louisot, P. and Martin, A. (1995) *FEBS Lett*, **359**, 169–172.

Teichberg, V.I., Sliman, I., Beitsch, D.D. and Resheff, G. (1975) *Proc. Natl .Acad. Sci. USA*, **72**, 1383–1387.

Timpl, R. and Brown, J.C. (1994) *Matrix Biol.*, **14**, 275–281.

Timpl, R., Rohde, H., Robey, P.G., Rennard, S.I., Foidart, J.M. and Martin, G.R. (1979) *J. Biol. Chem.*, **254**, 9933–9937.

Türeci, O., Schmitt, H., Fadle, N., Pfreundschuh, M. and Sahin, U. (1997) *J. Biol. Chem.*, **272**, 6416–6422.

van den Brûle, F. A., Bellahcène, A., Jackers, P., Liu, F.-T., Sobel, M.E. and Castronovo, V. (1997a) *Int. J. Oncol.*, **11**, 261–264.

van den Brûle, F.A., Berchuck, A., Bast, R.C., Liu, F.T., Gillet, C., Sobel, M.E. and Castronovo, V. (1994a) *Eur. J. Cancer*, **30A**, 1096–1099.

van den Brûle, F.A., Buicu, C., Baldet, M., Sobel, M.E., Cooper, D.N., Marschal, P. and Castronovo, V. (1995a) *Biochem. Biophys. Res. Commun.*, **209**, 760–767.

van den Brûle, F.A., Buicu, C., Berchuck, A., Bast, R.C., Deprez, M., Liu, F.T., Cooper, D.N., Pieters, C., Sobel, M.E. and Castronovo, V. (1996) *Hum. Pathol.*, **27**, 1185–1191.

van den Brûle, F.A., Buicu, C., Sobel, M.E., Liu, F.T. and Castronovo, V. (1995b) *Neoplasma*, **42**, 215–219.

van den Brûle, F.A., Engel, J., Stetler-Stevenson, W.G., Liu, F.T., Sobel, M.E. and Castronovo, V. (1992) *Int. J. Cancer*, **52**, 653–657.

van den Brûle, F.A., Fernandez, P.L., Buicu, C., Liu, F.T., Jackers, P., Lambotte, R. and Castronovo, V. (1997b) *Dev. Dyn.*, **209**, 399–405.

van den Brûle, F.A., Liu, F.-T. and Castronovo, V. (1998) *Cell Adh Comm*, **5**, 425–435.

van den Brûle, F. A., Price, J., Sobel, M. E., Lambotte, R. and Castronovo, V. (1994b) *Biochem. Biophys. Res. Comm.*, **201**, 399–393.

Vlassara, H., Li, Y.M., Imani, F., Wojciechowicz, D., Yang, Z., Liu, F.T. and Cerami, A. (1995) *Mol Med*, **1**, 634–646.

Voss, P.G., Tsay, Y.G. and Wang, J.L. (1994) *Glycoconj. J.*, **11**, 353–362.

Vracko, R. (1974) *Am J Pathol*, **77**, 314–346.

Vyakarnam, A., Dagher, S.F., Wang, J.L. and Patterson, R.J. (1997) *Mol. Cell Biol.*, **17**, 4730–4737.

Wada, J. and Kanwar, Y.S. (1997) *J. Biol. Chem.*, **272**, 6078–6086.

Wada, J., Ota, K., Kumar, A., Wallner, E.I. and Kanwar, Y.S. (1997) *J. Clin. Invest.*, **99**, 2452–2461.

Wang, L., Inohara, H., Pienta, K.J. and Raz, A. (1995) *Biochem. Biophys. Res. Commun.*, **217**, 292–303.

Wasano, K. and Hirakawa, Y. (1997) *J. Histochem. Cytochem.*, **45**, 275–283.

Weitlauf, H.M. and Knisley, K.A. (1992) *Biol. Reprod.*, **46**, 811–816.

Wells, V. and Mallucci, L. (1983) *J. Cell Physiol.*, **117**, 148–154.

Wells, V. and Mallucci, L. (1992) *Biochim. Biophys. Acta*, **1121**, 239–244.

Wells, V. and Malucci, L. (1991) *Cell*, **64**, 91–97.

Whitney, P.L., Powell, J.T. and Sanford, G.L. (1986) *Biochem. J.*, **238**, 683–689.

Wilson, T.J.G., Firth, M.N., Powell, J.T. and Harrison, F.L. (1989) *Biochem. J.*, **261**, 847–852.

Woo, H.J., Shaw, L.M., Messier, J.M. and Mercurio, A.M. (1990) *J. Biol. Chem.*, **265**, 7097–7099.

Woynarowska, B., Dimitroff, C.J., Sharma, M., Matta, K.L. and Bernacki, R.J. (1996) *Glycoconj. J.*, **13**, 663–674.

Xu, X. C., el-Naggar, A.K. and Lotan, R. (1995) *Am. J. Pathol.*, **147**, 815–822.

Yamaoka, A., Kuwabara, I., Frigeri, L. G. and Liu, F. T. (1995) *J. Immunol.*, **154**, 3479–3487.

Yamaoka, K., Hirrai, R., Tsugita, A. and Mitsui, H. (1984) *J. Cell Physiol.*, **119**, 307–314.

Yamaoka, K., Ingendoh, A., Tsubuki, S., Nagai, Y. and Sanai, Y. (1996) *J. Biochem. (Tokyo)*, **119**, 878–886.

Yamaoka, K., Ohno, S., Kawasaki, H. and Suzuki, K. (1991) *Biochemical and Biophysical Research Communications*, **179**, 272–279.

Yang, R. Y., Hsu, D. K. and Liu, F. T. (1996) *Proc Natl Acad Sci USA*, **93**, 6737–6742.

Zhou, Q. and Cummings, R.D. (1990) *Arch. Biochem. Biophys.*, **281**, 27–35.

Zhou, Q. and Cummings, R.D. (1993) *Arch. Biochem. Biophys.*, **300**, 6–17.

7. LECTINS AS DIAGNOSTIC TOOLS AND ENDOGENOUS TARGETS IN AIDS AND PRION DISEASES

HEINZ C. SCHRÖDER[1], WERNER E.G. MÜLLER[1] AND JOHN M.S. FORREST[2]

[1]*Institut für Physiologische Chemie, Abteilung Angewandte Molekularbiologie, Universität, Duesbergweg 6, 55099 Mainz, Germany*
[2]*Scottish Crop Research Institute, Invergowrie, Dundee DD2 5DA, United Kingdom*

INTRODUCTION

Based on their ability to specifically to recognize monosaccharides or larger segments within oligosaccharide side chains of glycoconjugates such as cellular or pathogen-specific glycoproteins, lectins are considered as useful tools in biomedical research, clinical diagnosis and also therapy. Most of these lectins come from plants. Animal lectins are a rather novel group of lectins acting as physiological receptors for cellular glycoproteins and gangliosides. In this chapter the potential role of animal and plant lectins in diagnosis and pathogenesis of AIDS and prion diseases is discussed.

LECTINS FOR DETECTION OF HIV AND PRION PROTEINS

Recognition of Human Immunodeficiency Virus Glycoproteins

Collectins

Among components of the mammalian innate immune system, the collectins (including mannan binding proteins), which are mainly mannose specific lectins, are involved in the recognition of diverse pathogens such as fungi, bacteria and some viruses (Epstein *et al.*, 1996). They bind to glycans which are themselves almost ubiquitous on hosts and pathogens, but the secret of their ability to distinguish non self from self lies in recognition of a pattern of glycans which may be invariant across a wide range of pathogens, but absent from self. In addition, galactose and sialic acid which are usually present on mammalian glycoproteins as the penultimate and ultimate sugars, cannot be accommodated by the carbohydrate recognition domain of mannan binding proteins (Epstein *et al.*, 1996, Medzhitov and Janeway, 1997). These abilities certainly qualify them for use *in vitro* to capture and quantify some pathogens. However, it is debatable whether harvesting the collectins from serum

---

[1] Abbreviations: CBP, carbohydrate-binding protein; CJD, Creutzfeldt-Jakob disease; ELISA, enzyme-linked immunosorbent assay; GNA, *Galanthus nivalis* agglutinin; gp120 (gp125), glycoprotein of M_r 120 kDa (M_r 125 kDa); HIV, human immunodeficiency virus; hnRNP, heterogeneous nuclear ribonucleoprotein complex; NPA, *Narcissus pseudonarcissus* agglutinin; PrP^C, cellular prion protein; PrP^Sc, scrapie prion protein; RRE, Rev-responsive element; TAR, *trans*-acting response element; WGA, wheat germ agglutinin.

would be economical, or even desirable as a potential source of blood-borne infections. Transgenic production of collectins may be a possibility.

Plant lectins

The envelope glycoproteins of the human immunodeficiency viruses 1 and 2 (HIV-1 and HIV-2),[1] gp120 and gp125 (glycoprotein of M_r 120 kDa and M_r 125 kDa, respectively), are heavily glycosylated and are bound not only by the collectins (Medzhitov and Janeway, 1997) but also by some plant lectins (Robinson *et al.*, 1987). One of these, concanavalin A, which can be extracted relatively cheaply and in abundance binds glycoproteins on the surface of uninfected cells as well as the virus. While this has not prevented its use to trap HIV envelope glycoproteins for subsequent quantification of anti-gp120 antibodies in human serum (Robinson *et al.*, 1990), more selective plant lectins were clearly preferable for such a task. *Narcissus pseudonarcissus* agglutinin (NPA) and *Galanthus nivalis* agglutinin (GNA) largely fulfil this criterion, as they have only been reported to bind to a very limited range of human glycoproteins, including α_2-macroglobulin and one unknown glycoprotein in human serum (Van Leuven *et al.*, 1993) (but also to the surface of cells under non-physiological temperatures, Hammar *et al.*, 1995). gp120 is a particularly appropriate target, because it appears in cell cultures, and most probably also *in vivo*, much earlier and in greater quantities than the commonly targeted HIV protein, p24. Ushijima *et al.* (1992) demonstrated with NPA bound to the solid phase and GNA as the detector, that quantification was valid for the range 0.6 to 20,000 ng gp120/ml. In practice, serum components did not interfere with binding of gp120 to the lectins (Ushijima *et al.*, 1992).

Discrimination between different pathogens

Interestingly, the characteristic arrays of glycans recognised by NPA and GNA may be the same as those recognised by the collectins (Forrest *et al.*, 1996). They are now known to be shared by some other viruses, including feline immunodeficiency virus, simian immunodeficiency virus, human cytomegalovirus (Balzarini *et al.*, 1991), rabies, rubella (Marchetti *et al.*, 1995), and indeed some fungi and bacteria, so the original suggestion that the NPA/GNA enzyme-linked immunosorbent assay (ELISA) should be used for specific detection may be invalid. However, lectins of the *Amaryllidaceae* exist as multiple isoforms which vary quantitatively between cultivars, plant parts and seasons (van Damme *et al.*, 1992). The microheterogeneity of individual lectins is perhaps unlikely to affect binding to simple sugar haptens, but where complexes are formed with other glycoproteins, small differences in composition distant from the lectin binding site may conceivably confer pathogen-specific recognition. Further work on the isolectins will reveal whether they can indeed discriminate between different pathogens with the same glycan patterns. In the meantime, lectins can be used to trap glycoproteins for subsequent probing with a range of pathogen-specific antibodies. For example, a positive agglutination test could reveal the possible presence of a range of pathogens, and individual pathogen(s) could be subsequently identified by a panel of specific antibodies.

Recognition of Prion Glycoproteins

Plant lectins

The same attributes may make some lectins suitable for differentiating between 'strains' of prion protein. In the absence of evidence for an accessory molecule such as a nucleic acid, 'strains' are primarily considered to be associated with post-translational modifications of the prion protein, especially its two potential asparagine-linked glycosylation sites. Quantitative or qualitative changes may well be revealed by differential lectin binding. Hamster PrP 27-30, produced by limited proteolysis of the scrapie prion protein, PrP^{Sc}, was specifically bound by wheat germ agglutinin (WGA) and *Lens culinaris* agglutinin as demonstrated by blocking with the respective sugar haptens. Treatment with sialidase abolished the binding of WGA, but enhanced that of *Ricinus communis* agglutinin (Sklaviadis *et al.*, 1986). These findings are consistent with the presence of complex oligosaccharides bearing terminal sialic acids, penultimate galactoses, and fucose residues attached to the innermost *N*-acetyl glucosamine (Haraguchi *et al.*, 1989). These structures should not be recognised by the collectins. It is not known whether glycans of PrP^{Sc} have the same structures as those of PrP^{C} (normal cellular isoform of prion protein), and there are difficulties in obtaining sufficient quantities of PrP^{C} for such studies. However, the absence of an immune response suggests that both isoforms of prion are recognised as self. The same question can be asked about the endogenous lectins - do they recognise PrP^{Sc} as abnormal, and is its distribution within the cell affected?

Factors possibly influencing lectin recognition

Even in the absence of glycosylation of prion protein, changes in conformation probably occur, as the molecules acquire resistance to proteases (Taraboulos *et al.*, 1990). Conformational change could also affect lectin binding by altering the orientation rather than the structure of glycans. Furthermore, when the prion protein exists as a fibrillar precipitate, this may present quite a different target from an intermediate form which is soluble, but has some resistance to proteinase K.

   Briefly, it is currently proposed that PrP^{Sc} is generated by dimerisation of normal and abnormal prions, with subsequent unfolding and refolding of normal prions in the abnormal conformation, leading to a depletion of normal prion protein and an increase of the abnormal form. Other possible sources of conformational change have received little or no attention. Protein splicing (Perler *et al.*, 1997) or its intermediate steps could, at the very least, lead to conformational change, or even hybrid molecules, although no evidence has yet been found for a role in formation of abnormal prions (Stahl *et al.*, 1993). Interestingly, there is an intein-like sequence created by the point mutation D178N which is linked to either familial Creutzfeldt-Jakob disease (CJD) or fatal familial insomnia, depending on which amino acid is present at codon 129 (Prusiner 1994). The putative intein begins at C22 and ends with a characteristic C terminal splice junction sequence — HNC — at 178.

   Western blot analysis of PrP^{Sc} which has undergone limited proteolysis and separation on polyacrylamide gels reveals three bands which correspond to proteins glycosylated at both sites, one site only, or none at all. The mobility and proportions

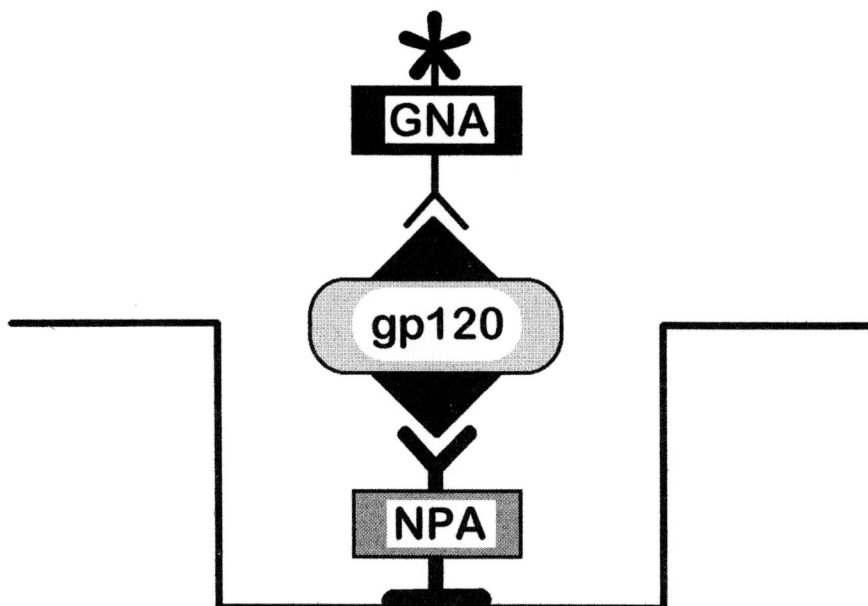

Figure 1   Scheme of lectin ELISA.

of these bands vary between 'strains' of prion, and may be the result of differential susceptibility of the glycosylated fragments to degradation by proteinase K. Collinge *et al.* (1996) found that the 'fingerprint' of the so-called 'new variant'- vCJD which has affected young people in the UK more closely resembled that of BSE rather than sporadic or iatrogenic CJD, each of which has its own distinctive pattern. Somerville and Ritchie (1990) separated PrP^{Sc} from scrapie-associated fibrils untreated with proteinase K on two-dimensional gels to reveal six major spots. They found no major differences in binding of fourteen plant lectins to three 'strains' of scrapie from mouse brains, although there were relative differences in the yields of each spot.

Future developments

During the past decade, much has been discovered about plant and animal lectins and prions, and it may be worth re-examining the possibility of a lectin ELISA (Figure 1). Among the endogenous lectins CBP35, also known as galectin 3, has been shown to be associated with partially purified prion protein, as well as in complexes with PrP mRNA and PrP (Schröder *et al.*, 1994; see also below). Because of its ability to bind both carbohydrate and RNA at separate sites, CBP35 may form complexes which present opportunities for specific recognition via PrP mRNA or differences in the prion proteins themselves. There is also the possibility of examining what may be a unique relationship between the components of the complex with plant lectins or other probes.

INTERACTION OF ENDOGENOUS LECTINS WITH HIV AND PRION PROTEINS

Nuclear Lectins (Carbohydrate-binding Proteins)

So far six major nuclear lectins (carbohydrate binding proteins, CBPs) from mammalian cells and tissues have been purified and characterized: CBP14, CBP22, CBP33, CBP35, CBP67 and CBP70 (for a review, see Hubert and Sève, 1994). A few glycosylated ligands of these lectins have been identified (Felin *et al.*, 1997). CBPs are preferentially localized in the cell nucleus, especially in the extranucleolar region (Sève *et al.*, 1985, 1986; Hubert *et al.*, 1985; Laing and Wang, 1988; Facy *et al.*, 1990), but they are found also in the cytoplasm, on the cell surface and in the extracellular matrix (Ho and Springer, 1982; Cherayil *et al.*, 1989). The pattern of nuclear lectins depends on the proliferative state of the cells (Moutsatsos *et al.*, 1986, 1987; Bourgeois *et al.*, 1987; Hubert *et al.*, 1989). Undifferentiated cells were found to bind significantly lower amounts of neoglycoprotein (synthetic glycoproteins) than differentiated cells (Facy *et al.*, 1990).

Only one nuclear lectin, CBP35 (galectin 3, also known as Mac2, L29, or IgE-binding protein; Wang *et al.*, 1991, Hirabayashi and Kasai, 1993), has been characterized in greater detail; see below (Moutsatsos *et al.*, 1987; Agrwal *et al.*, 1993; Knibbs *et al.*, 1993). CBP67 is a glucose-specific nuclear CBP from rat liver with a molecular mass of 67 kDa; it is present in nuclear ribonucleoprotein (RNP) complexes but absent in polysomal RNP complexes (Schröder *et al.*, 1992). Another glucose-specific nuclear lectin, CBP70 (molecular mass, 70 kDa), has been isolated and purified from the human tumoral cell line HL60; this lectin also recognises *N*-acetyl glucosamine (Sève *et al.*, 1993, 1994). A further glucose-specific nuclear CBP, CBP33, that has been purified from rat liver, is induced following exposure to immobilization stress (Lauc *et al.*, 1994). A *N*-acetylglucosamine-binding protein, CBP22, has been identified and purified from HL60 cell nuclei (Felin *et al.*, 1994). The lectin CBP14 that is probably also associated with RNP complexes, has been purified from rat liver nuclei (Cuperlovic *et al.*, 1995).

Carbohydrate-binding Protein 35

The carbohydrate-binding protein CBP35 (molecular mass, ~35 kDa), also termed galectin 3, specifically binds to ß-galactoside residues (Laing and Wang 1988, Roff and Wang, 1983). CBP35 has been identified as a protein of the hnRNP complex (Laing and Wang, 1988). The binding of CBP35 to β-galactoside-containing glycoconjugates does not depend on the presence of divalent cations; therefore CBP35 belongs to the class of S-type lectins (Drickamer, 1988). This lectin has been cloned and found to consist of two domains, a *C*-terminal carbohydrate recognition domain that is homologous to ß-galactoside-specific lectins and a *N*-terminal proline/glycine rich domain homologous to proteins of the hnRNP complex (Figure 2; Jia and Wang, 1988).

Evidence has been presented that CBP35 is involved in the cellular stress response, in the formation of spliceosomal complexes and in the transport of mRNA from nucleus to cytoplasm (Wang *et al.*, 1992). CBP35 is able to bind to the glucose-specific lectins, CBP67 (Lauc *et al.*, 1993) and CBP70 (Sève *et al.*, 1993); Figure 2.

CBP35 (dimerization)

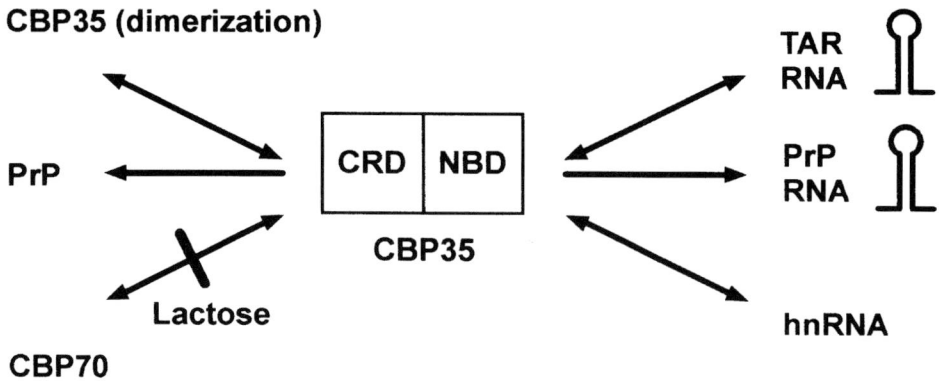

CBP70

Figure 2   Protein-protein and protein-nucleic acid interactions of CBP35, containing both a carbohydrate-recognition domain (CRD) and a nucleic acid binding domain (NBD), with CBP35 (dimerization), PrP, CBP70, TAR- and PrP-RNA stem-loops and hnRNA.

The CBP35–CBP67 association is induced under stress conditions (Lauc *et al.*, 1993). The protein-protein interaction between CBP70 and CBP35 is disrupted when lactose binds to CBP35 (Figure 2; Sève *et al.*, 1994). The nucleocytoplasmic transport of mRNA is an energy-dependent process (Schröder *et al.*, 1986, 1987). CBP35 is complexed with RNP both in the nucleus and in the cytoplasm (Wang *et al.*, 1992; Schröder *et al.*, 1995) and seems to be co-transported with the mRNA in the form of a RNP complex (Agrwal *et al.*, 1989; Lauc *et al.*, 1993).

CBP35 and HIV Pathogenesis

Changes in expression of CBP35 in HIV-1-infected cells

The expression of CBP35 transiently and strongly increases in the early period of infection of cells with HIV-1 (Schröder *et al.*, 1995). Molt-3 cells start to produce virus at day 2 following infection with HIV-1. Concomitantly with the onset of the expression of viral *tat* gene, a rise in the level of CBP35 mRNA was observed (Figure 3). The increase in CBP35 mRNA level results in an enhanced synthesis of CBP35. The mRNA levels of CBP35 in mock-infected cells and the mRNA levels of glyceraldehyde-3-phosphate dehydrogenase do not change markedly. The maximum increase in CBP35 mRNA level (at day 2) coincides with the integration of HIV-1 into the cell genome and the onset of production of the HIV-1 regulatory proteins Tat, Rev, and Nef. In the further course of infection, characterized by the synthesis and assembly of the viral structural proteins and the release of virus progeny, CBP35 mRNA level and activity decrease to levels below those found in uninfected cells (at day 5; Figure 3). The increase in the level of CBP35 transcripts at day 2 is most likely to be due to an increase in the rate of synthesis of the mRNA, while the decrease in CBP35 transcript level at day 5 may be caused by destabilization of the mRNA (Schröder *et al.*, 1995). Immunoblotting experiments revealed that CBP35 is present in the 40S hnRNP complex from HIV-1-infected Molt-3 cells (Schröder *et al.*, 1995).

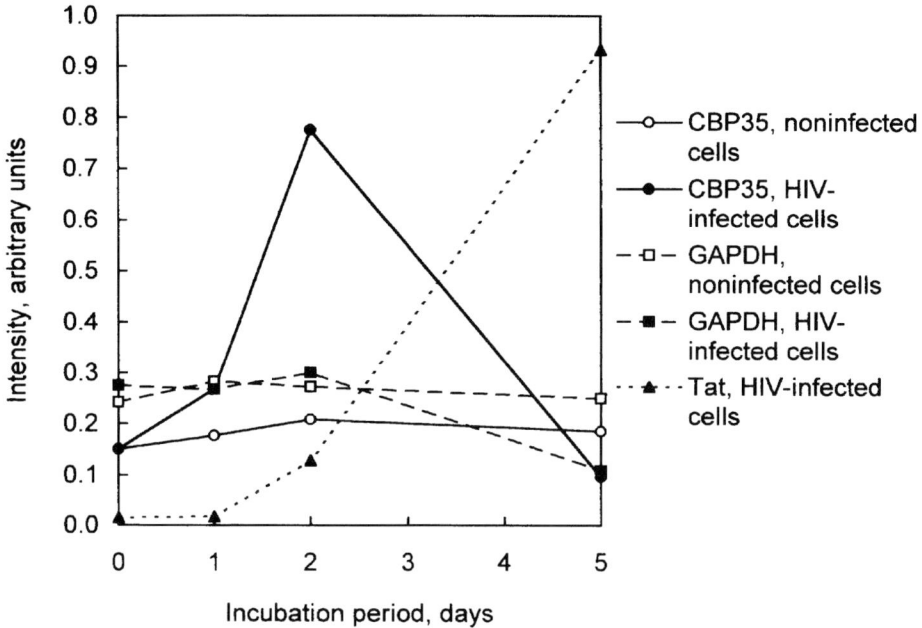

Figure 3   Changes of expression of CBP35 mRNA in the course of infection of Molt-3 cells with HIV-1. Levels of CBP35 RNA, glyceraldehyde-3-phosphate dehydrogenase (GAPDH) RNA (= control) and viral *tat* RNA were determined in mock-infected (= noninfected) and HIV-1-infected cells at days 0, 1, 2 and 5 post-infection. The RNA concentrations are given in arbitrary units. Results are mean values of triplicate determinations (SD $\leq$ 20%).

Possible role of CBP35 in HIV pathogensis

CBP35 has been found to bind to *in vitro* transcribed Rev-responsive element (RRE) RNA of HIV-1 (Bek *et al.*, 1995) as well as TAR RNA (Fig. 2). Binding of CBP35 to RRE RNA is blocked in the presence of Rev protein of HIV-1 (Bek *et al.*, 1995); this result is in line with the assumption that Rev prevents formation of hnRNP complexes, as well as spliceosome assembly, of nascent HIV transcripts, resulting in transport of the unspliced or incompletely spliced HIV-1 mRNA into the cytoplasm (Cullen *et al.*, 1991). The finding that the expression of CBP35 transiently increases after infection of cells with HIV-1 suggests that CBP35 may play a role during the initial stage of HIV infection, most likely during transport and/or splicing of HIV mRNA. Later CBP35 synthesis slows down and binding of CBP35 to the RRE of HIV mRNA is inhibited by Rev protein.

CBP35 and Prion Pathogenesis

Glycosylation of prion protein

Both prion protein isoforms and PrP 27-30 are glycosylated at one or both of their two *N*-glycosylation sites (Oesch *et al.*, 1985). The asparagine-linked oligosaccharides attached to these sites are of the complex type and consist of a mixture of bi-, tri-, and tetrantennary glycans (Endo *et al.*, 1989, Taraboulos *et al.*, 1990) with the following core; Manα1→6(GlcNAcβ1→4)(Manα1→3)Manβ1→4GlcNAcβ1→4 (Fucα1→6)GlcNAc, and the following outer chain moieties; Galβ1→4GlcNAcβ1→, Galβ1→4(Fucα1→3)GlcNAcβ1→, GlcNAcβ1→, Siaα2→3Galβ1→4GlcNAcβ1→, and Siaα2→6Galβ1→4GlcNAcβ1→. Over 400 different carbohydrate structures are possible by different combinations of these sugar chains and glycosylation of both *N*-glycosylation sites (Endo *et al.*, 1989). The heterogeneity of undigested PrP^C and PrP^{Sc} may be caused at least partially by differential glycosylation of the protein. In addition, both prion protein isoforms and PrP 27-30 possess a glycosylphosphatidylinositol anchor covalently attached to the *C*-terminus of the protein (Stahl *et al.*, 1987, Baldwin *et al.*, 1990). The PrP carbohydrates may constitute up to 30% of the total molecular mass of PrP 27-30.

Possible interaction of prion protein with endogenous lectins

The oligosaccharides in both prion protein isoforms, PrP^{Sc} and PrP^C, may represent potential binding sites for cellular CBPs (Figure 2). CBP35 is present also on the cell surface (Ho and Springer, 1982; Cherayil *et al.*, 1989). Therefore, this lectin might interact with PrP^{Sc} not only intracellularly but also act as cellular receptor for extracellular PrP^{Sc}. The identification of proteins on the cell surface that bind PrP^{Sc} is important not only for the understanding of the mechanism of PrP^{Sc} propagation but also for the development of strategies for prevention and therapy of prion diseases. Binding of PrP^{Sc} to cellular receptors may also be involved in the mechanism of induction of apoptosis by this protein. PrP^{Sc} and the PrP 106-126 peptide, corresponding to amino acid 106-126 of the deduced amino acid sequence of human PrP cDNA, have been shown to act neurotoxically *in vitro* (Müller *et al.*, 1993, Forloni et al. 1993). Future studies have to clarify the potential role of cellular lectins in the process of infection of cells with PrP^{Sc}/PrP 27-30.

Interaction of CBP35 with prion protein mRNA

The PrP mRNA contains three predictable hairpin structures at nt +170, +194, and +218 within the coding region (Wills and Hughes, 1990). These hairpin structures contain the pentanucleotide CUGGG in the loop, and a uridine- and adenine bulge in the stem, similar to the *cis*-acting TAR sequence within the long terminal repeat of HIV-1 RNA (Okamoto *et al.*, 1986). Indeed, the *trans*-activating Tat protein that is known to bind at the uridine bulge in the TAR RNA stem, was also found to be able to bind to the PrP RNA stem-loop (Müller *et al.*, 1992).

UV cross-linking/RNase protection and Northwestern blotting experiments revealed that a number of cellular proteins binds to the PrP RNA stem-loop structure (Schröder *et al.*, 1994; Scheffer *et al.*, 1995). The molecular masses of PrP RNA- and HIV-1 TAR RNA-binding proteins were found to be virtually identical. CBP35 could be detected among those proteins which associate with the PrP RNA stem-loop structure (Fig. 2; Schröder *et al.*, 1994).

PrP^C was also identified in the RNP complexes formed between PrP RNA and proteins present in brain extracts, but no evidence was obtained that this protein directly binds to the RNA (Schröder *et al.*, 1994). It was proposed that the association of PrP^C with the RNP complex formed with PrP RNA and cellular proteins is due to an interaction of PrP^C with CBP35 present in this complex. Interestingly, nucleic acid binding proteins have also been detected in purified prions (Sklaviadis *et al.*, 1993). However, any speculation that a cellular RNA-binding protein such as CBP35 may be an essential component of infectious prion particles through binding to a cellular or pathogen-specific nucleic acid constituent is worthless, as long as it has not been shown that such a nucleic acid is required for prion replication. There are also no hints that PrP^{Sc} affects expression of PrP gene. No marked differences in the steady-state PrP mRNA levels, PrP-specific transcriptional activity and half-life of PrP mRNA in uninfected and PrP^C-infected N$_2$a cells have been found (Pfeifer *et al.*, 1993).

ACKNOWLEDGEMENTS

This work was supported by a grant from the Commission of the European Communities (FAIR5-CT97-3315). Funding from Scottish Office Agriculture, Environment and Fisheries Department (SOAEFD) is also acknowledged.

REFERENCES

Agrwal, N., Wang, J.L. and Voss, P.G. (1989) Carbohydrate-binding protein 35. Levels of transcription and mRNA accumulation in quiescent and proliferating cells. *J. Biol. Chem.*, **264**, 17236–17242.

Agrwal, N., Sun, Q., Wang, S. and Wang, J.L. (1993) Carbohydrate-binding protein 35. I. Properties of the recombinant polypeptide and the individuality of the domains. *J. Biol. Chem.*, **268**, 14932–14939.

Baldwin, M.A., Stahl, N., Reinders, L.G., Gibson, B.W., Prusiner, S.B. and Burlingame, A.L. (1990) Permethylation and tandem mass spectrometry of oligosaccharides having free hexosamine: analysis of the glycoinositol phospholipid anchor glycan from the scrapie prion protein. *Anal. Biochem.*, **191**, 174–182.

Balzarini, J., Schols, D., Neyts, J., Van Damme, E., Peumans, W. and De Clercq, E. (1991) α-(1–3)- and α-(1–6)-D-mannose-specific plant lectins are markedly inhibitory to human immunodeficiency virus and cytomegalovirus *in vitro*. *Antimicrob. Agents Chemother.*, **35**, 410–416.

Bek, A., Müller, W.E.G., Sève, A.-P., Kavsan, V. and Schröder, H.C. (1995) Rev protein suppression of complex formation between nuclear proteins and Rev-responsive element-containing RNA of human immunodeficiency virus-1. *Int. J. Biochem. Cell Biol.*, **27**, 1317–1329.

Borchelt, D.R., Scott, M., Taraboulos, A., Stahl, N. and Prusiner, S.B. (1990) Scrapie and cellular prion proteins differ in their kinetics of synthesis and topology in cultured cells. *J. Cell Biol.*, **110**, 743–752.

Bourgeois, C.A., Sève, A.P., Monsigny, M. and Hubert, J. (1987) Detection of sugar-binding sites in the fibrillar and the granular components of the nucleolus: an experimental study in cultured mammalian cells. *Exp. Cell Res.*, **172**, 365–376

Cherayil, B.J., Weiner, S.J. and Pillai, S. (1989) The Mac-2 antigen is a galactose specific lectin that binds IgG. *J. Exp. Med.*, **170**, 1959–1972.

Collinge, J., Sidle, K.C.L., Meads, J., Ironside, J. and Hill, A.F. (1996) Molecular analysis of prion strain variation and the aetiology of 'new variant' CJD. *Nature*, **383**, 685–690.

Cullen, B.R., Malim, M.H. (1991) The HIV-1 Rev protein: prototype of a novel class of eukaryotic post-transcriptional regulators. *TIBS*, **16**, 346–350.

Cuperlovic, M., Jankovic, M., Pfeifer, K. and Müller, W.E.G. (1995) Isolation and characterization of a β-galactoside-binding protein (14 kD) from rat liver nuclei. *Cell. Physiol. Biochem.*, **5**, 33–44.

Drickamer, K. (1988) Two distinct classes of carbohydrate-recognition domains in animal lectins. *J. Biol. Chem.* **263**, 9557–9560.

Endo, T., Groth, D., Prusiner, S.B. and Kobata, A. (1989) Diversity of oligosaccharide structures linked to asparagines of the scrapie prion protein. *Biochemistry*, **28**, 8380–8388.

Epstein, J., Eichbaum, Q., Sheriff, S. and Ezekowitz, R.A.B. (1996) The collectins in innate immunity. *Curr. Opin. Immunol.*, **8**, 29–35.

Facy, P., Sève, A.-P., Hubert, M., Monsigny, M. and Hubert, J. (1990) Analysis of nuclear sugar-binding components in undifferentiated and *in vitro* differentiated human promyelocytic leukemia cells (HL60). *Exp. Cell Res.*, **190**, 151–160.

Felin, M., Doyennette-Moyne, M.A., Hadj-Sahraoui, Y., Aubery, M., Hubert, J. and Sève, A.P. (1994) Identification of two nuclear N-acetylglucosamine-binding proteins. *J. Cell. Biochem.*, **56**, 527–535.

Felin, M., Doyennette-Moyne, M.-A., Rousseau, C., Schröder H.C. and Sève A.-P. (1997) Characterization of a putative nuclear ligand of 82 kDa for the N-acetylglucosamine binding protein CBP70. *Glycobiology*, **7**, 23–29.

Forloni, G., Angeretti, N., Chiesa, R., Monzani, E., Salmona, M., Bugiani, O. and Tagliavini, F. (1993) Neurotoxicity of a prion protein fragment. *Nature*, **362**, 543–546.

Forrest, J.M.S., Stewart, D.. and Müller, W.E.G. (1996) Lectins of the Amaryllidaceae and their potential uses. *Annual Report of the Scottish Crop Research Institute for 1995*. Scotish Crop Research Institute, Dundee, 89–91.

Hammar, L., Hirsch, I., Machado, A.A., De Mareuil, J., Baillon, J.G., Bolmont, C. and Chermann, J.-C. (1995) Lectin-mediated effects on HIV Type 1 infection *in vitro*. *AIDS Res. Hum. Retroviruses*, **11**, 87–95.

Haraguchi, T., Fisher, S., Olofsson, S., Endo, T., Groth, D., Tarentino, A., Borchelt, D.R., Teplow, D., Hood, L., Burlingame, A., Lycke, E., Kobata, A. and Prusiner, S.B. (1989) Asparagine-linked glycosylation of the scrapie and cellular prion proteins. *Arch. Biochem. Biophys.*, **274**, 1–13.

Hirabayashi, J. and Kasai, K. (1993) The family of metazoan metal-independent β-galactoside-binding lectins: structure, function and molecular evolution. *Glycobiology*, **3**, 297–304.

Ho, M.K. and Springer, T.A. (1982) Mac-2, a novel 32,000 M_r mouse macrophage subpopulation-specific antigen defined by monoclonal antibodies. *J. Immunol.*, **128**, 1221–1228.

Hubert, J., Sève, A.P., Bouvier, D., Masson, C., Bouteille, M. and Monsigny, M. (1985) *In situ* ultrastructural localization of sugar-binding sites in lizard granulosa cell nuclei. *Biol. Cell* **55** (1985) 15–20.

Hubert, J., Sève, A.P., Facy, P. and Monsigny, M. (1989) Are nuclear lectins and nuclear glycoproteins involved in the modulation of nuclear functions? *Cell Different. Dev.*, **27**, 69–81.

Hubert, J. and Sève, A,-P. (1994) Nuclear Lectins. *Lectins: Biology, Biochemistry, Clinical Biochemistry*, **10**, 220–226.

Jia, S. and Wang, J.L. (1988) Carbohydrate binding protein 35. Complementary DNA sequence reveals homology with proteins of the heterogeneous nuclear RNP. *J. Biol. Chem.*, **263**, 6009–6011.

Knibbs, R.N., Agrwal, N., Wang, J.L. and Goldstein, I.J. (1993) Carbohydrate-binding protein 35. II. Analysis of the interaction of the recombinant polypeptide with saccharides. *J. Biol. Chem.*, **268**, 14940–14947.

Laing, J.G. and Wang, J.L. (1988) Identification of carbohydrate binding protein 35 in heterogeneous nuclear ribonucleoprotein complex. *Biochemistry*, **27**, 5329–5334.

Lauc, G., Sève, A.-P., Hubert, J., Flögel-Mrsic, M., Müller, W.E.G. and H.C. Schröder (1993) HnRNP CBP35. CBP67 interaction during stress response and ageing. *Mech. Ageing Dev.*, **70**, 227–237.

Lauc, G., Flögel, M., Diehl-Seifert, B., Schröder, H.C. and Müller, W.E.G. (1994) Identification and purification of a stress-associated nuclear carbohydrate binding protein (M_r 33,000) from rat liver by application of a new photoreactive carbohydrate probe. *Glycoconjugate J.*, **11**, 541–549.

Marchetti, M., Mastromarino, P., Rieti, S., Seganti, L. and Orsi, N. (1995) Inhibition of herpes simplex, rabies and rubella viruses by lectins with different specificities. *Res. Virol.*, **146**, 211–215.

Medzhitov, R. and Janeway, C.A. Jr (1997) Innate immunity: the virtues of a nonclonal system of recognition. *Cell*, **91**, 295–298.

Moutsatsos, I.K., Davis, J.M., Schindler, M. and Wang, J.L. (1986) Endogenous lectins from cultured cells: subcellular localization of carbohydrate-binding protein 35 in 3T3 fibroblasts. *J. Cell Biol.* **102**, 477–483

Moutsatsos, I.K., Wade, M., Schindler, M. and Wang, J.L. (1987) Endogenous lectins from cultured cells: nuclear localization of carbohydrate-binding protein 35 in proliferative 3T3 fibroblasts. *Proc. Natl. Acad. Sci. USA*, **84**, 6452–6456.

Müller, W.E.G., Pfeifer, K., Forrest, J., Rytik, P.G., Eremin, V.F., Popov, S.A. and Schröder, H.C. (1992) Accumulation of transcripts coding for prion protein in human astrocytes during infection with human immunodeficiency virus. *Biochim. Biophys. Acta*, **1139**, 32–40.

Müller, W.E.G., Ushijima, H., Schröder, H.C., Forrest, J., Schatton, W.F.H., Rytik, P.G. and Heffner-Lauc, M. (1993) Cytoprotective effect of NMDA receptor antagonists on prion protein (Prion[Sc])-induced toxicity in rat cortical cell cultures. *Eur. J. Pharmacol. (Molec. Pharmacol. Section)*, **246**, 261–267.

Oesch, B., Westaway, D., Wälchli, M., McKinley, M.P., Kent, S.B.H., Aebersold, R., Barry, R.A., Tempst, P., Teplow, D.B., Hood, L.E., Prusiner, S.B. and Weisssmann, C. (1985) A cellular gene encodes scrapie PrP 27–30 protein. *Cell*, **40**, 735–746.

Okamoto, T. and Wong-Staal, F. (1986). Demonstration of virus-specific transcriptional activator(s) in cells infected with HTLV-III by an *in vitro* cell-free system. *Cell*, **47**, 29–35.

Perler, F.B., Olsen, G.J. and Adam, E. (1997) Compilation and analysis of intein sequences. *Nucleic Acids Res.*, **25**,1087–1093

Pfeifer, K., Bachmann, M., Schröder, H.C., Forrest, J. and Müller, W.E.G. (1993) Kinetics of expression of prion protein in uninfected and scrapie-infected N_2a mouse neuroblastoma cells. *Cell Biochem. Funct.*, **11**, 1–11.

Prusiner, S.B. (1994) Molecular biology and genetics of prion diseases. *Philosoph. Transact. Royal Soc. London* Series B, **434**, 447–463.

Robinson, W.E. Jr., Montefiori, D.C. and Mitchell, W.M. (1987) Evidence that mannosyl residues are involved in human immunodeficiency virus type 1 (HIV-1) pathogenesis. *AIDS Res. Hum. Retroviruses*, **3**, 265–282.

Robinson, J.E., Holton, D., Liu, D., McMurdo, H., Murciano, A. and Gohd, R. (1990) A novel enzyme-linked immunosorbent assay (ELISA) for the detection of antibodies to HIV-1 envelope glycoptoteins based on immobilization of viral glycoproteins in microtitre wells coated with concanavalin A. *J. Immunol. Methods*, **132**, 63–71.

Roff, C.F. and Wang, J.L. (1983) Endogenous lectins from cultured cells. *J. Biol. Chem.*, **258**, 10657–10663.

Scheffer, U., Okamoto, T., Forrest, J.M.S., Rytik, P.G., Müller, W.E.G. and Schröder, H.C. (1995) Interaction of 68–kDa TAR RNA-binding protein and other cellular proteins with prion protein-RNA stem-loop. *J. NeuroVirol.*, **1**, 391–398.

Schröder, H.C., Rottmann, M., Bachmann, M. and Müller, W.E.G. (1986) Purification and characterization of the major NTPase from rat liver nuclear envelopes. *J. Biol. Chem.*, **261**, 663–668.

Schröder, H.C., Bachmann, M., Diehl-Seifert, B. and Müller, W.E.G. (1987) Transport of mRNA from nucleus to cytoplasm. *Prog. Nucleic Acid Res. Molec. Biol.*, 34, 89–142.

Schröder, H.C., Facy, P., Monsigny, M., Pfeifer, K., Bek, A. and Müller, W.E.G. (1992) Purification of a glucose-binding protein from rat liver nuclei. Evidence for a role in targeting of nuclear mRNP to nuclear pore complex. *Eur. J. Biochem.*, **205**, 1017–1025.

Schröder, H.C., Scheffer, U., Forrest, J.M.S., Sève, A.P., Rytik, P.G. and Müller, W.E.G. (1994) Association of scrapie prion protein and prion protein-RNA stem-loop with nuclear carbohydrate-binding protein 35 and other RNA-binding proteins. *Neurodegeneration*, **3**, 177–189.

Schröder, H.C., Ushijima, H., Theis, C., Sève, A.-P., Hubert, J. and Müller, W.E.G. (1995) Expression of nuclear lectin carbohydrate-binding protein 35 in human immunodeficiency virus type 1–infected Molt-3 cells. *J. Acq. Immune Defic. Synd.*, **9**, 340–348.

Sève, A.P., Hubert, J., Bouvier, D., Bouteille, M. and Monsigny, M. (1985) Detection of sugar-binding proteins in membrane-depleted nuclei. *Exp. Cell Res.*, **157**, 533–538.

Sève, A.P., Hubert, J., Bouvier, D., Bourgeois, C., Midoux, P., Roche, A.C. and Monsigny, M. (1986) Analysis of sugar-binding sites in mammalian cells nuclei by quantitative flow microfluorometry. *Proc. Natl. Acad. Sci. USA*, **83**, 5997–6001.

Sève, A.P., Felin, M., Doyennette-Moyne, M.A., Sahraoui, T., Aubery, M. and Hubert, J. (1993) Evidence for a lactose-mediated association between two nuclear carbohydrate-binding proteins. *Glycobiology*, **3**, 23–30.

Sève, A.P., Hadj-Sahaoui, Y., Felin, M., Doyennette-Moyne, M.-A., Aubery, M. and Hubert, J. (1994) Evidence that lactose binding to CBP35 disrupts its interaction with CBP70 in isolated HL60 cell nuclei. *Exp. Cell. Res.*, **213**, 191–197.

Sklaviadis, T., Manuelidis, L. and Manuelidis, E.E. (1986) Characterisation of major peptides in Creutzfeldt-Jakob disease and scrapie. *Proc. Natl. Acad. Sci. USA*, **83**, 6146–6150.

Sklaviadis, T., Akowitz, A., Manuelidis, E.E. and Manuelidis, L. (1993) Nucleic acid binding proteins in highly purified Creutzfeldt-Jakob disease preparations. *Proc. Natl. Acad. Sci. USA*, **90**, 5713–5717.

Somerville, R.A. and Ritchie, L.A. (1990) Differential glycosylation of the protein (PrP) forming scrapie-associated fibrils. *J. Gen. Virol.*, **71**, 833–839.

Stahl, N., Borchelt, D.R., Hsiao, K. and Prusiner, S.B. (1987) Scrapie prion protein contains a phosphatidylinositol glycolipid. *Cell*, **51**, 229–240.

Stahl, N., Baldwin, M.A., Teplow, D.B., Hood, L., Gibson, B.W., Burlingame, A.L. and Prusiner, S.B. (1993) Structural studies of scrapie prion protein using mass spectrometry and amino acid sequencing. *Biochemistry*, **32**, 1991–2002.

Taraboulos, A., Rogers, M., Borchelt, D.R., McKinley, M.P., Scott, M., Serban, D. and Prusiner, S.B. (1990) Acquisition of protease resistance by prion proteins in scrapie-infected cells does not require asparagine-linked glycosylation. *Proc. Natl. Acad. Sci. USA*, **87**, 8262–8266.

Taraboulos, A., Serban, D. and Prusiner, S.B. (1990) Scrapie prion proteins accumulate in the cytoplasm of persistently infected cultured cells. *J. Cell Biol.*, **110**, 2117–2132.

Ushijima, H., Mangel, A., Forrest, J.M.S., Kunisada, T., Matthes, E. and Müller, W.E.G. (1992) ELISA for the quantification of HIV-1 Env glycoprotein 120: some practical applications. *AIDS-Forschung*, **4**,185–189.

Van Damme, E.J.M., Goldstein, I.J., Vercammen, G., Vuylsteke, J. and Peumans, W. (1992) Lectins of the *Amaryllidaceae* are encoded by multigene families which show extensive homology. *Physiologia Plantarum*, **86**, 245–252.

Van Leuven, F., Torrekens, S., Van Damme, E., Peumans, W. and Van den Berghe, H. (1993) Mannose-specific lectins bind alpha-2–macroglobulin and an unknown protein from human plasma. *Protein Sci.*, **2**, 255–263.

Wang, J.L., Laing, J.G. and Anderson, R.L. (1991) Lectins in the cell nucleus. *Glycobiology*, **1**, 243–252.

Wang, J.L., Werner, E.A., Laing, J.G. and Patterson, R.J. (1992) Nuclear and cytoplasmic localization of a lectin-ribonucleoprotein complex. *Biochem. Soc. Transact.*, **20**, 269–274.

Wills, P.R. and Hughes, A.L. (1990) Stem loops in HIV and prion protein mRNAs. *J. Acquir. Immune Def. Syndr.*, **3**, 95–97.

8. PROTEIN-CARBOHYDRATE INTERACTIONS IN THE ATTACHMENT OF ENTEROTOXIGENIC *ESCHERICHIA COLI* TO THE INTESTINAL MUCOSA

EDILBERT VAN DRIESSCHE, FRANÇOISE DE CUPERE and
SONIA BEECKMANS

Laboratorium voor Scheikunde der Proteïnen, Instituut voor Moleculaire Biologie en Biotechnologie, Vrije Universiteit Brussel, Paardenstraat 65, B-1640 Sint-Genesius-Rode, Belgium

INTRODUCTION

Escherichia coli is a well-known normal inhabitant of the large intestine of mammals and birds. This bacterium, originally isolated in 1885 from faeces by the German paediatrician Theodore Escherich, is however also the causative agent of a diversity of diseases in man and his livestock, such as diarrhoea, urinary tract infections, cystitis, pyelonephritis, meningitis, peritonitis, septicaemia or gram-negative pneumonia. From all these diseases, diarrhoea was one of the first to be investigated, and the relation between outbreaks of diarrhoea and *E. coli* infections was already established since the end of the last century (for a historical review, see Robins-Browne, 1987).

In developed countries, coligenic diarrhoea in man is only of minor importance, and when it occurs it can mostly easily be controlled. In developing countries however, diarrhoea caused by any type of agent remains a life-threatening disorder, especially in neonates and children, and therefore it is not surprising that meticulous epidemiological investigations are still being performed all over the world (Krogfelt, 1995; Guth *et al.*, 1994; Estevez Touzard *et al.*, 1993; Lindblom *et al.*, 1995; Muñoz *et al.*, 1996; Nagy *et al.*, 1996; Nirdnoy *et al.*, 1997; Otto *et al.*, 1993; Sarinho *et al.*, 1993; Serichantalergs *et al.*, 1997; Subekti *et al.*, 1993; Takeda *et al.*, 1997; Vergara *et al.*, 1996; Wolk *et al.*, 1995, 1997; Molbak *et al.*, 1994; Oyofo *et al.*, 1995, 1997).

In its attempts to achieve rationalisation, intensification and optimisation, modern husbandry practice in Western countries has ironically enough also created the optimal conditions for the fast spreading of various diseases. In the case of diarrhoea, especially neonates and animals just after weaning are very susceptible to infection. Antibiotics used to be a successful tool to suppress this kind of infections, but the intensive use and over-consumption has resulted in the selection of strains that display resistance to any known antibiotic, a problem that emerged firstly in veterinary medicine. Nowadays, multi-resistance is taking alarming proportions also in human medicine, and consequently alternatives to antibiotics are urgently needed. As will be discussed later in this chapter, the new insights into the attachment of bacteria to host tissues has resulted in possible alternative strategies to interfere with the attachment process. In this chapter, we will only consider the attachment mechanism of enterotoxigenic *E. coli* (ETEC) mediated by surface lectins, although whatever *E. coli* strain considered, enterotoxigenic, enteroinvasive (EIEC), enteropathogenic

(EPEC) or enterohaemorrhagic (EHEC), the attachment to the intestinal mucosa is always an initial and essential step in pathogenesis.

Most ETEC strains express at their surface long proteinaceous filaments, known as fimbriae. Diarrhoea provoked by these strains is the result of the production of heat-stable (ST) and/or heat-labile (LT) enterotoxins (Gyles, 1992; Spangler, 1992). These toxins provoke the secretion of water and electrolytes in the small intestinal lumen. When not treated rapidly and adequately, the infected host will face severe dehydration eventually resulting in death. Therefore it is not surprising that, over the last years, many investigators have focused their research on fimbriae-mediated attachment of ETEC, as well as on strategies to prevent adherence. Consequently many excellent reviews have been published on these topics, that, together with the information provided in this chapter, should give a comprehensive picture of the subject (Gaastra and Svennerholm, 1996; Van Driessche and Beeckmans, 1993; Van Driessche *et al.*, 1995; Lintermans *et al.*, 1995; Mol and Oudega, 1996; Mouricout, 1991, 1997; Smyth *et al.*, 1996; Pohl *et al.*, 1992, 1995; Krogfelt, 1991, 1995; De Graaf, 1988, 1990; Moon, 1990).

FIMBRIAL LECTINS MEDIATE ATTACHMENT OF ETEC TO MUCUS AND BRUSHBORDERS OF ENTEROCYTES

Already in the beginning of this century it had been observed that some *E. coli* strains agglutinate human and/or animal erythrocytes. This agglutination was later shown by Collier and De Miranda (1955) to be specifically inhibited by mannose and some mannose derivatives. Duguid and Gillies (1957) found mannose-sensitive agglutination to be correlated with the expression by the bacterial cells of long thread-like structures, indicating that the latter might be responsible for binding to erythrocytes and causing them to become agglutinated. Likewise, the same authors found that fimbriated *E. coli* can bind to intestinal cells.

Although the pioneering studies just mentioned pointed to the expression by *E. coli* of mannose-sensitive surface lectins in the form of long threads that recognise glycoconjugates at the surface of eukaryotic cells, it were Ofek and his co-workers (1977, 1978) who unequivocally proved that this hypothesis was correct. Their conclusion was based on a series of original and rather simple experiments. They found that, after destruction of oligosaccharides of epithelial cells by sodium metaperiodate, these cells failed to bind *E. coli*. Similarly, incubation of epithelial cells with the mannose/glucose specific plant lectin ConA resulted in inhibition of binding of *E. coli* that provoke mannose-specific haemagglutination. Ofek and his co-workers could also show that yeast mannan is a strong inhibitor of attachment of these *E. coli* strains to epithelial cells and that yeast cells themselves are agglutinated by *E. coli* expressing mannose-sensitive fimbriae. Although these experiments conclusively showed that the receptors are mannose-containing glycoconjugates expressed at the surface of epithelial cells, they only pointed to the implication of mannose-specific lectins on the bacterial surface being involved in the attachment process. Final proof that fimbriae, isolated in a high purified form bind to epithelial cells in a mannose-inhibitable way and provoke mannose-sensitive haemagglutination was provided by Ofek *et al.* (1977) and Salit and Gotschlich (1977).

Based on their haemagglutination properties, fimbrial lectins are generally classified as mannose-sensitive (type-1) and mannose-resistant. Type-1 fimbriae are expressed by both commensal and enterotoxigenic *E. coli* and are easily recognised in the electron microscope as 0.2–1 micron long rigid filaments that protrude from the bacterial surface. *E. coli* cells expressing these fimbriae agglutinate a wide variety of erythrocytes, other animal cells and yeast cells, and this agglutination is inhibitable by mannose. Type-1 fimbriae are expressed within a broad temperature range.

Mannose-resistant haemagglutination on the other hand points to the presence of what are called host-specific fimbriae which are responsible for host and tissue tropism of enterotoxigenic *E. coli* strains. These fimbriae are not expressed below 18°C. Unlike type-1 or common fimbriae, the host-specific fimbriae agglutinate or bind to a restricted variety of eukaryotic cells. Today many host-specific fimbriae and the genes involved in their expression have been characterised (see Table 1).

Although haemagglutination is the most straightforward and easiest test for the detection of surface lectins on *E. coli*, some lectins may not be detected by this method. For example for 987P and CS31A lectins, no erythrocytes are known to be agglutinated by *E. coli* expressing these antigens. Also, haemagglutination does not allow to discriminate between fimbrial and non-fimbrial lectins. This difference can however be readily made by electron microscopy after negative staining of the *E. coli* cells with uranyl acetate (see Van Driessche *et al.*, 1995, for a review). Electron microscopy can also be used for the unequivocal identification of surface adhesins when monospecific antisera against known and highly purified adhesins are available. However, if specific antibodies are available, a simple slide agglutination test is easier to be performed when a newly isolated *E. coli* strain is assayed for the expression of known surface lectins. When a new *E. coli* strain is to be investigated, several techniques can thus be used for the detection and identification of surface lectins i.e. haemagglutination, attachment to isolated villi, isolated enterocytes or cell lines such as Caco-2, attachment to glycoproteins covalently immobilised on a solid support (Van Driessche *et al.*, 1995). Attachment studies on villi or enterocytes require fixation, a process that might affect the lectin receptors and prevent them from interacting with lectins. An easy to use *in vitro* attachment system using Eupergit-C beads which can be easily derivatised with glycoproteins (including solubilised brushborder membranes or mucus) has been developed by Van Driessche *et al.* (1988). These beads can be stored for years at 4°C and allow the investigation of attachment as well as attachment-inhibition. For example, Eupergit-C-glycoprotein beads were successfully used to investigate the expression by *E. coli* strains of F17 fimbriae for which initially no erythrocytes were known that were recognised by these fimbriae (Lintermans *et al.*, 1991). Similarly, with Eupergit-glycoprotein beads we were able to demonstrate the carbohydrate-binding heterogeneity of F17 fimbriae expressed by different *E. coli* strains (Van Driessche *et al.*, 1988).

BIOSYNTHESIS OF FIMBRIAL LECTINS

Upon isolation, fimbrial lectins mostly display one single band on SDS-gels (Van Driessche *et al.*, 1993). Consequently it was originally believed that fimbriae are just polymers of one single subunit. Genetic analysis of several fimbrial systems revealed

Table 1   Characteristics of fimbriae of enterotoxigenic *E. coli*

Fimbriae		Natural Host	Morphology	Molecular Mass Major Subunit (Dalton)	Gene Localization	Erythrocytes Agglutinated		Inhibiting Sugars
Type 1	(F1)	no host specificity, common fimbriae	rigid, Ø 7 nm	17,000	chromosome	guinea pig	MS	mannose, mannosides
K88 (ab, ac, ad)	(F4)	pig	flexible, Ø 2.1 nm	23,000–27,000	plasmid	guinea pig, chicken	MR	galactosides
987P		pig	rigid, Ø 7 nm	20,000	plasmid	none	MR	not known
F18	(F107)	postweaned pig	rather rigid, Ø 3-4 nm	15,000	not known	none	MR	not known
K99	(F5)	pig, lamb, calf	flexible, Ø 5 nm	18,500	plasmid	horse, sheep	MR	sialic acid
F41		pig, lamb, calf	flexible, Ø 3.2 nm	29,500	chromosome	human, guinea pig, horse, sheep	MR	GalNAc
F17 **		calf	flexible	19,500	chromosome	bovine	MR	GlcNAc
F111		calf	flexible	17,500	not known	bovine	MR	GlcNAc
CS31A		calf	flexible, Ø 2 nm	30,000	plasmid	not known		not known
CFA/I	(F2)	human	rigid, Ø 7 nm	15,000	plasmid	human, bovine, chicken	MR	sialic acid
CFA/II	(F3)	human			plasmid	bovine, chicken	MR	not known
CS 1			rigid, Ø 7 nm	16,800				
CS 2			rigid, Ø 7 nm	15,300				
CS 3			flexible, Ø 2 nm	15,000				
CFA/III		human	rigid, Ø 7 nm	16,000–18,000	plasmid	human, bovine	MR	not known

Table 1 Continued.

Fimbriae	Natural Host	Morphology	Molecular Mass Major Subunit (Dalton)	Gene Localization	Erythrocytes Agglutinated	MR	Inhibiting Sugars
CFA/IV	human			plasmid	human, bovine	MR	not known
CS 4		rigid, Ø 6-7 nm	17,000				
CS 5		rigid, Ø 5-6 nm	21,000				
CS 6		fine fibrillar	14,500				
Longus	human	rod-like, Ø 7 nm, L>20μm, polar	22,000	plasmid	not known		not known

Legend:
For detailed information on the structure of the fimbriae mentioned in this table, and on the organisation of their gene clusters, the reader is referred to:De Graaf (1986, 1990), De Graaf and Mooi (1986), Gaastra and De Graaf (1982), Gaastra and Svennerholm (1996), Girón et al. (1997), Imberechts et al. (1993, 1997b), Krogfelt (1991), Lintermans et al. (1995), Mol and Oudega (1996), Moon (1990), Mouricout (1991, 1997), Smyth et al. (1996), Van Driessche et al. (1995).

Abbreviations used: MS mannose sensitive; MR mannose resistant; Ø diameter; L length; GalNAc N-acetylgalactosamine; GlcNAc N-acetylglucosamine.

**Fimbriae which are highly homologous to F17 fimbriae from enterotoxigenic E. coli have been shown to be expressed by human uropathogenic E. coli (G-fimbriae) (Rhen et al., 1986; Saarela et al., 1995, 1996; Martin et al., 1997), bovine septicaemia E. coli strains (20K-fimbriae) (Bertin et al., 1996; Martin et al., 1997) and ovine cytotoxic necrotizing factor type 2 toxin producing E. coli strains (F17b) (El Mazouari et al., 1994). 20K-Fimbriae were recently shown to be identical to G-fimbriae and were renamed F17c (Martin et al., 1997).

Besides the fimbriae described in this table, several putative colonization factors (PCF) on ETEC strains have been described but fall outside the scope of this publication: PCFO149 (Knutton et al., 1987), PCFO159 (Tacket et al., 1987), PCFO166 (McConnell et al., 1989), CS17 (McConnell et al., 1990), PCFO9 (Heuzenroeder et al., 1990), PCFO20 (Viboud et al., 1993), PCFO2 (Ricci et al., 1997).

however that this picture is a very strong oversimplification of the situation, and that the biosynthesis and expression of fimbriae requires a complex and meticulously regulated interplay of several genes and the polypeptides they encode (Lintermans *et al.*, 1988, 1991; Bertels *et al.*, 1991; De Graaf, 1990; Hultgren *et al.*, 1993; Smyth *et al.*, 1996; Mouricout, 1997).

From the information available today it can be concluded that the gene clusters directing the expression of fimbriae contain genes that encode for:

1. A large outer membrane "pore" protein which is implicated in the translocation of fimbrial subunits across the outer membrane, and also serves as a mould upon which fimbriae polymerisation occurs.
2. Multifunctional periplasmic proteins involved in stabilising non-polymerised fimbrial subunits and in transporting subunits from the inner to the outer membrane. These proteins obviously act as chaperones (see e.g. Edwards *et al.*, 1996; Mol *et al.*, 1996a,b).
3. The major fimbrial subunits that build up the "body" of the fimbriae. This subunit is the most prominent polypeptide seen on SDS-gels after electrophoresis of purified fimbriae. In some fimbrial systems such as K88, K99, CFA I and others, the major subunits display carbohydrate-binding activity.
4. Minor fimbrial subunits which may be involved in initiation and termination of subunit polymerisation, in regulation of the extent of fimbriation, in determining the length of fimbriae, in carbohydrate-binding, etc..

For updated reviews on the biogenesis of fimbriae and their structure, the reader is referred to recent review papers by Van Driessche *et al.* (1995), Smyth *et al.* (1996) and Mouricout (1997).

As mentioned earlier in this chapter, the expression of fimbriae may depend on the growth temperature, but also the composition of the growth medium may be of utmost importance. This is the reason why upon examining newly isolated strains for fimbriae production, different growth media should be tested, and bacteria grown in different circumstances should be examined by electron microscopy after negative staining.

HOW TO PREVENT THE COLONISATION OF THE SMALL INTESTINE BY *E. COLI*?

Since it is now firmly established that attachment of ETEC or other pathogens to the mucosa is an initial but essential step in pathogenesis, strategies to prevent attachment have been and are still being developed. Several approaches can be distinguished:

1. Blocking of the intestinal receptors by for example plant lectins or *E. coli* adhesins;
2. Blocking of the carbohydrate-binding sites using receptor-analogues i.e. glycoproteins and/or oligosaccharides;
3. Vaccination of dams during pregnancy to protect the offspring thanks to secretory antibodies in colostrum and milk;

4. Administration of sublethal dose of antibiotics that interfere with the expression of functional fimbriae;
5. Temporal modification of intestinal receptors for bacterial adhesins.

Some examples of the application of these strategies will be discussed below.

A. Sublethal Dose of Antibiotics

In vitro studies convincingly showed that sublethal dose of antibiotics can interfere with the expression of functional fimbriae, inhibit fimbriae expression, or change the shape, hydrophobicity and motility of *E. coli* in such a way that attachment to the mucosa is hampered (Chopra and Linton, 1986; Schifferli and Beachy, 1988a,b; Zhanel and Nicolle, 1992; Breines and Burnham, 1994, 1995; Loubeyre *et al.*, 1993; Sonstein and Burnham, 1993). Experiments carried out in the lab of the authors of this chapter (unpublished results) revealed that the effects of antibiotics such as novobiocin, polymyxin B, furazolidone and erythromycin strongly depend on the *E. coli* strain used as well as on the fimbrial system under investigation. These antibiotics are currently used in the treatment of neonatal diarrhoea in calves, and are known not to be resorbed nor broken down in the small intestine of calves. In view of the general consensus that developed over the last years on the need to reduce the use of antibiotics, this strategy will not further be discussed here.

B. Prevention by Antibodies: Fimbriae as Vaccine Candidates

Vaccination of dams during pregnancy has in many cases resulted in high titers of specific antibodies in colostrum and milk. Since fimbriae are known to be very strong immunogens, and easy to prepare in high amounts (Van Driessche *et al.*, 1993), they can be considered as good vaccine candidates. Already in the 1970's Rutter *et al.* (1976) described the antibacterial activity of colostrum of sows vaccinated with K88 fimbriae, and they could show that this activity is due to the presence of anti-adhesive K88 antibodies. It should be kept in mind however that when an *E. coli* cell expresses different types of fimbriae, for example K99, F41 and FY, it may be necessary to include each fimbrial type in the vaccine in order to get antibodies to each of them in the colostrum (Contrepois and Girardeau, 1985; Runnels *et al.*, 1987). These observations clearly demonstrate the urgent need for searching new types of adhesins in the *E. coli* population, so that they can be included to extend existing vaccines and to increase their protective effect.

Successful protection of new-born lambs (Sojka *et al.*, 1978), piglets (Nagy *et al.*, 1986) and calves (Acres *et al.*, 1979) by colostral antibodies raised by vaccination of dams with K99 fimbriae during pregnancy have been reported. Similarly, colostral antibodies against 987P fimbriae (Lösch *et al.*, 1986) passively protected piglets against infection by 987P positive *E. coli*.

As pointed out by Moon and Bunn (1993), the success of fimbrial vaccines given parenterally to pregnant cattle, sheep and ovine to protect suckling new-borns against ETEC infections, is due to the following:

1. most ETEC infections occur soon after birth, when colostral and milk antibody titers are high;

2. only a limited number of fimbrial *E. coli* types are of importance in farm animals;
3. fimbriae are good immunogens, being present at the bacterial surface;
4. fimbriae are implicated in an initial but essential step in the development of disease.

Although it might be expected that, because of the intensive use of fimbrial vaccines, a rapid selection would occur in favour of previously rather infrequent occurring fimbrial antigens, this does not seem to be the case (Moon and Bunn, 1993).

The prophylactic effect of various inactivated *E. coli* vaccines in the control of pig colibacillosis has been investigated under large-scale farm conditions by Osek *et al.* (1995). These investigators immunised 2472 pregnant sows with eight different vaccines containing *E. coli* fimbrial adhesins and adjuvants. Upon considering the general health status of the new-born pigs, the percentage of piglets with diarrhoea and dead piglets, as well as the mean body weight gain of weaned piglets, it was found that vaccination had an overall positive effect, although the vaccines tested differed in their protective effect. Best results were obtained when pregnant sows were immunised with a vaccine containing K88, K99 and 987P fimbriae and the B-subunit of the LT-enterotoxin. From the examples given above it is clear that new-borns can be passively protected from colonisation of the small intestine by maternal antibodies secreted in colostrum and milk. However, upon weaning, a new period of high sensitivity to pathogenic *E. coli* starts as a result of the lack of maternal antibodies, change of diet, stress at weaning, new environmental conditions etc..

Two approaches can be considered to treat or prevent postweaning diarrhoea immunologically, i.e. by passive immunisation or by active immunisation in the attempt to raise mucosal antibodies.

Protection of piglets against postweaning diarrhoea was described by Alexa *et al.* (1995) who used a combination of parental and oral vaccination against ETEC expressing F18 fimbriae. When piglets received an intramuscular injection the day before weaning and were treated perorally with a live but non-toxic *E. coli* strain expressing the same F18 fimbriae one day after weaning, they were protected against a subsequent challenge with the virulent strain. Field trials confirmed the protective effect of this immunisation scheme. Sarrazin and Bertschinger (1997) showed that intestinal colonisation of pigs by live *E. coli* expressing F18 fimbriae resulted in significant increased levels of anti-fimbrial antibodies, especially IgA, both in serum and in intestinal wash fluids. These authors also concluded that F18 fimbriae are important candidates for vaccination of pigs against oedema disease and postweaning diarrhoea.

Bianchi *et al.* (1996) on the other hand reported that parental immunisation of mice or piglets with *E. coli* expressing F4 fimbriae or with purified F4 fimbriae is ineffective in inducing protective immunity at the mucosal level against a subsequent challenge with live bacteria. These authors reported that this immunisation procedure might even be detrimental by inducing a state of suppression. Oral vaccination with live bacteria expressing F4 fimbriae induces an enteric immune response, while killed bacteria were without effect.

Using rabbits as a model, Reid *et al.* (1993) observed that the colonisation factor CFA/II isolated from ETEC, when incorporated in biodegradable polymer capsules, is immunogenic when administered intradermally. After vaccination, Peyer's patch cells responded by lymphocyte proliferation to *in vitro* challenge with CFA/II, and B-cells secreting specific anti-CFA/II antibodies were detected in the spleen of vaccinated animals. In human volunteers, the studies of Tacket *et al.* (1994) showed that CFA/II encapsulated in biodegradable microspheres stimulate the secretion of s-IgA anti-CFA/II in the jejunal fluid. Some volunteers were found to be protected against challenge with a pathogenic strain expressing CFA/II antigens.

Because of the lag period between vaccination and the production of protective antibodies in the intestinal mucosa, it is clear that passive immunisation is the method of choice just after weaning. Several reports have been published describing the successful use of hen egg yolk antibodies which can easily be raised upon immunisation of laying hens (Heller *et al.*, 1990). Of course, these antibodies can be used as well in the neonatal period in situations where dams have not been vaccinated, or produce inadequate amounts of colostrum or colostrum containing too low titers of specific antibodies. For example, Yokoyama *et al.* (1992) could passively protect neonatal colostrum-deprived piglets against fatal enteric colibacillosis with powder preparations of specific antibodies obtained by spray-drying the water-soluble fraction of egg yolks from hens immunised against K88, K99 and 987P fimbriae. The investigations of O'Farrelly *et al.* (1992) showed that also rabbits could be protected from developing diarrhoea when they were fed egg yolk from hens previously immunised with *E. coli* that produced heat-labile enterotoxin and colonisation factor antigen-I. Independently, Imberechts *et al.* (1997a) and Yokoyama *et al.* (1997) described the protective effect of hen egg yolk containing specific antibodies against F18 fimbriae. These authors also showed that the yolk antibodies interfere with the attachment of bacteria to the intestinal mucosa. The prophylactic effect of hen egg yolk antibodies to K88, K99, 987P and rotavirus has also been investigated by Erhard *et al.* (1996). When egg yolk containing specific antibodies was included in a 5% feed ration, a significant decrease in the number of piglets with diarrhoea as well as a decrease in the severity of the symptoms was observed when compared to animals fed on a diet with egg yolk of non-immunised hens or without egg yolk supplementation.

An original approach to oral immunisation was described by Haq *et al.* (1995). These investigators succeeded in producing transgenic tobacco and potato plants that expressed the natural ligand-binding LT-B subunits (the B-subunit of the heat-labile enterotoxin behaves as a lectin). It was shown that mice fed transgenic potato tubers got orally immunised and produced specific toxin-neutralising antibodies both in their serum and at the level of the mucosa.

C. Intestinal Mucosa Receptors for Adhesins of Enterotoxigenic *E. coli*: "Receptor-analogue Therapy"

The susceptibility of animals and humans towards colonisation of the intestinal mucosa depends on the presence of receptors in either or both mucus and brushborder membranes. Since the fimbrial adhesins are recognised to be lectins, by definition the corresponding receptors should be glycoconjugates i.e. glycoproteins or glycolipids. The variation of oligosaccharide composition and structure with age might thus explain why susceptibility often depends on the age of animals. Similarly, tissue-specific glycosylation explains the tissue tropism displayed by pathogens. For several fimbrial systems, the intestinal receptors have been isolated and characterised. Receptors for K99 fimbriae in the piglet small intestine have been shown by Ono *et al.* (1989) and Teneberg *et al.* (1990, 1993) to be the gangliosides NeuGc-GM3, NeuGc-GM2, NeuGc-CD1a and NeuAc-5PG. Non-acid glycolipids were shown not to display receptor activity. A good correlation was observed by Yuyama *et al.* (1993) between the postnatal changes the NeuGc-GM3 content and the susceptibility of piglets to *E. coli* K99 infection. The K99-receptor was shown to be maximally expressed at birth and gradually decreases with age. Glycolipids on horse erythrocytes have been shown by Smit *et al.* (1984) to act as receptors for K99 fimbriae. Mucin glycopeptides of the pig small intestine were shown by Lindahl and Carlstedt (1990) to display K99 receptor activity. Receptor activity of these glycopeptides was found to be destroyed upon desialylation confirming the sialic acid specificity of K99 fimbriae.

Genetic differences in susceptibility between piglets was shown by Seignole *et al.* (1991) to reside in differences in the glycolipid composition of the enterocyte membrane. Also in calves mucus covering the intestinal epithelium was shown by Mouricout and Julien (1987) to contain receptors for K99 fimbrial adhesins. They were shown to be glycoproteins which upon desialylation loose their receptor activity.

Susceptibility or resistance to intestinal colonisation by *E. coli* does not only depend on the presence or absence of specific receptors in mucus and/or enterocyte membranes recognised by the bacterial adhesins. For example, Dean *et al.* (1989) found that colonisation of the small intestine of piglets by *E. coli* expressing 987P fimbriae is age-dependent. Both colonisation and the incidence and severity of diarrhoea were found to be highest in neonatal animals, while older animals are not colonised and do not develop diarrhoea upon infection. Nevertheless, *in vitro*, fimbriated 987P *E. coli* attach to intestinal epithelial brushborders of both neonatal and older piglets, indicating that sensitivity or resistance is not directly correlated to the presence of 987P receptors. More detailed investigations revealed that 987P receptors on brushborder membranes of both neonatal and resistant animals are glycoproteins with a molecular mass between 33 and 40 kDa. However the mucus receptors in older pigs proved to be low molecular weight glycoproteins, trace amounts of which are only detectable in neonates. These results conclusively show that resistance to *E. coli* 987P resides in the presence of attachment blockers secreted in the small intestinal mucus layer in older piglets.

Except from glycoproteins, Khan *et al.* (1996) showed that brushborders of piglets also contain two glycolipids that act as receptor for 987P adhesins. Similar to type-1 fimbriae, 987P fimbriae contain adhesive subunits not only at their tip but also at various positions along the fimbriae (Khan *et al.*, 1996).

The susceptibility of piglets to infection by *E. coli* expressing K88 fimbriae was shown by Sellwood (1980) to depend on the presence of K88 receptors in brushborder membranes. Sensitive animals possess receptors while resistant animals lack K88-specific receptors. Depending on the binding properties of the K88 serotypes (ab, ac, ad), different pig phenotypes can be distinguished, i.e. phenotypes A-E. Several investigations have shown that brushborder membranes as well as mucus receptors for K88 are glycoproteins (Willemsen and De Graaf, 1992; Metcalfe *et al.*, 1991; Erickson *et al.*, 1992; Seignole *et al.*, 1994).

Edfors-Lilja *et al.* (1995) reported that K88 receptor loci are located on chromosome 13. A 74 kDa O-glycosylated glycoprotein, present in mucus and epithelial cells of piglets was identified by Grange and Mouricout (1996) as mucosal transferrin, which was furthermore shown to be more abundant in adhesive intestines than in non-adhesive ones. From these observations, Grange *et al.* (1997) suggested that the susceptibility/resistance phenotype could be related to iron metabolism in the intestinal tract. Similar to the K88 receptor loci, the transferrin locus is located on chromosome 13 (Guérin *et al.*, 1993).

Recently, Mouricout *et al.* (1995) proposed a "three receptor model" to explain the observed adhesion types in piglets. Moreover these authors suggested the existence of multi-F17 mucosal receptor complexes in new-born calf intestines, the density of which might be age-dependent, and might also depend on the location in the intestinal tract. Calf intestinal receptors for F17 fimbriae have been characterised by Mouricout *et al.* (1995) and were shown to be glycoproteins.

Investigations of Sanchez *et al.* (1993a) revealed that F17 receptors are present in both the intestinal mucus and brushborder membranes all along the small intestine. *In vitro* studies showed that these receptors are expressed at least until the age of 4 months. Although F17 and F111 fimbriae are highly homologous, the receptors recognised in the intestinal mucus and brushborder membranes might be different. As will be described below, the administration of receptor analogues to block the adhesin carbohydrate-binding sites has in many cases been shown to be successful in preventing attachment of ETEC to the intestinal wall and consequently the development of diarrhoea. It can thus be hoped that by unravelling the structure of intestinal receptors, tailor-made drugs can be produced as adhesion blockers.

The usefulness of receptor analogues to prevent attachment was already shown by Aronson *et al.* in 1979. These investigators reported that methyl-α,D-mannopyranoside, a sugar shown in *in vitro* studies to inhibit attachment of some *E. coli* strains to human buccal epithelial cells, also *in vivo* prevent the colonisation of the urinary tract of mice. Similarly, Neeser *et al.* (1986) used short oligomannoside-type glycopeptides of ovalbumin and oligomannoside-type glycopeptides derived from legume storage proteins to inhibit the agglutination of erythrocytes caused by *E. coli* expressing type-1 fimbriae. It were however the studies of Mouricout and co-workers that boosted the idea of using receptor analogues as therapeutic agents. In 1986, Mouricout *et al.* showed that agglutination of sheep erythrocytes by ETEC expressing K99 and F41 fimbriae can be inhibited by bovine glycoprotein glycans, and that these preparations protect colostrum deprived new-born calves against lethal doses of K99-positive *E. coli* (Mouricout and Julien, 1987; Mouricout *et al.*, 1990). *In vitro*, Sanchez *et al.* (1993a,b) showed that glycoproteins such as fetuin, ovomucoid, submaxillary gland mucin, hen egg white glycoproteins, as well as plasma glycoproteins (especially those of cow) are strong inhibitors of the attachment of *E. coli* expressing F17 fimbriae to mucus and brushborder membranes prepared from different parts of calve intestines. The inhibitory effect of cow plasma preparations was shown by Sanchez *et al.* (1993a) not to be due to specific antibodies to surface components of the *E. coli* strains used. Subsequent *in vivo* studies of Nollet *et al.* (1996) revealed that non-immune plasma powder protected new-born and colostrum deprived calves from developing diarrhoea upon infection with pathogens expressing K99 or F17 fimbriae. Deprez *et al.* (1996) showed that plasma from swine without detectable antibodies against *E. coli* strain (O141 K85abF18ac SLTII v+) can be used to prevent the colonisation of the gut of piglets by this strain. These authors further showed that the amount of plasma powder required can be drastically reduced when using plasma powder of slaughter pigs vaccinated with purified F18ab fimbriae.

Several investigations have demonstrated that milk is a rich source of glycoproteins and oligosaccharides that can be used as receptor analogues to competitively inhibit the adhesion of *E. coli* to the gut. Already in 1983, Wadström and co-workers (1983a,b) reported milk as a source of oligosaccharide structures which mimic intestinal receptors for *E. coli* surface adhesins. Schroten *et al.* (1992) found milk fat globules to be protective against enteric infection. *In vitro*, S-fimbriated *E. coli* could be prevented from binding to human buccal cells by human milk fat globule membranes. The active principle was shown to reside in the mucus glycoproteins of the membrane. Since milk fat globule membranes, isolated from stools, are still agglutinated by S-fimbriated *E. coli*, Schroten *et al.* (1992) postulated that milk fat globules or their derived membranes might be protective all along the intestinal tract. Subsequently, the same authors (Schroten *et al.*, 1993) showed that carbohydrate residues of secreted mucus of human skim milk are able to inhibit bacterial adhesion to mucosal surfaces. They also found a remarkable difference between the inhibitory effect of colostrum and mature milk, and thus concluded that the inhibitory potential depends on the lactation period. Except from fat globules, milk contains several other components with anti-bacterial activity such as lysozyme, the lactoperoxidase system, macrophages and lymphocytes, as well as secretory IgA.

The protective action of IgA is not solely due to its specific antigen-binding properties, but also to its covalently linked oligosaccharide chains that will bind for example mannose-specific lectins from type-1 fimbriae of *E. coli* (Wold *et al.*, 1990, 1994). As such, IgA does not only display antibacterial activity as a specific antibody but also as a glycoprotein, that can prevent mucosal colonisation by saturating the carbohydrate binding sites of surface lectins of *E. coli* and other pathogens. Carbonare *et al.* (1995) found that IgA in human colostrum and milk is responsible for the inhibition of invasion of Hep-2 cells by enteroinvasive *E. coli* (EIEC).

Except from IgA, other non-specific defence factors against enterotoxigenic *E. coli* infections have been discovered. For example, Giugliano *et al.* (1995) identified lactoferrin and free secretory components (fsc) in human milk to be inhibitors of *E. coli* CFA/I adhesion, as monitored by haemagglutination. At least in the case of type-1 fimbriae, Teraguchi *et al.* (1996) showed that inhibition of haemagglutination is due to the glycan part of bovine lactoferrin. Except from inhibitory bacterial attachment, some glycoconjugates from milk may also interfere with binding of enterotoxins to their intestinal receptors. Shida *et al.* (1994) evidenced that two glycoproteins of the protease-peptone fraction of bovine milk bind to the heat-labile enterotoxin, and that binding was mediated by the carbohydrate part of the molecules which were identified as α-lactalbumin and β-lactoglobulin.

Instead of blocking the receptor-binding sites of bacterial surface lectins, colonisation of the small intestine might principally also be prevented by blocking the receptors by for example plant lectins or isolated bacterial lectins. It is well known that most plant lectins are rather resistant to proteolytic breakdown in the gastro-intestinal tract and reach the small intestine in an active form, where they can bind to mucus and the enterocytes (Pusztai *et al.*, 1993a,b). As a result, lectins can occupy the fimbrial intestinal receptors and as such prevent colonisation. The mannose-specific lectin from the bulbs of *Galanthus nivalis* has been shown by Pusztai *et al.* (1993c) to prevent the overgrowth of the small intestine by type-1 fimbriated *E. coli*. It should be kept in mind however that many plant lectins have adverse effects on the physiology of the small bowel, and are consequently considered to be an important group of anti-nutritional factors (ANF). Whether this approach has a future in practice remains to be proven, and much more research will be needed before it can be applied successfully.

Finally a few reports have been published on the temporal modification of intestinal receptors for bacterial lectins. For example, Mynott *et al.* (1991) showed that intestinal colonisation in rabbits by a diarrhoegenic *E. coli* strain expressing the human colonisation factor CFA/I could be prevented by the administration of an enteric-coated protease preparation 18 hours prior to infection. The same authors (Mynott *et al.*, 1996) investigated recently the effect of oral administration of bromelain, a proteolytic extract from pineapple stems, on the colonisation of pig small intestine by *E. coli* K88+. They observed a dose-dependent inhibition of the attachment of K88+ cells to the small intestinal wall. After treatment, the attachment of *E. coli* K88 was found to be negligible and at the same level of attachment of K88 cells to piglets that are genetically resistant to *E. coli* K88 infection. Most important, no adverse effects of the bromelain treatment was noticed and, consequently, the authors concluded that the administration of bromelain can inhibit ETEC receptor activity and is useful to prevent *E. coli* K88 induced diarrhoea.

CONCLUSION

All information available till now clearly demonstrates the importance of protein-carbohydrate interactions in the adhesion process which allows micro-organisms such as pathogenic *E. coli* to colonise mucosal surfaces. Although in this chapter emphasis was given to the binding of enterotoxigenic *E. coli* to the intestinal mucosa, it should be kept in mind that similar attachment mechanisms govern the colonisation of mucosae such as those of the urinary and respiratory tract. Similarly, *E. coli* is not an exception to the more general rule that micro-organisms produce surface-bound adhesive carbohydrate-binding molecules that specifically recognise glycosylated receptors on host tissues. The overwhelming evidence available that attachment of micro-organisms to mucosal surfaces is an initial but essential event in all kinds of pathological processes, has stimulated many research groups in investigating ways to prevent attachment in a specific way. Antibiotics have long been considered as "magic bullets" that can be used at any time to combat microbial and fungal infections in man and his livestock. However, during recent years, mankind has to face the alarming reality that the unrestricted use of antibiotics, both in the prophylaxis and treatment of infectious diseases, resulted in the selection of resistant and even multiresistant strains of pathogens, which cannot be eliminated anymore with existing antibiotics. In spite of the many adverse effects that are today ascribed to the often uncontrolled use of antibiotics, it is only fair to mention and keep in mind that, since their discovery by Sir Alexander Fleming, these compounds saved the lives of many hundreds of thousands of citizens, often in the most dramatic episodes of our civilisation. Obviously, in our attempts to expel microbial infectious diseases, antibiotics have proven to be a strong weapon allowing us to win a few battles but not the universally proclaimed war against pathogenic micro-organisms.

Without any doubt, this cool observation has stimulated research to finding means of blocking microbial attachment to host tissues. Different avenues have been explored to achieve this goal. From the information summarised in this chapter, the immunological approach seems to be the most successful and easiest one in preventing intestinal colonisation, at least by *E. coli*. In many cases, passive immunisation by maternal antibodies directed against the adhesive surface compounds of *E. coli* proved to be protective in the case of neonatal diarrhoea. As mentioned above, immunisation programmes do not seem to stimulate propagation in the population of surface antigens not included in vaccines. Unfortunately, the protective effect of antibodies delivered to new-borns through colostrum and milk seems to be critically dependent on the immunisation scheme and on the titer of specific antibodies present. Also, maternal immunity does not last long and is not adequate at later stages of life such as the weaning and postweaning period. In these cases however, passive immunisation with antibodies raised in laying hens proves to be an alternative. Of course, this strategy can also be applied in the neonatal phase of life.

Except from specific antibodies directed against *E. coli* surface compounds, colostrum and milk were shown over the last years to contain glycosylated macromolecules that block the carbohydrate-binding sites of bacterial adhesins. Also part of the protective effect of IgA is now ascribed to its glycoprotein nature and not necessarily solely due to its specific antigen binding potential. The successful protective effects against neonatal and postweaning diarrhoea by receptor analogues that block the carbohydrate-binding sites of surface lectins clearly demonstrate that these compounds might gain in importance in the future. A major advantage of the administration of receptor analogues to prevent pathogens from attaching to the small intestinal wall is without any doubt the resistance of the active part of these molecules to breakdown or major modification in this environment. It may even be hoped that, as more detailed structural data on receptors for surface lectins of pathogens become available, tailor-made analogues can be produced with an increased affinity for their binding partners. Oligosaccharides as anti-infective substances might prove to be of general use in all cases where carbohydrate-binding proteins are involved in the attachment of pathogens to host tissues (Zopf and Roth, 1996). These authors further argued that, since oligosaccharides are not bactericidal, resistance to them is unlikely to develop.

Finally, blocking intestinal receptors by for example plant lectins or even bacterium-derived fimbriae or adhesins is principally possible, and has been achieved in a limited number of reports. However, this approach seems to us to be of no practical use in view of the huge amounts of receptors-blockers that have to be used in order to achieve constant saturation of the receptors. Moreover, it should be kept in mind that binding of lectins to the gut may result in a number of adverse effects, and can profoundly affect the metabolism of the intestine. It can also be argued that binding of extravenously administered lectins, being often glycoproteins themselves, can act as "neo-receptors" and as such they can create new types of attachment sites at the level of the intestinal wall. Finally, binding of lectins, aimed at saturating intestinal receptors for bacterial adhesins, will critically depend on the glycosylation of these receptors which is known to change in an age-dependent way.

Although temporal enzymatic modification of intestinal receptor-sites might be an attractive alternative, also this approach seems to us of marginal applicability. Indeed, macromolecules identified by us as "receptors" for bacterial adhesins fulfil of course other functions in the intestine, and consequently, even temporal modification might have serious consequences on the normal functioning of the intestine.

In conclusion: all information available until now seems to justify the argument that either an immunologic approach or an approach aiming at saturating the carbohydrate binding sites of surface lectins are of practical use in preventing the attachment of microbial pathogens such as *E. coli* in the small intestine. At present, the successful protection against intestinal colonisation achieved by these approaches has been amply reported in animal model systems.

Passive immunisation and a "receptor-analogue therapy" in animal husbandry is already being applied at limited scale and will without any doubt be increasingly used in the future. It may be hoped that the new insights in the attachment process, and the strategies developed to prevent adherence will contribute to the treatment of diarrhoea in man as well. Up to date, in developing countries diarrhoea still remains one of the most frequent causes of death during childhood. It is also clear that the results of the intensive research performed over the last decades on attachment mechanisms of enterotoxigenic *E. coli*, their adhesins and the intestinal receptors they recognise will boost research efforts to unravel the basic mechanisms that govern other specific cellular recognition phenomena that finally cumulate in disease.

ACKNOWLEDGEMENTS

The Belgian "Ministerie van Landbouw en Middenstand", VEOS n.v. (Zwevezele, Belgium), and the Flemish FWO are gratefully acknowledged for financial support. S.B. is Senior Research Associate of the FWO.

REFERENCES

Acres, S.D., Isaacson, R.E., Babink, L.A. and Kapitany, R.A. (1979) Immunization of calves against enterotoxigenic colibacillosis by vaccinating dams with purified K99 antigen and whole cell bacterins. *Infect. Immunity*, **25**, 121–126.

Alexa, P., Salajka, E., Salajkova, Z. and Machova, A. (1995) Combined parenteral and oral immunization against enterotoxigenic *Escherichia coli* diarrhea in weaned piglets. *Vet. Med. (Praha)*, **40**, 365–370.

Aronson, M., Medalia, O., Schori, L., Mirelman, D., Sharon, N. and Ofek, I. (1979) Prevention of colonization of the urinary tract of mice with *Escherichia coli* by blocking of bacterial adherence with methyl-a,D-mannopyranoside. *J. Infect. Dis.*, **139**, 329–332.

Bertels, A., De Greve, H. and Lintermans, P. (1991) Function and genetics of fimbrial and nonfimbrial lectins from *Escherichia coli*. in: *"Lectin Reviews"*, vol. 1 (D.C. Kilpatrick, E. Van Driessche, T.C. BØg-Hansen, eds.), pp. 53–67 (Sigma Chem. Comp., St. Louis MO, USA).

Bertin, Y., Girardeau, J-P., Darfeuille-Michaud, A., andd Contrepois, M. (1996) Characterization of 20K fimbria, a new adhesin of septicemic and diarrhea-associated *Escherichia coli* strains, that belongs to a family of adhesins with N-acetyl-D-glucosamine recognition. *Infect. Immunity*, **64**, 332–342.

Bianchi, A.T., Scholten, J.W., van Zijderveld, A.M., van Zijderveld, F.G. and Bokhout, B.A. (1996) Parenteral vaccination of mice and piglets with F4+ *Escherichia coli* suppresses the enteric anti-F4 response upon oral infection. *Vaccine*, **14**, 199–206.

Breines, D.M. and Burnham, J.C. (1994) Modulation of *Escherichia coli* type 1 fimbrial expression and adherence to uroepithelial cells following exposure of logarithmic phase cells to quinolones at subinhibitory concentrations. *J. Antimicrob. Chemother.*, **34**, 205–221.

Breines, D.M. and Burnham, J.C. (1995) The effects of quinolones on the adherence of type-1 fimbriated *Escherichia coli* to mannosylated agarose beads. *J. Antimicrob. Chemother.*, **36**, 911–925.

Carbonare, S.B., Silva, M.L., Trabulsi, L.R. and Carneiro-Sampaio, M.M. (1995) Inhibition of HEp-2 cell invasion by enteroinvasive Escherichia coli by human colostrum IgA. *Int. Arch. Allergy Immunol.*, **108**, 113–118.

Chopra, I. and Linton, A. (1986) The antibacterial effects of low concentrations of antibiotics. Adv. Microbial Physiol., **28**, 211–259.

Collier, W.A. and De Miranda, J.C. (1955) Microbial Serology. *Antonie van Leeuwenhoek*, **21**, 133–140.

Contrepois, M.G. and Girardeau, J.P. (1985) Additive protective effects of colostral antipili antibodies in calves experimentally infected with enterotoxigenic *Escherichia coli*. *Infect. Immunity*, **50**, 947–949.

Dean, E.A., Whipp, S.C. and Moon, H.W. (1989) Age-specific colonization of porcine intestinal epithelium by 987P-piliated enterotoxigenic *Escherichia coli*. *Infect. Immunity*, **57**, 82–87.

De Graaf, F.K. (1986) The fimbrial lectins of E. coli. in: "Lectins: Biology, Biochemistry, Clininal Biochemistry" vol. 5 (T.C. BØg-Hansen, E. Van Driessche, eds.) pp. 285–296 (W. deGruyter, Berlin).

De Graaf, F.K. (1988) Fimbrial structures of enterotoxigenic *E. coli*. *Antonie van Leeuwenhoek*, **54**, 395–404.

De Graaf, F.K. (1990) Genetics of adhesive fimbriae of intestinal *Escherichia coli*. *Curr. Topics Microbiol. Immunol.*, **151**, 29–53.

De Graaf, F.K. and Mooi, F.R. (1986) The fimbrial adhesins of *Escherichia coli*. *Adv. Microb. Physiol.*, **28**, 65–143.

Deprez, P., Nollet, H., Van Driessche, E and Muylle, E. (1996) The use of swine plasma components as adhesin inhibitors in the protection of piglets against Escherichia coli enterotoxemia. Proceedings of the 14th IVPS Congress, Bologna, Italy, pp. 276.

Duguid, J.P. and Gillies, R.R. (1957) Fimbriae and adhesive properties in dysentery bacilli. *J. Path. Bacteriol.*, **74**, 397–411.

Edfors-Lilja, I., Gustafsson, U., Duval-Iflah, Y., Ellergren, H., Johansson, M., Juneja, R.K., Marklund, L. and Andersson, L. (1995) The porcine intestinal receptor for Escherichia coli K88ab, K88ac: regional localization on chromosome 13 and influence of IgG response to the K88 antigen. *Anim. Genet.*, **26**, 237–242.

Edwards, R.A., Cao, J. and Schifferli, D.M. (1996) Identification of major and minor chaperone proteins involved in the export of 987P fimbriae. *J. Bacteriology* 178, 3426–3433.

El Mazouari, K., Oswald, E., Hernalsteens, J-P., Lintermans, P. and De Greve, H. (1994) F17–Like fimbriae from an invasive Escherichia coli strain producing cytotoxic necrotizing factor type 2 toxin. *Infect. Immunity* 62, 2633–2638.

Erhard, M.H., Bergmann, J., Renner, M., Hofmann, A. and Heinritzi, K. (1996) Prophylactic effect of specific egg yolk antibodies in diarrhea caused by Escherichia coli K88 (F4) in weaned piglets. *Zentralbl. Veterinarmed. A*, **43**, 217–223.

Erickson, A.K., Willgohs, J.A., McFarland, S.Y., Benfield, D.A. and Francis, D.H. (1992) Identification of two porcine brushborder glycoproteins that bind the K88ac adhesin of Escherichia coli and correlation of these glycoproteins with the adhesive phenotype. *Infect. Immunity*, **60**, 983–988.

Estevez Touzard, M., Diaz Gonzalez, M., Monte Boada, R.J., Toledo Rodriguez, I. and Ramon Bravo, J. (1993) The infectious etiology of acute diarrheal diseases in the Republic of Cuba. *Rev. Cubana Med. Trop.*, **45**, 139–145.

Gaastra, W. and De Graaf, F.K. (1982) Host-specific fimbrial adhesins of noninvasive enterotoxigenic Escherichia coli strains. *Microbiol. Rev.*, **46**, 129–161.

Gaastra, W. and Svennerholm, A.M. (1996) Colonization factors of human enterotoxigenic Escherichia coli (ETEC). *Trends Microbiol.*, 4,444–452.

Girón, J.A., Gómez-Duarte, O.G., Jarvis, K.G. and Kaper, J.B. (1997) Longus pilus of enterotoxigenic Escherichia coli and its relatedness to other type-4 pili – a minireview. *Gene*, **192**, 39–43.

Grange, P. and Mouricout, M.A. (1996) Transferrin associated with the porcine intestinal mucosa is a receptor specific for K88ab fimbriae of *Escherichia coli*. *Infect. Immunity* 64, 606–610.

Grange, P., Vedrine, B. and Mouricout M. (1997) Adhesion of K88ab fimbriated E. coli in piglet small intestines in relation with iron transport molecules. *Adv. Exp. Med. Biol.*, **412**, 357–361.

Guérin, G., Duval-Iflah, Y., Bonneau, M., Bertaud, M., Guillaume, P. and Ollivier, L. (1993) Evidence for linkage between K88ab, K88ac intestinal receptors to Escherichia coli and transferrin loci in pigs. *Anim. Genet.*, **24**, 393–396.

Giugliano, L.G., Ribeiro, S.T., Vainstein, M.H. and Ulhoa, C.J. (1995) Free secretory component and lactoferrin of human milk inhibit the adhesion of ETEC. *J. Med. Microbiol.*, **42**, 3–9.

Guth, B.E., Aguiar, E.G., Griffin, P.M., Ramos, S.R. and Gomes, T.A. (1994) Prevalence of colonization factor antigens (CFAs) and adherence to HeLa cells in enterotoxigenic Escherichia coli isolated from feces of children in Sao Paulo. *Microbiol. Immunol.*, **38**, 695–701.

Gyles, C.L. (1992) Escherichia coli cytotoxins and enterotoxins. *Can. J. Microbiol.*, **38**, 734–746.

Haq, T.A., Mason, H.S., Clements, J.D. and Arntzen, C.J. (1995) Oral immunization with a recombinant bacterial antigen produced in transgenic plants. *Science*, **268**, 714–716.

Heller, E.D., Leitner, G., Drabkin, N. and Melamed, D. (1990) Passive immunization of chicks against Escherichia coli. *Avian Pathol.*, **19**, 345–354.

Heuzenroeder, M.W., Elliot, T.R., Thomas, C.S., Halter, H.R. and Manning, P.A. (1990) A new fimbrial type (PCFO9) on enterotoxigenic Escherichia coli O9:H_LT+ isolated from a case of infant diarrhea in Central Australia. *FEMS Microbiol. Lett.*, **66**, 55–60.

Hultgren, S.J., Abraham, S., Caparon, M., Falk, P., St. Geme, J.W.III. and Normark, S. (1993) Pilus and nonpilus bacterial adhesins: assembly and function in cell recognition. *Cell*, **73**, 887–901.

Imberechts, H., Deprez, P., Lintermans, P. and Broes, A. (1993) La maladie de l'oedème du porcelet. *Rec. Méd. Vét.*, **169**, 665–674.

Imberechts, H., Deprez, P., Van Driessche, E. and Pohl, P. (1997a) Chicken egg yolk antibodies against F18ab fimbriae of Escherichia coli inhibit shedding of F18 positive E. coli by experimentally infected pigs. *Vet. Microbiol.*, **54**, 329–341.

Imberechts, H., Bertschinger, H.U., Nagy, B., Deprez, P. and Pohl, P. (1997b) Fimbrial colonisation factors F18ab and F18ac of Escherichia coli isolated from pigs with postweaning diarrhea and edema disease. *Adv. Exp. Med. Biol.*, **412**, 175–211.

Khan, A.S., Johnston, N.C., Goldfine, H. and Schifferli, D.M. (1996) Porcine 987P glycolipid receptors on intestinal brush borders and their cognate bacterial ligands. *Infect. Immunity*, **64**, 3688–3693.

Knutton, S., Lloyd, D.R. and McNeish, A.S. (1987) Identification of a new fimbrial structure in enterotoxigenic Escherichia coli (ETEC) serotype O149:H28 which adheres to human intestinal mucosa: a potentially new humen ETEC colonization factor. *Infect. Immunity* 55, 86–92.

Krogfelt, K.A. (1991) Bacterial adhesion: genetics, biogenesis and role in pathogenesis of fimbrial adhesins of *Escherichia coli*. *Rev. Infect. Dis.*, **13**, 721–735.

Krogfelt, K.A. (1995) Adhesin-dependent isolation and characterization of bacteria from their natural environment. *Methods Enzymology*, **253**, 50–53.

Lindahl, M. and Carlstedt, I. (1990) Binding of K99 fimbriae of enterotoxigenic Escherichia coli to pig small intestinal mucin glycopeptides. *J. Gen. Microbiol.*, **136**, 1609–1614.

Lindblom, G.B., Ahren, C., Changalucha, J., Gabone, R., Kaijser, B., Nilsson, L.A., Sjogren, E., Svennerholm, A.M. and Temu, M. (1995) Campylobacter jejuni and enterotoxigenic Escherichia coli (ETEC) in faeces from children and adults in Tanzania. *Scand. J. Infect. Dis.*, **27**, 589–593.

Lintermans, P., Pohl, P., Deboeck, F., Bertels, A., Schlicker, C., Vandekerckhove, J., Van Damme, J., Van Montagu, M. and De Greve, H. (1988) Isolation and nucleotide sequence of the F17-A gene encoding the structural protein of the F17 fimbriae in bovine enterotoxigenic *Escherichia coli*. *Infect. Immunity*, **56**, 1475–1484.

Lintermans, P.F., Bertels, A., Schlicker, C., Deboeck, F., Charlier, G., Pohl, P., Norgren, M., Normark, S., Van Montagu, M. and De Greve, H. (1991) Identification, characterization and nucleotide sequence of the F17-G gene, which determines receptor binding of Escherichia coli F17 fimbriae. *J. Bacteriol.*, **173**, 3366–3373.

Lintermans, P., Bertels, A., Van Driessche, E. and De Greve, H. (1995) Identification of the F17 gene cluster and development of adhesion blockers and vaccine components. in: "Lectins: Biomedical Perspectives" (A. Pusztai, S. Bardocz, eds.) pp. 235–292 (Taylor & Francis, London).

Lösch, U., Schranner, I., Wanke, R. and Jürgens, L. (1986) The chicken egg, an antibody source. *J. Vet. Med.*, **33**, 609–619.

Loubeyre, C., Desnottes, J.F. and Moreau, N. (1993) Influence of sub-inhibitory concentrations of antibacterials on the surface properties and adhesion of *Escherichia coli*. *J. Antimicrob. Chemother.*, **31**, 37–45.

Martin, C., Rousset, E. and De Greve, H. (1997) Human uropathogenic and bovine septicaemic Escherichia coli strains carry an identical F17–related adhesin. *Res. Microbiol.*, **148**, 55–64.

McConnell, M.M., Chart, H., Field, A.M., Hibberd, M. and Rowe, B. (1989) Characterization of a putative colonization factor (PCFO166) of enterotoxigenic Escherichia coli strain. *Infect. Immunity 135*, 1135–1144.

McConnell, M.M., Hibberd, M., Field, A.M., Chart, H. and Rowe, B. (1990) Characterization of a new putative colonization factor (CS17) from a human enterotoxigenic Escherichia coli of serotype O114:H22 which produces only heat-labile enterotoxin. *J. Infect. Dis.*, **161**, 343–347.

Metcalfe, J.W., Krogfelt, K.A., Krivan, H.C., Cohen, P.S. and Laux, D.C. (1991) Characterization and identification of a porcine small intestine mucus receptor for the K88ab fimbrial adhesin. *Infect. Immunity*, **59**, 91–96.

Mol, O. and Oudega, B. (1996) Molecular and structural aspects of fimbriae biosynthesis and assembly in *Escherichia coli*. *FEMS Microbiol. Rev.*, **19**, 25–52.

Mol, O., Fokkema, H. and Oudega, B. (1996a) The Escherichia coli K99 periplasmic chaperone FanE is a monomeric protein. *FEMS Microbiol. Lett.*, **138**, 185–189.

Mol, O., Oudhuis, W.C., Fokkema, H. and Oudega, B. (1996b) The N-terminal b-barrel domain of the Escherichia coli K88 periplasmic chaperone FaeE determines fimbrial subunit recognition and dimerization. *Mol. Microbiol.*, **22**, 379–388.

Molbak, K., Wested, N., Hojlyng, N., Scheutz, F., Gottschau, A., Aaby, P. and Da Silva, A.P. (1994) The etiology of early childhood diarrhea: a community study from Guinea-Bissau. *J. Infect. Dis.*, **169**, 581–587.

Moon, H.W. (1990) Colonisation factor antigens of enterotoxigenic Escherichia coli in animals. *Curr. Topics Microbiol. Immunol.*, **151**, 147–165.

Moon, H.W. and Bunn, T.O. (1993) Vaccines for preventing enterotoxigenic Escherichia coli infections in farm animals. *Vaccine* 11, 213–220.

Mouricout, M. (1991) Swine and cattle enterotoxigenic Escherichia coli-mediated diarrhoea. Development of therapies based on inhibition of bacteria-host interactions. *Eur. J. Epidemiol.*, **7**, 588–604.

Mouricout, M. (1997) Interactions between the enteric pathogen and the host. An assortment of bacterial lectins and a set of glycoconjugate receptors. *Adv. Exp. Med. Biol.*, **412**, 109–123.

Mouricout, M.A. and Julien, R.A. (1987) Pilus-mediated binding of bovine enterotoxigenic Escherichia coli to calf small intestinal mucins. *Infect. Immunity*, **55**, 1216–1223.

Mouricout, M., Petit, J.M., Carias, J.R. and Julien, R. (1990) Glycoprotein glycans that inhibit adhesion of Escherichia coli mediated by K99 fimbriae: treatment of experimental colibacillosis. *Infect. Immunity*, **58**, 98–106.

Mouricout, M., Milhavet, M., Durié, C. and Grange, P. (1995) Characterization of glycoprotein glycan receptors for Escherichia coli F17 fimbrial lectin. *Microb. Pathogenesis 18*, 297–306.

Muñoz, M., Alvarez, M., Lanza, I. and Carmenes, P. (1996) Role of enteric pathogens in the aetiology of neonatal diarrhoea in lambs and goat kids in Spain. *Epidemiol. Infect.*, **117**, 203–211.

Mynott, T.L., Chandler, D.S. and Luke, R.K.J. (1991) Efficacy of enteric-coated protease in preventing attachment of enterotoxigenic Escherichia coli and diarrheal disease in the RITARD model. *Infect. Immunity*, **59**, 3708–3714.

Mynott, T.L., Luke, R.K. and Chandler, D.S. (1996) Oral administration of protease inhibits enterotoxigenic Escherichia coli receptor activity in piglet small intestine. *Gut*, **38**, 28–32.

Nagy, L.K., Mackenzie, T., Pickard, D.J. and Dougan, G. (1986) Effects of immune colostrum on the expression of a K88 plasmid encoded determinant: role of plasmid stability and influence of phenotypic expression of K88 fimbriae. *J. Gen. Microbiol.*, **132**, 2497–2503.

Nagy, B., Nagy, G., Meder, M. and Mocsari, E. (1996) Enterotoxigenic Escherichia coli, rotavirus, porcine epidemic diarrhoea virus, adenovirus and calici-like virus in porcine postweaning diarrhoea in Hungary. *Acta Vet. Hung.*, **44**, 9–19.

Neeser, J.R., Koellreutter, B. and Wuersch, P. (1986) Oligomannoside-type glycopeptides inhibiting adhesion of Escherichia coli strains mediated by type 1 pili: preparation of potent inhibitors from plant glycoproteins. *Infect. Immunity*, **52**, 428–436.

Nirdnoy, W., Serichantalergs, O., Cravioto, A., LeBron, C., Wolf, M., Hoge, C.W., Svennerholm, A.M., Taylor, D.N. and Echeverria, P. (1997) Distribution of colonization factor antigens among enterotoxigenic Escherichia coli strains isolated from patients with diarrhea in Nepal, Indonesia, Peru and Thailand. *J. Clin. Microbiol.*, **35**, 527–530.

Nollet, H., Deprez, P., Muylle, E. and Van Driessche, E. (1996) The use of non-immune plasma powder in the profylaxis of neonatal E.T.E.C. diarrhoea in calves. Proceedings of the 19th World Buiatrics Congress, Edinburgh, pp. 117–120.

O'Farrelly, C., Branton, D. and Wanke, C.A. (1992) Oral ingestion of egg yolk immunoglobulin from hens immunized with an enterotoxigenic Escherichia coli strain prevents diarrhea in rabbits challenged with the same strain. *Infect. Immunity*, **60**, 2593–2597.

Ofek, I., Mirelman, D. and Sharon, N. (1977) Adherence of Escherichia coli to human mucosal cells mediated by mannose receptors. *Nature (London)*, **265**, 623–625.

Ofek, I., Beachey, E.H. and Sharon, N. (1978) Surface sugars of animal cells as determinants of recognition in bacterial adherence. *Trends Biochem. Sci.*, **3**, 159–160.

Ono, E., Abe, K., Nakazawa, M. and Naiki, M. (1989) Ganglioside epitope recognized by K99 fimbriae from enterotoxigenic *Escherichia coli*. *Infect. Immunity*, **57**, 907–911.

Osek, J., Truszczynski, M., Tarasiuk, K. and Pejsak, Z. (1995) Evaluation of different vaccines to control of pig colibacillosis under large-scale farm conditions. *Comp. Immunol. Microbiol. Infect. Dis.*, **18**, 1–8.

Otto, G., Sandberg, T., Marklund, B.I., Ulleryd, P. and Svanborg, C. (1993) Virulence factors and pap genotype in Escherichia coli isolates from women with acute pyelonephritis, with or without bacteremia. *Clin. Infect. Dis.*, **17**, 448–456.

Oyofo, B.A., El-Etr, S.H., Wasfy, M.O., Peruski, L., Kay, B., Mansour, M., Campbell, J.R. Svennerholm, A.M., Churilla, A.M. and Murphy, J.R. (1995) Colonization factors of enterotoxigenic E. coli (ETEC) from residents of Northern Egypt. *Microbiol. Res.*, **150**, 429–436.

Oyofo, B.A., Peruski, L.F., Ismail, T.F., El-Etr, S.H., Churilla, A.M., Wasfy, M.O., Petruccelli, B.P. and Gabriel, M.E. (1997) Enteropathogens associated with diarrhea among military personnel during Operation Bright Star 96, in Alexandria, *Egypt. Mil. Med.*, **162**, 396–400.

Pohl, P., Van Muylem, K., Lintermans, P., Imberechts, H. and Mainil, J. (1992) Adhésines connues et inconnues chez les Escherichia coli isolées de cas de diarrhée du veau. *Ann. Méd. Vét.*, **136**, 479–481.

Pohl, P., Contrepois, M., Imberechts, H., Van Muylem, K., Jacquemin, E., Oswald, E. and Mainil, J. (1995) Recherches de structures d'attachement et de facteurs de virulence chez des Escherichia coli adhérentes aux villosités intestinales du veau mais qui ne produisent pas les adhésines F5 (K99), F17 ni Att111 *in vitro*. *Ann. Méd. Vét.*, **139**, 421–425.

Pusztai, A., Grant, G. and Bardocs, S. (1993a) Lectins as metabolic and growth signals: an overview. in: "Lectins: Biology, Biochemistry, Clinical Biochemistry", vol. 8 (E. Van Driessche, H. Franz, S. Beeckmans, U. Pfüller, A. Kallikorm, T.C. BØg-Hansen, eds.), pp. 245–249 (Textop, Hellerup, DK).

Pusztai, A., Ewen, S.W.B., Carvalho, A. de F.F.U., Grant, G., Stewart, J.C. and Bardocs, S. (1993b) Lectins as metabolic signals for the gut and body. in: "Lectins: Biology, Biochemistry, Clinical Biochemistry", vol. 8 (E. Van Driessche, H. Franz, S. Beeckmans, U. Pfüller, A. Kallikorm, T.C. BØg-Hansen, eds.), pp. 250–257 (Textop, Hellerup, DK).

Pusztai, A., Grant, G., Spencer, R.J., Duguid, T.J., Brown, D.S., Ewen, S.W., Peumans, W.J., Van Damme, E.J.M. and Bardocz, S. (1993c) Kidney bean lectin-induced Escherichia coli overgrowth in the small intestine is blocked by GNA, a mannose-specific lectin. *J. Appl. Bacteriol.*, **75**, 360–368.

Reid, R.H., Boedeker, E.C., McQueen, C.E., Davis, D., Tseng, L.Y., Kodak, J., Sau, K., Wilhelmsen, C.L., Nellore, R. and Dalal, P. (1993) Preclinical evaluation of microencapsulated CFA/II oral vaccine against enterotoxigenic *E. coli*. *Vaccine*, **11**, 159–167.

Rhen, M., Klemm, P. and Korhonen, T.K. (1986) Identification of two new hemagglutinins of Escherichia coli, N-acetyl-D-glucosamine-specific fimbriae and a blood group M-specific agglutinin, by cloning the corresponding genes in Escherichia coli K-12. *J. Bacteriology*, **168**, 1234–1242.

Ricci, L.C., De Faria, F.P., Porto, P.S.S., De Oliveira, E.M.G. and Pestana de Castro, A.F. (1997) A new fimbrial putative colonization factor (PCFO2) in human enterotoxigenic Escherichia coli isolated in Brazil. *Res. Microbiol.*, **148**, 65–69.

Robins-Browne, R.M. (1987) Traditional enteropathogenic Escherichia coli of infantile diarrhea. *Rev. Infect. Dis.*, **9**, 28–53.

Runnels, P.L., Moseley, S.L. and Moon, H.W. (1987) F41 pili as protective antigens of enterotoxigenic Escherichia coli that produce F41, K99, or both pilus antigens. *Infect. Immunity*, **55**, 555–558.

Rutter, J.M., Jones, G.W., Brown, G.T.H., Burrows, M.R. and Luther, P.D. (1976) Antibacterial activity in colostrum and milk associated with protection of piglets against enteric disease caused by K88-positive *Escherichia coli*. *Infect. Immunity*, **13**, 667–676.

Saarela, S., Taira, S., Nurmiaho-Lassila, E-L., Makkonen, A. and Rhen, M. (1995) The Escherichia coli G-fimbrial lectin protein participates both in fimbrial biogenesis and in recognition of the receptor N-acetyl-D-glucosamine. *J. Bacteriology 177*, 1477–1484.

Saarela, S., Westerlund-Wikström, B., Rhen, M. and Korhonen, T.K. (1996) The GafD protein of the G (F17) fimbrial complex confers adhesiveness of Escherichia coli to laminin. *Infect. Immunity*, **64**, 2857–2860.

Salit, I.E. and Gotschlich, E.C. (1977) Hemagglutination by purified type 1 Escherichia coli pili. *J. Exper. Med.*, **146**, 1169–1181.

Sanchez, R., Kanarek, L., Koninkx, J., Hendriks, H., Lintermans, P., Bertels, A., Charlier, G. and Van Driessche, E. (1993a) Inhibition of adhesion of enterotoxigenic Escherichia coli cells expressing F17 fimbriae to small intestinal mucus and brush-border membranes of young calves. *Microb. Pathogenesis*, **15**, 407–419.

Sanchez, R., Hendriks, H., Koninckx, J., Lintermans, P., Kanarek, L. and Van Driessche, E. (1993b) Studies on attachment-inhibitors of Escherichia coli strains expressing F17 fimbriae. in: "Lectins: Biology, Biochemistry, Clinical Biochemistry", vol. 8 (E. Van Driessche, H. Franz, S. Beeckmans, U. Pfüller, A. Kallikorm, T.C. BØg-Hansen, eds.), pp. 223–228 (Textop, Hellerup, DK).

Sarinho, S.W., Da Silva, G.A., Magalhaes, M. and Carvalho, M.R. (1993) A study of the importance of the enterotoxigenic E. coli in children with accute diarrhoea in Recife, Brazil. *J. Trop. Pediatr.*, **39**, 304–306.

Sarrazin, E. and Bertschinger, H.U. (1997) Role of fimbriae F18 for actively acquired immunity against porcine enterotoxigenic *Escherichia coli*. *Vet. Microbiol.*, **54**, 133–144.

Schifferli, D.M. and Beachey, E.H. (1988a) Bacterial adhesion: modulation by antibiotics which perturb protein synthesis. *Antimicrob. Agents Chemother.*, **32**, 1603–1608.

Schifferli, D.M. and Beachey, E.H. (1988b) Bacterial adhesion: modulation by antibiotics with primary targets other than protein synthesis. *Antimicrob. Agents Chemother.*, **32**, 1609–1613.

Schroten, H., Hanisch, F.G., Plogmann, R., Hacker, J., Uhlenbruck, G., Nobis-Bosch, R. and Wahn, V. (1992) Inhibition of adhesion of S-fimbriated Escherichia coli to buccal epithelial cells by human milk fat globule membrane components: a novel aspect of the protective function of mucins in the nonimmunoglobulin fraction. *Infect. Immunity*, **60**, 2893–2899.

Schroten, H., Plogmann, R., Hanisch, F.G., Hacker, J., Nobis-Bosch, R. and Wahn, V. (1993) Inhibition of adhesion of S-fimbriated E. coli to buccal epithelial cells by human skim milk is predominantly mediated by mucins and depends on the period of lactation. *Acta Paediatr.*, **82**, 6–11.

Seignole, D., Mouricout, M., Duval-Iflah, Y., Quintard, B. and Julien, R. (1991) Adhesion of K99 fimbriated Escherichia coli to pig intestinal epithelium: correlation of adhesive and non-adhesive phenotypes with the sialoglycolipid content. *J. Gen. Microbiol.*, **137**, 1591–1601.

Seignole, D., Grange, P., Duval-Iflah, Y. and Mouricout, M. (1994) Characterization of O-glycan moieties of the 210 and 240 kDa pig intestinal receptors for Escherichia coli K88ac fimbriae. *Microbiology*, **140**, 2467–2473.

Sellwood, R. (1980) The interaction of the K88 antigen with porcine intestinal epithelial cell brush borders. *Biochim. Biophys. Acta*, **632**, 326–335.

Serichantalergs, O., Nirdnoy, W., Cravioto, A., LeBron, C., Wolf, M., Svennerholm, A.M., Shlim, D., Hoge, C.W. and Echeverria, P. (1997) Coli surface antigens associated with enterotoxigenic Escherichia coli strains isolated from persons with traveler's diarrhea in Asia. *J. Clin. Microbiol.*, **35**, 1639–1641.

Shida, K., Takamizawa, K., Nagaoka, M., Osawa, T. and Tsuji, T. (1994) Enterotoxin-binding glycoproteins in a protease-peptone fraction of heated bovine milk. *J. Dairy Sci.*, **77**, 930–939.

Smit, H., Gaastra, W., Kamerling, J.P., Vliegenthart, J.F.G. and De Graaf, F.K. (1984) Isolation and structural characterization of the equine erythrocyte receptor for enterotoxigenic Escherichia coli K99 fimbrial adhesin. *Infect. Immunity*, **46**, 578–584.

Smyth, C.J., Marron, M.B., Twohig, J.M. and Smith, S.G. (1996) Fimbrial adhesins: similarities and variations in structure and biogenesis. *FEMS Immunol. Med. Microbiol.*, **16**, 127–139.

Sojka, W.J., Wray, C. and Morris, J.A. (1978) Passive protection of lambs against experimental enteric colibacillosis by colostral transfer of antibodies from K99 vaccinated ewes. *J. Med. Microbiol.*, **11**, 493–499.

Sonstein, S.A. and Burnham, J.C. (1993) Effect of low concentrations of quinolone antibiotics on bacterial virulence mechanisms. *Diagn. Microbiol. Infect. Dis.*, **16**, 277–289.

Spangler, B.D. (1992) Structure and function of cholera toxin and the related Escherichia coli heat-labile enterotoxin. *Microb. Rev.*, **56**, 622–647.

Subekti, D., Lesmana, M., Komalarini, S., Tjaniadi, P., Burr, D. and Pazzaglia, G. (1993) Enterotoxigenic Escherichia coli and other causes of infectious pediatric diarrheas in Jakarta, Indonesia. *Southeast Asian J. Trop. Med. Public Health*, **24**, 420–424.

Tacket, C.O., Maneval, D.R. and Levine, M.M. (1987) Purification, morphology and genetics of a new fimbrial putative colonization factor of enterotoxigenic Escherichia coli O159:H4. *Infect. Immunity*, **55**, 1063–1069.

Tacket, C.O., Reid, R.H., Boedeker, E.C., Losonsky, G., Nataro, J.P., Bhagat, H. and Edelman, R. (1994) Enteral immunization and challenge of volunteers given enterotoxigenic E. coli CFA/II encapsulated in biodegradable microspheres. *Vaccine*, **12**, 1270–1274.

Takeda, T., Yoshino, K., Uchida, H., Matsuda, E. and Yamagata, M. (1997) Epidemiological study of hemolytic uremic syndrome associated with enterohemorrhagic Escherichia coli in Japan. *Nippon Rinsho*, **55**, 693–699.

Teneberg, S., Willemsen, P.T.J., De Graaf, F.K. and Karlsson, K.A. (1990) Receptor-active glycolipids of epithelial cells of the small intestine of young and adult pigs in relation to susceptibility to infection with Escherichia coli K99. *FEBS Letters*, **263**, 10–14.

Teneberg, S., Willemsen, P.T.J., De Graaf, F.K. and Karlsson, K.A. (1993) Calf small intestine receptors for K99 fimbriated enterotoxigenic *Escherichia coli*. *FEMS Microbiol. Lett.*, **109**, 107–112.

Teraguchi, S., Shin, K., Fukuwatari, Y. and Shimamura, S. (1996) Glycans of bovine lactoferrin function as receptors for the type 1 fimbrial lectin of *Escherichia coli*. *Infect. Immunity*, **64**, 1075–1077.

Van Driessche, E. and Beeckmans, S. (1993) Fimbrial lectins from enterotoxigenic E. coli: a mini-review. in: "Lectins: Biology, Biochemistry, Clinical Biochemistry", vol. 8 (E. Van Driessche, H. Franz, S. Beeckmans, U. Pfüller, A. Kallikorm, T.C. BØg-Hansen, eds.), pp. 207–216 (Textop, Hellerup, DK).

Van Driessche, E., Schoup, J., Charlier, G., Lintermans, P., Beeckmans, S., Zeeuws, R., Pohl, P. and Kanarek, L. (1988) The attachment of E. coli to intestinal calf villi and Eupergit-C-glycoprotein beads. in: "Lectins: Biology, Biochemistry, Clinical Biochemistry", vol. 6 (T.C. BØg-Hansen, D.L.J. Freed, eds.), pp. 55–62 (Sigma Chem. Comp., St. Louis, MO, USA).

Van Driessche, E., Sanchez, R., Beeckmans, S., De Cupere, F., Charlier, G., Pohl, P., Lintermans, P. and Kanarek, L. (1993) A general procedure for the purification of fimbrial lectins from Escherichia coli. in: "Lectins and Glycobiology", (H.J. Gabius, S. Gabius, eds.), pp. 47–54 (Springer-verlag, Heidelberg).

Van Driessche, E., Sanchez, R., Dieussaert, I., Kanarek, L., Lintermans, P. and Beeckmans, S. (1995) Enterotoxigenic fimbrial Escherichia coli lectins and their receptors: targets for probiotic treatment of diarrhoea. in: "Lectins: Biomedical Perspectives" (A. Pusztai, S. Bardocz, eds.) pp. 235–292 (Taylor & Francis, London).

Vergara, M., Quiroga, M., Grenon, S., Pegels, E., Oviedo, P., Deschutter, J., Rivas, M., Binsztein, N. and Claramount, R. (1996) Prospective study of enteropathogens in two communities of Misiones, Argentina. *Rev. Inst. Med. Trop. Sao Paulo*, **38**, 337–347.

Viboud, G.I., Binsztein, N. and Svennerholm, A.M. (1993) A new fimbrial putative colonization factor, PFC020, in human enterotoxigenic *Escherichia coli*. *Infect. Immunity*, **61**, 5190–5197.

Wadström, T., Trust, T.J. and Brooks, D.E. (1983a) Bacterial surface lectins. in: "Lectins: Biology, Biochemistry, Clinical Biochemistry" vol. 3 (T.C. BØg-Hansen, G.A. Spengler, eds.) pp. 479–494 (W. deGruyter, Berlin).

Wadström, T., Faris, A., Lindahl, M., Lönnerdahl, B. and Ersson, B. (1983b) Mannose-specific lectins on enterotoxigenic E. coli recognizing different glycoconjugates. in: "Lectins: Biology, Biochemistry, Clinical Biochemistry" vol. 3 (T.C. BØg-Hansen, G.A. Spengler, eds.) pp. 503–510 (W. deGruyter, Berlin).

Willemsen, P.T.J. and De Graaf, F.K. (1992) Age and serotype dependent binding of K88 fimbriae to porcine intestinal receptors. *Microb. Pathogenesis*, **12**, 367–375.

Wold, A.E., Mestecky, J., Tomana, M., Kobata, A., Ohbayashi, H., Endo, T. and Svanborg Eden, C. (1990) Secretory immunoglobulin A carries oligosaccharide receptors for Escherichia coli type 1 fimbrial lectin. *Infect. Immunity*, **58**, 3073–3077.

Wold, A.E., Motas, C., Svanborg, C. and Mestecky, J. (1994) Lectin receptors on IgA isotypes. *Scand. J. Immunology*, **39**, 195–201.

Wolk, M., Ohad, E., Shafran, R., Safir, S., Cohen, Y., Wiklund, G. and Svennerholm, A.M. (1995) Epidemiological aspects of enterotoxigenic Escherichia coli diarrhoea in infants in the Jerusalem area. *Public Health Rev.*, **23**, 25–33.

Wolk, M., Ohad, E., Shafran, R., Schmid, I. and Jarjoui, E. (1997) Enterotoxigenic Escherichia coli

(ETEC) in hospitalised Arab infants from Judea area-west bank, Israel. *Public Health*, **111**, 11–17.

Yokoyama, H., Peralta, R.C., Diaz, R., Sendo, S., Ikemori, Y. and Kodama, Y. (1992) Passive protective effect of chicken egg yolk immunoglobulins against experimental enterotoxigenic Escherichia coli infection in neonatal piglets. *Infect. Immunity*, **60**, 998–1007.

Yokoyama, H., Hashi, T., Umeda, K., Icatlo, F.C., Kuroki, M., Ikemori, Y. and Kodama, Y. (1997) Effect of oral egg antibody in experimental F18+ Escherichia coli infection in weaned pigs. *J. Vet. Med. Sci.*, **59**, 917–921.

Yuyama, Y., Yoshimatsu, K., Ono, E., Saito, M. and Naiki, M. (1993) Postnatal change of pig intestinal ganglioside bound by Escherichia coli with K99 fimbriae. *J. Biochem.*, **113**, 488–492.

Zhanel, G.G. and Nicolle, L.E. (1992) Effect of subinhibitory antimicrobial concentrations (sub-MICs) on in-vitro bacterial adherence to uroepithelial cells. *J. Antimicrob. Chemother.*, **29**, 617–627.

Zopf, D. and Roth, S. (1996) Oligosaccharide anti-infective agents. *The Lancet*, 347, 1017–1021.

9. LECTIN-CARBOHYDRATE INTERACTIONS IN BACTERIAL PATHOGENESIS

MICHÈLE MOURICOUT and BRUNO VÉDRINE

Biotechnologie, Faculté des Sciences, 123 Avenue Albert Thomas, 87060, Limoges, France

INTRODUCTION

A broad spectrum of bacteria have been associated with a wide variety of diseases, including enteropathies, septicemia, œdema disease, pulmonary infections, uropathologies, meningitis. Among infections, diarrhea is an important and common problem in human health and in animal production. The disease is worldwide in its distribution despite some enteric infections may appear to be on the decline in some parts of the world. Intestinal infections are frequently caused by *E. coli* and, strains from several categories of diarrhoeagenic *E. coli* are distinguished. For example, enterotoxigenic *E. coli* species (ETEC) are associated with travellers diarrhea in humans and neonatal diarrhea in young animals, enteropathogenic (EPEC) found in infants and rabbits, enteroinvasive *E. coli* (EIEC) associated with dysentery syndromes, verotoxin-producing *E. coli* (VTEC) linked to hemorrhagic colitis and hemolytic uremic syndrome (Lingwood, 1996). This syndrome is also a feature of *Shigella dysenteriae* infections (Fontaine *et al.*, 1988).

Many bacterial pathogens share common mechanisms of interaction with their host cells. Distincts steps are essential for transmission of bacteria to the host, the adhesion to target tissue, the survival at the selected site, the colonization of the site and, under special circumstances, the invasion into cells.

BIOLOGICAL SIGNIFICANCE OF BACTERIAL ADHESION

The adhesion of pathogenic bacteria to host cells appears to be essential and is often a critical first stage in infectious process, for successful multiplication in situ and host colonization. Bacteria use multiple strategies to exploit the host environment. Mammalian polysaccharides can serve as receptors for carbohydrate-binding adhesins of distinct bacteria. In some cases, bacteria as enteropathogenic and enterohemorraghic *Escherichia coli* cause attaching-effacing lesions in the intestinal mucosa of humans and various animals. Sequence of infectious events involves initial recognition of epithelial cells, adhesion, exploitation of many cell functions, including signal transduction, cytoskeletal rearrangements, vacuolar trafficking (Agin *et al.*, 1997; Donnenberg *et al.*, 1992; Finlay *et al.*, 1997).

Although lectins are not the only means developed by bacteria to adhere to targets, the wide distribution of fimbrial or afimbrial adhesins acting as lectins, among the different oral, enteric, extraintestinal bacterial species, suggest that they serve as an important function for the bacteria and may be important virulence factors in bacterial pathogenicity.

The biosynthesis of adhesins is a relatively conserved process controlled by gene clusters which encode regulatory proteins, chaperones, structural and adhesive proteins (Ofek *et al.*, 1994; Mol *et al.* 1997). The ETEC fimbrial adhesins contain lectin domains usually at the top of appendices and/or along adhesive structures. The fimbrial lectins recognize their receptor only after being assembled in their fimbrial quaternary structure. At present, 20 different colonization factors in human ETEC, have been described (Gaastra *et al.* 1996; Di Martino *et al.*, 1997). *E. coli* has been found to express several different types of fimbriae in animals (Mouricout, 1997). Furthermore, nonpathogenic colibacilli, which are normal intestinal flora, possess many of the genes necessary for interactions with host cells and, are predisposed to become pathogens upon acquisition of virulence gene cluster (Groisman *et al.*, 1997). For example, P fimbriae virulence factors in uropathologies are encoded by genes normally present in commensal *E. coli* strains of colon and many of the genes implicated in *Salmonella* virulence are also present in nonpathogenic *E. coli* (Groisman *et al.*, 1997).

A bacterium frequently expresses several lectins at its surface that interact with distinct host glycoconjugates. Bacterial lectins are discriminated in their carbohydrate recognition through the primary sugar specificity. However, the primary specificity does not explain the overall selectivity (Table 1).

The multistep process of infection needs in first the recognition of carbohydrates which determine species specificities and tropism for particular host and tissues. One of difficulties which had been encountered was the identification of molecules involved in bacterial adhesion in complex biological fluids, cell membranes and tissues. Lectins mediate the bacterial binding to host glycoconjugates either tightly membrane bound or integral part of cell membranes. Hence, number of cell surface compounds, glycoproteins, glycolipids, proteins, oligosaccharides, proteoglycans can serve as receptors for bacteria (Karlsson, 1995; Mouricout, 1997; Rostand *et al.*, 1997).

Main constituents of mucous layer are the first sites of adhesion of pathogens that colonize the oral and intestinal tissues. Glycoconjugates must be available and their function depends on their molecular conformation at the surface of cells. The carbohydrate chains and protein or lipid moieties directly or indirectly involved in the bacteria-host interactions present an extreme variability. In digestive, urinary and repiratory tracts, tissues are bathed in mucus. Mucus which acts as a protective barrier, can be matrix for bacterial multiplication. Consequently, pathogens must cross the constantly flowing mucus and adhere to the underlying cells.

The host is, of course an obligate participant. The involvment of the individual variation in receptor repertoire (due to the age, genetic basis of receptor expression and targeting of bacterium), can influence the susceptibility/resistance to infection in a given population. Since the susceptibility to infection is usually a complex genetic trait, the approach has been to detect animals that differ in their susceptibility or resistance to a pathogen. The studies of genetic variation and the inheritance of susceptibility and resistance characters on controlled breedings, allows the assignment of host-resistance/susceptibility gene(s) to specific chromosome(s).

In this review, we focus on some examples which will be drawn primarily from adhesins of diarrheagenic *E. coli*. When examined from the viewpoint of the bacterial virulence, the underlying principles of the recognition process may be applied equally well to other organisms and other diseases.

Table 1 Carbohydrate Specificities of Bacterial Lectins.

	target	adhesion
Campylobacter jejuni	intestines	sialylglycopeptides (unpublished data)
Escherichia coli	intestines	mannosides (type 1 lectin)
		NeuGc (F5)
		NeuAc in glyco-lipids and -proteins (CFA)
		Galβ1-4GlcNAc (F4ab)
		Galβ1-3GalNAc, Fuc α1-2Gal β13GlcNAc (F4ac)
		GlcNAc (F17)
		SO₃Gal β1-, Gal β1-4GlcNAc (F6)
	urinay tract	Galα1-4Gal (Pap lectin binds globoside convex side)
	brain tissues	NeuAc α2-3Gal β1-3
Helicobacter pylori	stomach	SO₃-Gal, sialylated Lewis a&b
		SO₃-Lewis a
Haemophilus influenza	lung	GM1, GM2, GM3, GD1a gangliosides
Klebsellia pneumoniae	lung	mannose (type 1 lectin)
Neisseria gonorrhoeae		Lewis a, proteoglycans
P. aeruginosa	buccal, tracheal cells	asialoganglioside Gal α1-4Gal
	corneal cells	A, H group blood substances
Salmonella	intestinal cells	mannosides
Shigella	intestinal cells	mannosides
Streptococcus sanguis	saliva	NeuAc α2-3Gal β1-4
Vibrio cholerae	intestinal cells	fucosides

Gal, galactose; GlcNAc N-acetylglucosamine; NeuAc, NeuGc, acetylated and glycolylated neuraminic acids
References: Dean-Nystrom, 1995; Demuth *et al.*, 1990; Fakid *et al.*, 1997; Gaastra *et al.*, 1996; Grange *et al.*, 1996a; Khan *et al.*, 1996; Mouricout *et al.*, 1995; review Mouricout, 1997; Piotrowski *et al.*, 1991; Rostand *et al.*, 1997; Snellings *et al.*, 1997; Van Alphen *et al.*, 1991; Veerman *et al.*, 1997.

Role of Lectins (Type 1 Fimbrial) in the Targeting

Mannose specific type 1 fimbriae are found in most species, *as Salmonella, Klebsellia pneumoniae,* fimbriated species of the *Enterobacteriaceae* family, which colonize different host and sites. The type 1 fimbriae confer to pathogenic *E. coli* a selective advantage in colonizing hosts (Bloch *et al.*, 1992).

Recently it has become clear that type 1 fimbriae, responsible of mannose- and mannoside-binding activity, exhibit several different variants, due to the allelic variation of the gene for the lectin subunit FimH. Type 1 lectins of phenotype MN recognize only substrates rich in exposed mannose residues such as yeast mannans. Another phenotype (MF) binds to yeast mannans and complex-type oligosaccharides such as in human plasma fibronectin (Sokurenko *et al.*, 1994, 1995). A third variant

(MFP) also binds to peptides devoid of saccharide moieties. While all FimH subunits mediate adhesion via trimannosyl residues, certain variants are capable of mediating adhesion via monomannosyl residues and strongly target uroepithelial cells. This recent observation could explain the predominance of the high MN-adhesive phenotype among urinary tract infections (Sokurenko *et al.*, 1997).

Although closely related genetically, *E. coli* and *Shigella* differ in both pathogenicity and host and tissue specificities. *Shigella* express type 1 fimbrial adhesins which bind mannose (Snellings *et al.*, 1997) but infect only the colonic epithelial cells of humans and higher primates to cause bacillary disentery. *Salmonella* species, associated with consumption of contamined food products or water, preferentially associate with M cells within Peyer's patches of the distal ileum, penetrate the intestinal epithelium in membrane-bound vacuoles from which they spread into host tissues. They induce ruffles on the cell membranes at the site of entry. Depending on the host, the bacteria may remain in the lamina propia of the epithelium or the bacteria may be transported by macrophages to liver and spleen.

Host-bacteria Interactions May Involve Both Lectins and Lipo-oligosaccharides (LOS) and Lipopolysaccharides (LPS) on Bacterial Cells

Pseudomonas aeruginosa is a human pathogen cause of morbidity and mortality for subjects suffering from congenital or acquired immunodeficiency. It also produces destructive corneal infection in humans. This bacterium may damage almost every tissue and organ. For invasiveness and keratitis to occur, its must first adhere to host tissues. *Pseudomonas aeruginosa*-host interactions are multifactorial and are presented here as a prototype of distinct receptor-ligand interactions, both by adhesins that recognize host's glycoconjugates, and by lipopolysaccharide (LPS) that binds host's proteinic receptors.

Pseudomonas aeruginosa colonizes mucus of respiratory tracts by the mediation of two groups of lectins: the galactose/mannose- and fucose-binding lectins (PA-I and PA-II, respectively) which can, in addition, affect ciliary beat frequency. PA-I preferentially interacts with several receptors containing terminal galactose such as Pk, P1, B antigens (Gilboa-Garber 1994). Furthermore, LPS of *Pseudomonas aeruginosa* binds galectin-3 isolated from human corneal epithelium (Gupta *et al.*, 1997).

The lectin PA-I has a preference for the α-anomer of galactose at the non-reducing end, in the following decreasing order Galα1-6, Galα1-4, Galα1-3. In the binding established between the PA-I lectin and Galα1-4Gal motif, hydrophobic interactions are important for recognition (Chen *et al.*, 1997). PA-II lectin exhibits a very high affinity for fucose (Ka: 1.5×10^6 M^{-1}) and it is grouped with lectins which may exhibit H blood group specificity (Adam *et al.*, 1997; Gilboa-Garber 1994). The lectins contribute to the virulence by enabling the targeting and homing of bacteria on cells and the action of cytotoxic compounds as microbial glycosidases, proteases, hemolysins.

The affinity of *Pseudomonas aeruginosa* lectins is also exhibited by some lectins of E. coli strains, which cause urinary tract infections in humans. Pyelonephritis mediated by P-fimbriated *E. coli* involve the receptor Galα1-4Gal, expressed in glycolipids and not in glycoproteins (Yang *et al.*, 1994). This selection might afford the bacteria an advantage when ascending the ureter, enabling them to stick to true membrane receptors without binding to glycoproteins which are subsequently eluted (Karlsson, 1995). Gal α1-4Gal- is also recognized by Streptococcus suis, a pathogen for pig and humans (cause of meningitis). However, P-fimbriated E. coli and Streptococcus recognize different epitopes in the same disaccaharide (Haataja et al., 1994). The conformation of the Gal α1-4Gal disaccharide, in relation to the plasma membrane, changes as the oligosaccharide sequence is extended (Strömberg et al., 1991).

Extracellular capsular oligosaccharides are widespread among pathogenic bacteria (Escherichia coli, Campylobacter jejuni, Helicobacter pylori, Pseudomonas aeruginosa, Salmonella typhi, Shigella flexneri). Among the large number of other potential virulence factors, LPS or LOS potentially contribute to pathogenesis. In this type of interactions, mammalian lectins mediate host-pathogen interactions by their specific recognition of bacterial carbohydrates (Mandrell et al., 1994) and the role of LPS/LOS in the adhesion to the host has been established (Jacques 1996; Paradis et al., 1994). Extracellular oligosaccharides are important in bacteria on mucosal surfaces whose strategy for survival in the host depends on evasion of phagocytosis (Amako et al., 1988; Roche et al., 1995; Whitfield 1988).

Asialoglycoprotein receptor (ASGP-R) and protein of 70 kDa (p70) serve as receptors for Neisseria gonorrhoeae lipooligosaccharide LOS which contains a terminal Galβ-GlcNAc residue (Porat et al., 1995a,b). Two independant LPS binding sites on galectin-3 confers direct binding to galactoside-containing polysaccharide chains of Klebsiella pneumoniae LPS and to inner core region of Salmonella minnesota R7 LPS which is only made of covalently bound lipid component lipid A, 2-keto-3-deoxy-octulosonic acid (KDO) and heptose and is devoid of galactoside-containing polysaccharides (Mey et al., 1996). By its binding to LPS, extracellular galectin could enhance the bacterial interaction with host cells such as macrophages and neutrophil leukocytes (Mandrell et al., 1994).

Phenotypic variation of capsular polysaccharides and lipopolysaccharide in Haemophilus influenza illustrate some of genetic mechanisms used by pathogenic bacteria to adapt to different host's microenvironments (Roche et al., 1995). So, the ability to generate phenotypic variety provides an adaptative mechanism to combat the polymorphism and immune repertoire of the host.

Bacteria express oligosaccharides on their cell surfaces that are also found in mammalian cells. O antigens of Helicobacter pylori, contain Lewis X determinant (Galβ1-4[Fucα1-3]GlcNAcβ-) synthesized by β1,3 fucosyltransferase and β1,4 galactosyltransferase in sequential addition of galactose and fucose in an analogous fashion to that found in humans (Chan et al., 1995).

BACTERIAL BINDING TO TARGET-HOST GLYCOCONJUGATES: A RECEPTOR REPERTOIRE

Although the role of the adhesins of human ETEC in the adhesion of bacteria to intestines is well documented, the structures involved in the binding have not been established in detail because the limited human material. The lectin-carbohydrate interactions are best exemplied in studies on animal enterobacterial species which produce lectins on fimbrial structures. Colibacillosis occurs most frequently in calves, lambs and piglets 1–3 days after birth and at weaning period. E. coli strains which cause postweaning diarrhea and œdema disease often differ from those that cause neonatal diarrhea. In domestic animals (pigs and cattle), lectins predominantly found in enterotoxigenic E. coli strains (ETEC) are F4 (K88, ab, ac, ad), F5 (K99), F6 (987P), F17, F18 and F41 fimbriae (Mouricout, 1991, 1997).

F5 Lectin

The role of glycolipid-receptor binding in ETEC pathogenesis has been clearly demonstrated for F5 fimbriated bacteria which cause diarrhea in young pigs and newborn calves but not in adult animals (Teneberg et al., 1990). F5 fimbrial lectin strongly binds to N-glycoloyl- sialoglycoconjugates (Mouricout et al., 1987; Kyogashima et al., 1989; Teneberg et al., 1990). Furthermore, F5 lectin detects fine age-related changes in N-glycolylation or N-acetylation of neuraminic acid (for which it presents more lower affinity). The distribution of N-glycolylated neuraminic acids on sialoglycoproteins and sialoglycolipids of intestinal cell membranes determine the host susceptibility.

The broad range of specificity of the bacterial lectins can explain in part the capacity of enterobacteria to cause severe diarrhea. The majority of pathogenic strains tested are capable of producing more than one kind of lectin. Bovine E. coli co-expressing the sialic acid- (F5), galactose-(F41), N-acetyl-glucosamine-(F17) lectins strongly adhere to intestines (Mouricout et al., 1987).

F17 Lectin

It has been observed that F17 fimbrial lectin binding significantly vary in mucins and membrane glycoconjugates along proximal and distal portions of calf small intestinal mucosa (Mouricout et al., 1995). Furthermore, F17 lectin did not recognizes terminal GlcNAc(α1-4) linked to galactose (our data), identified in duodenal-gland oligosaccharides (Van Halbeek et al., 1983). Recent works indicate that GafD lectin of the F17 fimbriae adhere to laminin carbohydrate and suggest a novel function for the F17 fimbria in binding to mammalian basement membranes (Saarela et al., 1996).

F6 Lectin

F6 (987P) fimbriae expressed by certain strains of enterotoxigenic E. coli also mediate the adhesion to small intestinal epithelial cells. The F6 lectin is structured in helical arrangement of three fimbrial proteins, the major subunit FasA and two minor subunits FasF and FasG. FasG located at the tip and at various position along the fimbriae, possess binding properties and recognizes both glycoproteins (Khan et al., 1994) and ceramide monohexoside and sulfatide with hydroxylated fatty acyl chains (Khan et al., 1996). Furthermore, FasA may participe to the ligand-receptor interaction by galactosylceramide with hydroxylated fatty acids and hydroxylated ceramide monohexoside. Neonatal pigs are susceptible to F6-mediated enterotoxigenic E. coli, but older pigs are resistant. Resistant pigs have functional F6 receptors in their brush borders as susceptible pigs. Fimbriated strains adhere to intestinal epithelium in susceptible pigs and, associate with intestinal mucus and debris, but not with the epithelium in resistant pigs (Dean-Nystrom, 1995). Author hypothezises that F6 receptors in the mucous bind and prevent F6 fimbriated strains from attaching to receptors on the epithelial cell surfaces.

F4ab, ac and ad Isolectins of Enterotoxigenic *Escherichia coli*

Enterotoxigenic *Escherichia coli* (ETEC) strains that express F4ab, ac and ad fimbriae, constitute the majority of the ETEC strains isolated from infected piglets. F4 fimbriated ETEC do not adhere to enterocytes of all pigs, and phenotypes are distinguished with respect to susceptibility to bacterial adhesion (Bijlsma *et al.*, 1985; Rapacz and Hasler-Rapacz, 1986).

No bacteria was found adhered to the intestinal brush border membranes obtained from resistant animals. Our results suggested that mucosal glycoproteins or glycolipids, could participate to the host protection against bacterial adhesion. Effectively, one 45 kDa receptive glycoprotein, abundant in the mucus, was detected only in the mucus of non adhesive pigs (unpublished data). Furthermore, we identified in intestines of all animals, independently the adhesion phenotypes, the precursors Glc-Cer and Gal-Cer and glycolipids of 5–6 carbohydrate residues with substitution in ceramide (our results). It has been observed that Gal-Cer and β1-linked galactosyl residues in glycosphingolipids are recognized by *E. coli* F4 fimbriae (Blomberg *et al.*, 1993, Payne *et al.*, 1993). So, lipid-bound galactosyl sequences which were present both in non adhesive and adhesive piglets in the mucus of small intestine are possible attachment sites. These receptors are present in the intestinal mucosa of adhesive phenotype animals but in small amount present in mucus, which might facilitate adhesion and subsequent penetration to the underlying epithelial cells. As the thin mucus of pigs of adhesive phenotypes may not retard bacterial colonization, animals must be also examined on the basis of the expression of the glycoproteins bound to the brush border membranes, which act as receptors for the bacteria.

Distinct set of receptors are specific for each of the F4 variants and that N- and O-glycoprotein membrane or membrane-bound receptors played a role in the adhesion of K88ab and K88ac fimbriae (Grange *et al.*, 1996a, Seignole *et al.*, 1994). F4ab lectin binds intestinal transferrin (Grange *et al.*, 1996a) which possess sialylated and fucosylated N-acetyllactosamine glycan chain. In pig intestines, we observed that F4ab receptors (Table 2) did not possess Lewis a, Galβ1-3 GlcNAcβ1-2[Fucα1-4]-R or Lewis β Galβ1-3 [Fucα1-2]GlcNAcβ1-2[Fucα1-4]-R sequences (the structure Lewis-type Gal β1-3/4 [Fuc α1-3/4] GlcNAc β1-, is produced by attaching a fucose residue to the penultimate sugar N-acetylglucosamine).

F4ac lectin binds membrane O-glycoproteins. Oligosaccharides of membrane O-glycoproteins may contain Galβ1-3-GalNAcα1-, Galβ1-3[GlcNAcβ1-6]-GalNAcα-, GlcNAcβ1-3-GalNAcα1- and GlcNAcβ1-3[GlcNAcβ1-6]-GalNAcα- core structures with additional backbones Galβ1-3GlcNAcβ1- and Galβ1-4 GlcNAcβ1-3-. Backbones are extended by A, B, H, and Lewis blood group substances also encountered in glycolipids of group A, H (Bäker *et al.*, 1997). F4ac lectin recognizes Galβ1-3GalNAcβ1, and Fucα1-2Galβ1-3/4GlcNAc-β1 on the brush borders of enterocytes (Seignole *et al.*, 1994). Interactions of F4ac lectin with intestinal brush borders might be, in part, related to the fucosylation (Seignole *et al.*, 1994). The presence of fucose residues in the complex galactoside receptors could be a common point between the variants ab and ac of F4 fimbriae. Sialoglycoproteins may act as

Table 2   Oligosaccharide Sequences in Receptors for *E. coli* F4ab and F4ac Lectins.

Sequences	F4ab	F4ac
	recognition	
SAα2-6Galβ1-4GlcNAcβ1-2GalMan3GlcNAcβ1-4GlcNAc[Fucα1-6]β1-N	++	–
Galβ1-4GlcNAcβ1-2GalMan3GlcNAcβ1-4GlcNAc[Fucα1-6]β1-N	+++	+
SAα2-6Galβ1-4GlcNAcβ1-2[Fucα1-3]GalMan3GlcNAcβ1-4GlcNAc[Fucα1-6]β1	+	+
Fucα1-2Galβ1-3GlcNAcβ1-3Galβ1-4Glc (H)	–	+++
Fucα1-2Galβ1-3[Fucα1-4]GlcNAcβ1-3Galβ1-4Glc (Le^b)	–	+++
Galβ1-3[Fucα1-4]GlcNAcβ1-3Galβ1-4Glc (Le^a)	–	?
Galβ1-4GlcNAc β1	+ +++	
Galβ1-4GlcNAcβ1-2[Fucα1-3]-R (Lewis X)	?	?
Galβ1-4[Fucα1-2]GlcNAcβ1-2[Fucα1-3]-R (Lewis Y)	?	?
NeuAcα2-6Gal/ GalNAc	–	–
Manα1-2/ 3/ 6Man	–	–

High and low intensities of recognition were noted (+++) and (+), respectively. Absence of recognition was (–).

receptor for the galactose-binding F4 lectin by the involvement of the polypeptide backbone in the definition of receptor sites (Table 2), neuraminic acid in α2–3/6 linkage to galactose is typical constituent of pig small-intestine glycoproteins.

Adhesion of F4ad lectin is not an all-or-none characteristic, low and high affinity receptors have been found (Hu *et al.*, 1993), this variant binds glycolipids (Grange, personal communication).

Additionally, it is interesting to notice that heat-labile enterotoxins (LTp) from strains of *E. coli* isolated from pig intestines bind to several blood-group A-active glycoproteins rather than H-active glycoproteins. The binding of LTp to A-active glycoproteins is inhibited more by fucose than by galactose (Balanzino *et al.*, 1994).

F4 Receptor Repertoire Defines Host Susceptibility

Studies in animals and humans have suggested that genetic factors influence the host's response to pathogens. The polymorphism that affect the response of individuals to the bacterial environment are candidate risk factors in multifactorial diseases. Host susceptibility to infections will be exemplified by the phenotypic variation of the glycoconjugate receptors in porcine species.

Porcine phenotypes are distinguished with respect to susceptibility or resistance to adhesion of F4 fimbriated ETEC to intestinal brush borders. At least five different phenotypes of piglets are distinguished using an in vitro enterocyte brush border adhesion test (Sellwood *et al.*, 1975). A genetic basis exists for these phenotypes (Sellwood *et al.*, 1975; Edfors-Lilja, 1991). The ability to bind F4-ETEC appears to be controlled by a dominant (susceptible) and recessive (resistant) alleles (Guérin *et al.*, 1993). Phenotyping and genetic linkage analysis are underway to find an association between markers and infection loci which contibute either to suceptibility or resistance. Resistance/susceptibility character is located on the chromosome 13 (Chowdhary *et al.*, 1993) and it has been observed a genetic linkage between the electrophoretic variants of serum transferrin and the locus encoding F4 receptors (Guérin *et al.*, 1993).

We examined the genetic basis of susceptibility through the TF gene encoding the polypeptide of TFintestinal receptor for F4ab. To determine allelic frequencies in various swine breeds, variants of the protein TF were firstly resolved, frequencies being not uniform among porcine populations (Mouricout *et al.*, 1997). 108 animals issued from 11 families and 14 litters were studied. It appeared that offsprings were informative for the linkage between an allelic form of Tf gene and the susceptibility character where receptors able to bind to F4ab fimbriated E. coli strains. RFLPs analysis showed that genotypes were inherited in a Mendelian fashion and no recombinants were observed. Adhesive phenotype is concomitantly associated to one particular allele TF, which was mainly present in adhesive animals. The presence of such allelic form might suggest that the gene product expressed in susceptible animals could differ from the glycoprotein found in the resistant ones.

Host Susceptibility to Bacterial Infection in Relation to Glycosyltransferase Activities

As F4 fimbriae are isolectins with different affinities for complex galactoside glycoconjugates, the pathway biosynthesis of receptors may be considered and differences in adhesion of F4 strains could reflect genetic differences (Table 3). However, resistance to infections is an end-point of many genetically controlled physiological processes.

Considering the susceptibility/resistance to F5 fimbriated ETEC, two receptor phenotypes have been found in porcine populations (Seignole *et al.*, 1991). The susceptibility/resistance character has been studied on an infant mouse diarrhoea model and appeared to be the consequence of differential glycosyltransferase activites in intestinal mucosa from resistant or susceptible animals (Grange *et al.*, 1996b). An elevated GM3-synthase activity is observed in the intestines of susceptible infant mice, at the age of high susceptibility.

Resistance/susceptibility of pig to infection by F18 E. coli is also genetically inheritcd (Bertschinger *et al.*, 1993). Resistance and susceptibility to colonization is controlled by two alleles, susceptibility being dominant over resistance. The locus for F18 receptor is located on chromosome 6 (Table 3), and is closely linked to blood group factors and erythrocyte enzyme system (Vögeli *et al.*, 1992). Two a(1,2) fucosyltransferas genes on porcine chromosome 6q11 are intimately linked to the blood group inhibitor (S) and *Escherichia coli* F18 receptor (ECF18R loci) (Meijerink *et al.*, 1997). Furthermore, stress is often associated with absence of F18 receptors.

The involvement of glycosylation machinery has been demonstrated for the interactions established between verotoxins and glycolipids (globotriaosyl ceramide Galα1-4Galβ1-4 glucosyl ceramide (Gb3) or globotetraosyl ceramide which contains an additional N-acetylgalactosamine residue linked in β1-3 at terminal position)

Table 3 Bacterial Receptors and Host's Genes involved in Animal Diseases.

Infectious agent	Receptors	locus	Chromosomal location
Escherichia coli F4	transferrin[a]	ECF4R	Pig 13q31b
Escherichia coli F5	gangliosides[c]	ECF5R	polygenic[d]
Escherichia coli F18		ECF18R	Pig 6q11[e]
Shigella toxin[d]	P histogroup system[f]	Rabbit	
	Gal α1-4Gal-containing glycolipids		
Salmonella	mannose receptor	Nramp (Bcg, Ity)	mouse 1[g]
	natural-resistance-macrophage		
	associated protein		

[a]Grange *et al.*, 1996a; [b]Guérin *et al.*, 1993; [c]Seignole *et al.*, 1991; [d]Duchet-Suchaux *et al.*, 1992; [e]Meijerink *et al.*, 1997; [f]Mobassaleh *et al.*, 1994; [g]review Blackwell 1996.

(Mobassaleh *et al.*, 1994). In rabbit, the site of pathology correlate with the presence of glycolipids receptors for toxins. The development of sensitivity to Shiga toxin coincides with the age-related appearance of Gb3 in the intestinal epithelial cells and increase α-galactosyl transferase and decrease α-galactosidase activities (Mobassaleh *et al.*, 1994). So these variations could define cell sensitivity, tissue tropism and explain the epidemiology of renal pathology which may follow VTEC infection.

CONCLUSION

The diversity of microorganisms reflects their abilities to occupy different targets, and the diversity in oligosaccharides represent a mosaic of potential sites for the adhesion of bacteria. Since infection is dynamic, bacteria must respond appropriately to changes in their environment. The involvement of lectin in bacterial adhesion is an essential process, because its confers resistance to cleanup by fluids, improvement of nutriment acquisition, increase in the delivery of toxins that causes tissue damage.

Mucins, prominent components of mucous secretions and mucus layers, play a protective part in preventing bacterial diseases (Amerongen *et al.*, 1995). Bacteria can be trapped on the mucous layers, subsequently cleared. Loss or decrese of the mucous layer could also be a factor that promotes adhesion or in some cases invasion. Cell invasion often involves triggering host cell transduction mechanisms to induce reorganisation of cytoskeleton at the site of attachment (De Ricke *et al.*, 1997; Palmer *et al.*, 1997).

Binding of bacteria to host cells is mediated through both lectin-carbohydrate, protein-glycoprotein or protein-glycolipid interactions. It appears that three-dimensional conformation is generally important for the binding. Bacterial lectins by their binding to carbohydrate of membrane proteins could lead to conformational changes in glycans and affect physiologic response of receptor glycoprotein or glycolipid. Bacterial binding to glycans of the cell surfaces, alters the state of glycosylation of the epithelium, may damage membranes, affects the turnover and loss of cepithelial cells. So, bacteria could create favorable niches by the modification of cellular differentiation.

Complementation of diet with antibiotics provoke an increase of bacterial multiple resistance and the risk of dissemination of one or more resistance plasmids. The emergence of new resistance mechanisms in bacteria raises concerns about the usefulness of preventive antibiotic therapies.

For these reasons, the development of preventive therapeutic strategies, specific, safe and effective, is crucial. The most effective way of countering *E. coli* multiplication without damage to the host or without selection of multiresistant strains, could be to inhibit the bacterial colonisation ability. The development of therapeutics against bacterial diseases requires a detailed understanding of the pathogenicity determinants and identification of host pathogen relationships which are relevant in an in vivo physiological situation. Characterization of the molecular basis of diseases by the identification of receptors is a real challenge for the control of the various types of pathogens and infections affecting youngs. Glycoconjugate receptor knowledge will facilitate not only the elucidation of host-pathogen interactions but will also provide new target molecules for modulating the host response. It could be possible to interfere with bacterial binding and thereby have an opportunity to develop anti-adhesive molecules for prevent or treat infections. Colonisation by pathogens can be prevented by different ways: by the saturation of fimbriae receptors by receptors analogs, by non-toxic plant lectins, by antibodies directed against fimbriae.

Understanding the molecular basis of these diseases (presence or absence of host receptors,...) is a real challenge. Site-specific recognition by bacterial colonization factors determines which tissue and host is targeted by each pathogen. Bacteria colonize the host by interactions established between their lectins and glycoconjugate receptors available on epithelial cells. Genetic variation, a prerequisite for successful selection, has been found in domestic animals exposed to bacterial infections. Works outline the necessity of advances in the understanding of the possible mechanism of genetic control of animal's resistance or susceptibility on the basis of the absence or presence of glycoconjugate receptors for pathogenic E. coli strains. Breeding for resistance to specific diseases can play a significant role alone or in combination with other control measures including disease eradication and vaccination. Development of disease resistance through animal selection is presently limited by insufficient understanding of bacteria-host interaction mechanisms, so better understanding can open new avenues to better assess the feasibility of increasing postweaning disease resistance. It will be important to determine the effects (positive or negative affects on health characters and performance traits), of a selection based on the resistance of piglets to neonatal colibacillosis and the resistance/susceptibility to postweaning infections.

ACKNOWLEDGEMENTS

Works performed in the Laboratory were supported by grants-in-aid from INRA and Limousin Region.

REFERENCES

Adam, E., Mitchell, A., Schumacher, D., Grant, G. and Schumacher, U. (1997) *Pseudomonas aeruginosa* II lectin stops human ciliary beating : therapeutic implications of fucose. *American Journal of Respiratory Critical Care Medicine*, **155**, 2102–2104.

Agin, T. and Wolf, M. (1997) Identification of a family of intimins common to *Escherichia coli* causing attaching-effacing lesions in rabbits, humans and swine. *Infection and Immunity*, **65**, 320–326.

Amako, K., Meno, Y. and Takade, A. (1988) Fine structures of the capsules of *Klebsellia pneumoniae* and *Escherichia coli* K1. *Journal of Bacteriology*, **170**, 4960–4962.

Amerongen, A., Bolscher, J. and Veerman, E. (1995) Salivary mucin: protective functions in relation to their diversity. *Glycobiology*, **5**, 733–740.

Bäker, A., Breimer, M., Samuelson, B. and Holgersson, J. (1997) Biochemical and enzymatic characterization of blood group ABH and related histo-group glycosphingolipids in the epithelial cells of porcine small intestine. *Glycobiology*, **7**, 943–953.

Balanzino, L., Barra, J.L., Montferran, C. and Cumar, F. (1994) Differential interaction of *Escherichia coli* heat-labile toxin and cholera toxin with pig intestinal brush border glycoproteins depending on their ABH and related blood group antigenic determinants. *Infection and Immunity*, **62**, 1460–1464.

Bertschinger, H, Stamm, M. and Vögeli, P. (1993) Inheritance of resistance to oedema disease in the pig: Experiments with an *Escherichia coli* strain expressing fimbriae 107. *Veterinary Microbiology*, **35**, 79–89.

Bijlsma, I. and Bouw, J. (1985). Inheritance of K88–mediated adhesion of *Escherichia coli* to jejunal brush borders in pigs : a genetic analysis. In B. Altura, A.M. Lefer and W. Schumer, (eds.), *Adhesion of K88–positive Escherichia coli to porcine intestinal: Immunological and genetic aspects*, Raven Press, New York, pp. 51–95.

Blackwell, J. (1996) Structure and function of the natural-resistance-associated macrophage protein (Nramp1), a candidate protein for infectious and autoimmune disease susceptibility. *Molecular medicine Today*, **2**, 205–211.

Bloch, C. and Orndorff, P. (1992) A key role for type 1 pili in enterobacterial communicability. *Molecular Microbiology*, **6**, 697–701.

Blomberg, L., Krivan, H.C., Cohen, P. and Conway, P.L. (1993) Piglet ileal mucouscontains protein and glycolipid (galactosylceramide) receptors specific for *Escherichia coli* K88 fimbriae. *Infection and Immunity*, **61**, 2526–2531.

Chan, N., Stangier, K., Sherburne, R., Taylor, D., Zhang, Y., Dovichi, N. and Palcic, M. (1995) The biosynthesis of Lewis X in *Helicobacter pylori*. *Glycobiology*, **5**, 683–688.

Chen, C.P., Song, S.C, Gilboa-Garber N. and Wu, A.M. (1997) The combining site of Gal-specific agglutinin (PA-IL) isolated from *Pseudomonas aeruginosa*. *Glycoconjugate Journal*, **14**, S62.

Chowdhary, B., Johansson, M., Chaudhary, R., Ellegren, H., Gu, F., Andersson, L. and Gustavson, I. (1993) In situ hybridization mapping and restriction fragment length polymorphism analysis of the porcine albumin (ALB) and transferrin (TF) in the pigs. *Animal Genetics*, **24**, 85–90.

Dean-Nystrom, E. (1995) Identification of intestinal receptors for enterotoxigenic *Escherichia coli*. *Methods in Enzymology*, **253**, 315–324.

Demuth, D., Golub, E. and Malamud, D. (1990) Streptococcal-host interactions : structural and functional analysis of a *Streptococcus sanguis* receptor for a human salivary glycoprotein. *Journal of Biological Chemistry*, **265**, 7120–7126.

De Rycke, J., Comtet, E., Chalareng, C., Boury, M., Tasca, C. and Milon, A. (1997) Enteropathogenic *Escherichia coli* O103 from rabbit elicits actin stress fibers and focal adhesions in Hela epithelial cells, cytopathic effects that are linked to an analog of the locus of enterocyte effacement. *Infection and Immunity*, **65**, 2555–65.

Di Martino, P., Girardeau, J.P., Der Vartanian, M., Joly, B. and Darfeuille-Michaud, A. (1997) The central variable V2 region of the CS31A major subunit is involved in the receptor-binding domain. *Infection and Immunity*, **65**, 609–616.

Donnenberg, M.S. and Kaper, J.B. (1992) Enteropathogenic *Escherichia coli*. *Infection and Immunity*, **60**, 3953–3961.

Duchet-Suchaux, M., Menanteau, P., Le Roux H., Elsen, J.M. and Lechopier, P. (1992) Genetic control of resistance to enterotoxigenic *Escherichia coli* in infant mice. *Microbial Pathogenesis*, **13**, 157–160.

Edfors-Lilja, J. (1991) *Escherichia coli* resistance in pigs. In J.B. Owen and R.F. Axford, (eds.), *Breeding for diseases resistance in farm animals*, C.A.B. Int., Wallingford, pp. 424–435.

Fakid, M., Murphy, T., Pattoli, M. and Berenson, C. (1997) Specific binding of *Haemophilus influenza* to minor gangliosides of human respiratory epithelial cells. *Infection and Immunity,* **65,** 1695–1700.

Finlay, B. and Cossart, P. (1997) Exploitation of mammalian host cell functions by bacterial pathogens. *Science,* **276,** 718–725.

Fontaine, A., Arondel, J. and Sansonetti, P.J. (1988) Role of Shiga toxin in the pathogenesis of bacillary disentery, studied by using a *Tox* - mutant of *Shigella dysenteria* 1*Infection and Immunity,* **56,** 3099–3109.

Gaastra, W. and Svennerholm, A.M. (1996) Colonization factors of human enterotoxigneic *Escherichia coli* (ETEC). *Trends in Microbiology,* **4,** 444–452.

Gilboa-Garber, N. (1994) *Pseudomonas aeruginosa* PA-I and PA-II lectins. In J. Beuth and G. Pulverer, (eds.), *Lectin blocking: new strategies for the prevention and therapy of tumor metastasis and infectious diseases,* Verlag, Stuttgart, pp.44–58.

Grange, P. and Mouricout, M. (1996a) Transferrin associated with the porcine intestinal mucosa is a receptor specific for K88ab fimbriae of *Escherichia coli. Infection and Immunity,* **64,** 606–610.

Grange P. and Mouricout, M. (1996b) Susceptibility of infant mice to F5 (K99) E. coli infection; differences in glycosyltransferase activities in intestinal mucosa of inbred CBA and DBA/2 strains. *Glycoconjugate Journal,* **13,** 45–52.

Groisman, E. and Ochman, H. (1997) How *Salmonella* became a pathogen. *Trends in Microbiology,* **5,** 343–349.

Guérin, G., Duval-Iflah, Y., Bonneau, M., Bertaud, M., Guillaume, P. and Ollivier, L. (1993) Evidence for linkage between K88ab, K88ac intestinal receptors to *Escherichia coli* and transferrin loci in pigs. *Animal Genetics,* **24,** 393–396.

Gupta, S.K., Masinick, S., Garrett, M. and Hazlett, L. (1997) *Pseudomonas aeruginosa.* liposaccharide binds galectin-3 and other human corneal epithelial proteins. *Infection and Immunity,* **65,** 2747–2753.

Haataja, S., Tikkanen, K., Nilsson, U., Magnusson, G., Karlsson, KA. and Finne, J. (1994) Oligosaccharide-receptor interaction of the Galα4Gal binding adhesin of *Streptococcus suis.* Combinating site architecture and characterization of two variant adhesin specificities. *The Journal of Biological Chemistry,* **269,** 27466–27472.

Hu, Z.L., Hasler-Rapacz, J., Huang, S.C. and Rapacz, J. (1993) Studies in Swine on inheritance and variation in expression of small intestinal receptors mediating adhesion of the K88 enteropathogenic *Escherichia coli* variants. *Journal of Heredity,* **84,** 157–165.

Jacques, M. (1996) Role of lipo-oligosaccharides and liposaccharides in bacterial adherence. *Trends in Microbiology,* **4,** 408–410.

Karlsson, K.A. (1995) Microbial recognition of target-cell glycoconjugates. *Current Opinion in Structural Biology* **5,** 622–635.

Khan, A. and Schifferli, D. (1994) A minor 987P protein different from the structural fimbrial subunit is the adhesin. *Infection and Immunity,* **62,** 4223–4243.

Khan, A., Johnston, N., Goldfine, H. and Schifferli, D. (1996) Porcine 987P glycolipid receptors on intestinal brush borders and their cognate bacterial ligands. *Infection and Immunity,* **62,** 4223–4243.

Kyogashima, M., Ginsburg, V. and Krivan, H. (1989) *Escherichia coli* K99 binds to N-glycolylsialoparagloboside and N-glycolyl-GM3 found in piglet small intestine. *Archives of Biochemistry and Biophysics,* **270,** 391–397.

Lingwood, C. (1996) Role of verotoxin receptors in pathogenesis. *Trends in Microbiology,* **4,** 147–153.

Mandrell, R., Apicella, M., Lindstedt, R. and Leffler, H. (1994) Possible interactions between animal lectins and bacterial carbohydrates. *Methods of Enzymology,* **236,** 231–241.

Meijerink, E., Fries, R., Vogeli, P., Masabanda, J., Wigger, G., Stricker, C., Neuenschwander, S., Bertschinger, Hu. and Stranzinger, G. (1997) Two a(1,2) fucosyltransferas genes on porcine chromosome 6q11 are closely linked to the blood group inhibitor (S) and *Escherichia coli* F18 receptor (*ECF18R*) loci. *Mammalian Genome,* **8,** 736–741.

Mey, A., Hmama, Z., Normier, G. and Revillard, J.P. (1996) The animal lectin galectin-3 interacts with bacterial liposaccharides via two independant sites. *Journal of Immunology,* **156,** 1572–1577.

Mobassaleh, M., Koul, O., Mishra, K., Mc Cluer R.H. and Keusch, G.T. (1994) Developmentally regulated Gb3 galactosyltransferase and alpha-galactosidase determine Shiga Toxin receptors in intestine. *American Journal of Physiology,* **267,** G618–G624.

Mol, O. and Oudega, B. (1997) Molecular and structural aspects of fimbriae biosynthesis and assembly in *Escherichia coli. FEMS Microbiology Reviews,* **19,** 25–52.

Mouricout, M. (1991) Swine and cattle enterotoxigenic *Escherichia coli* mediated diarrhea. Development of therapies based on inhibition of bacterial-host interactions. *European Journal of Epidemiology,* **7,** 588–604.

Mouricout M. (1997) Interactions between pathogens and the host: An assortment of bacterial lectins and a set of glycoconjugate receptors. *Advances in Experimental Medicine and Biology,* **412,** 109–123.

Mouricout, M. and Julien, R. (1987) Pilus-mediated binding of bovine enterotoxigenic *Escherichia coli* to calf small intestinal mucins. *Infection and Immunity,* **55,** 1216–1223.

Mouricout, M., Milhavet, M., Durié, C. and Grange, P. (1995) Characterization of glycoprotein glycan receptors for *Escherichia coli* F17 fimbrial lectin. *Microbial Pathogenesis,* **18,** 297–306.

Ofek, I., Madison, B. and Abraham, S. (1994) Bacterial lectins as adhesins. *Zentralblatt für Bakteriologie,* **S25,** 16–25.

Payne, D., O'Reilly, M. and Williamson, D. (1993) The K88 fimbrial adhesin of enterotoxigenic *Escherichia coli* binds to β1–linked galactosyl residues in glycosphingolipids. *Infection and Immunity,* **61,**. 3673–3677.

Palmer, L., Reilly, T., Utsalo, S. and Donneberg, M. (1997) Internalization of *Escherichia coli* by human renal epithelial cells is associated with tyrosine phosphorylation of specific host cell proteins. *Infection and Immunity,* **65,** 2570–2573.

Paradis, S.E., Dubreuil, D. Rioux, S., Gottschalk, M. and Jacques, M. (1994) High molecular-mass lipopolysaccharides are involved in Actinobacillus pleuropneumoniae adherence to porcine respiratory tract cells.*Infection and Immunity,* **62,** 3311–3319.

Piotrowski, J., Slomiany, A., Murty, V.L., Fekete, Z. and Slomiany, B.L. (1991) Inhibition of *Helicobacter pylori* by sulfated gastric mucin. *Biochemistry International,* **4,** 749–756.

Porat, N., Apicella, M.A. and Blake, M. (1995a) *Neisseria gonorrhoeae* utilizes and enhances the biosynthesis of the asialoglycoprotein receptor expressed on the surface of the hepatic HepG2 cell line. *Infection and Immunity,* **63,** 1498–1506.

Porat, N., Apicella, M.A. and Blake, M. (1995b) A lipooligosaccharide-binding site on HepG2 cells similar to the gonococcal opacity-associated surface protein Opa. *Infection and Immunity,* **63,** 2164–2172.

Rapacz, J. and Hasler-Rapacz, J. (1986) Polymorphism and inheritance of swine small intestinal receptors mediating adhesion of three serological variants of *Escherichia coli*-producing K88 pilus antigen. *Animal Genetics,* **17,** 305–321.

Roche, R.J. and Moxon, E. (1995) Phenotypic variation of carbohydrate surface antigens and the pathogenesis of *Haemophilus influenzae* infections. *Trends in Microbiology,* **3,** 304–309.

Rostand, K. and Esko, J. (1997) Microbial adherence to and invasion through proteoglycans. *Infection and Immunity,* **65,** 1–8.

Saarela, S., Westerlund-Wikström, B., Rhen, M. and Korhonen, T. (1996) The GafD protein of the G (F17) fimbrial complex confers adhesiveness of *Escherichia coli* to laminin. *Infection and Immunity,* **64,** 2857–2860.

Seignole, D., Mouricout, M., Duval-Iflah, Y., Quintard, B. and Julien, R. (1991) Adhesion of K99 fimbriated *Escherichia coli* to pig intestinal epithelium: correlation of adhesive and non-adhesive phenotypes with the sialoglycolipid content. *Journal of General Microbiology,* **137,** 1591–1601.

Seignole, D., Grange, P., Duval-Iflah, Y. and Mouricout, M. (1994) Characterization of O-glycan moieties of the 210 and 240 kDa pig intestinal receptors for *E. coli* K88ac fimbriae. *MicroBiology,* **140,** 2467–2473.

Sellwood, R., Gibbons, R., Jones, J.W. and Rutter, J.M. (1975) Adhesion of enteropathogenic *Escherichia coli* to pig intestinal brush borders: the existence of two pig phenotypes. *Journal of Medical Microbiology,* **8,** 405–411.

Sokurenko, E.V., Courtney, H., Ohman, D., Klemm, P. and Hasty, D. (1994) Fim H family of type 1 fimbrial adhesins: functional heterogeneity due to minor sequence variations among Fim H genes.*Journal of Bacteriology,* **176,** 748–755.

Sokurenko, E.V.,. Courtney, H., Maslow, J., Siitonen, A. and Hasty, D. (1995) *Journal of Bacteriology,* **177,** 3680–3686.

Sokurenko, E.V., Chesnokova, V., Doyle, R. and Hasty, D. (1997) Diversity of the *Escherichia coli* type 1 fimbrial lectin. *The Journal of Biological Chemistry,* **272,** 17880–17886.

Snellings, N.J., Tall, B.D. and Venkatesan, M.M. (1997) Characterization of Shigella type 1 fimbriae: expression, fim A sequence and phase variation. *Infection and Immunity,* **65,** 2462–2467.

Strömberg, N., Nyholm, P.G., Pascher, J. and Normak, S. (1991) Saccharide orientation at the cell surface affects glycolipid receptor function. *Proceedings of National Academy of Sciences, USA,* **88,** 9340–9344.

Teneberg, S., Willemsen, P., de Graaf, F.K. and Karlsson, K.A. (1990) Receptor-active glycolipids of epithelial cells of the small intestine of young and adult pigs in relation to susceptibility to infection with *Escherichia coli* K99. *FEBS Letters,* **263,** 333–339.

Van Alphen, L., Geelen-van den Broek, L. and Ryan, J.L. (1991) Blocking of fimbriae-mediated adherence of *Haemophilus influenza* by sialyl gangliosides. *Infection and Immunity,* **55,** 2355–2358.

Van Halbeek, H., Gerwig, G., Vliegenthart, J.G., Smits, H., Van Kerkhof, P. and Kramer, M. (1983). Terminal (α1–4)-linked N-acetylglucosamine: a characteristic constituent of duodenal-gland mucous glycoproteins in rat and pig. A high-resolution H-^{1}NMR study. *Biochemica et Biophysica Acta,* **747,** 107–116.

Veerman, E.C., Bank, C.M., Namavar, F., Appelmelk, B.J., Bolscher, J.G. and Nieuw Amerongen, A. (1997) Sulfated glycans on oral mucin as receptors for *Helicobacter pylori. Glycobiology,* **7,** 737–743.

Vögeli, P., Delacretz, H. , Kuhn, B., Stamm, M., Bertschinger, H. and Stranzinger, G. (1992) Association between the H blood group system and the GPI red cell enzyme system and the locus specifying receptors of an *Escherichia coli* strain expressing fimbriae F107. *Animal Genetics,* **23,** 93.

Whitfield, C. (1988) Bacterial extracellular polysaccharides. *Canadian Journal of Microbiology,* **34,** 415–420.

Yang, Z., Bergström, J. and Karlsson, K.A. (1994) Glycoproteins with Gal α4Gal are absent from human erythrocyte membranes, indicating that glycolipids are the sole carriers of blood group P activities. *The Journal of Biological Chemistry,* **269,** 14620–14624.

10. LECTIN-MEDIATED INTERACTION OF PARASITES WITH HOST CELLS

NAJMA BHAT AND HONORINE D. WARD

Division of Geographic Medicine and Infectious Diseases, Department of Medicine, New England Medical Center, Tufts University School of Medicine, Boston, MA 02111, USA

INTRODUCTION

Lectins are carbohydrate-binding proteins of non-immune origin which, among other functions, mediate cell-cell interactions by binding to complementary carbohydrate moieties (Sharon and Lis, 1989). Increasing structural and functional understanding of this group of proteins has led to the realization that in addition to their carbohydrate-binding activity, biological functions of lectins may also be mediated by interaction with non-carbohydrate ligands (Barondes, 1988). Cell-cell interactions mediated by lectins include those between parasites and host cells (Pereira, 1986; Ward, 1996). One of the hallmarks of host-parasite interactions is specificity, which may be manifested at the level of host species, tissue site, cell type or receptor (Pereira, 1986). This selectivity suggests that specific parasite molecules and complementary ligands on host cells mediate these interactions. The specificity of lectins together with the diversity of carbohydrate residues confers these molecules with recognition and discrimination properties, which enable them to contribute to the selectivity of the host cell-parasite interaction. Interactions between parasite and host cells are complex processes, which involve multiple receptors and complementary ligands on both cell types. Examples of such interactions include recognition, adhesion, invasion, cell signaling and cytotoxicity.

Parasite lectins may bind to carbohydrate residues on a number of cell-associated glycoconjugates, including glycoproteins or mucins. A number of parasite proteins have also been shown to mediate intercellular interactions by binding to glycosaminoglycans (GAGs), such as heparan sulphate or chondroitin sulphate present on proteoglycans on the cell surface or in the extracellular matrix. Proteoglycans consist of a protein core, with covalently attached GAG chains. GAGs are linear polysaccharides that consist of repeating disaccharide units (Jackson et al., 1991).

Lectins or carbohydrate-binding proteins have been described in a number of parasites, including protozoa. Protozoan parasites comprise a diverse group of unicellular, lower eukaryotes, which exist in multiple developmental stages, frequently in more than one host and often transmitted by insect vectors. They may have simple or complex life cycles involving sexual and/or asexual reproduction, which take place in extra or intracellular locations. Pathogenic protozoa cause a number of different diseases in humans, which are responsible for significant mortality and morbidity worldwide. This chapter will focus on lectins in pathogenic protozoa of human importance and their role in mediating host-parasite interactions. Many of these may not be "lectins" in the classical sense. However, the term "lectin," in this chapter is used in the broader sense to encompass proteins, whose biological functions may be mediated at least in part, by their interaction with specific carbohydrate residues (Barondes, 1988; Sharon and Lis, 1989).

LECTINS IN PATHOGENIC PROTOZOA OF HUMANS

Intestinal and Mucosal Protozoa

Entameba histolytica

The enteric protozoan *Entameba histolytica*, is the causative agent of amebiasis, a spectrum of diseases, which include amebic dysentery and amebic liver abscess. This parasite infects an estimated 500 million people worldwide, causing invasive disease in 50 million of them and resulting in up to 100,000 deaths annually (Walsh, 1986). Infection is initiated by ingestion of the cyst form of the parasite, which excysts in the small or large intestine to release trophozoites. Attachment of trophozoites to colonic epithelial cells leads to contact mediated-lysis of host cells, followed by invasion and in some cases, dissemination of the parasite to extraintestinal sites. Two lectins, one specific for Galactose/N-acetylgalactosamine (Gal/GalNAc) and the other for chitotriose have been identified in *E. histolytica*.

1. Gal/GalNAc-binding Lectin

The Gal/GalNAc-specific lectin of *Entameba histolytica* is one of the best characterized of all the parasite lectins. This protein has been convincingly shown to be a virulence determinant involved in a number of pathogenic mechanisms including adhesion, cytotoxicity and resistance to complement (Petri and Mann, 1993; McCoy *et al.*, 1994a; Petri, 1996). The lectin was first described by Ravdin *et al.*, who showed that adherence of trophozoites to CHO cells could be specifically inhibited by the monosaccharides, Gal and GalNAc (Ravdin, 1981). Inhibition of adherence was 1000 time more potent with terminal Gal-containing glycoproteins such as asialofetuin and asialoorosomucoid than Gal monomers (Petri *et al.*, 1987). The specificity of the lectin for Gal and GalNAc residues on host cells was confirmed by studies employing CHO cell glycosylation mutants defective in the production of various N- and O-linked oligosaccharides (Li *et al.*, 1988; Li *et al.*, 1989; Ravdin *et al.*, 1989; Ravdin and Murphy, 1992). Trophozoite attachment to mutant CHO cells, which express increased Gal residues at the non-reducing termini, was almost twice that to the wild type cells. Sialidase treatment of the wild type cells, leading to increased exposure of terminal Gal residues, resulted in enhanced attachment of trophozoites. There was minimal adherence to CHO mutants lacking lactosamine units on β1-6 branched N-linked carbohydrates. Binding of purified lectin to CHO cells was inhibited by α as well as β linked Gal-containing saccharides, suggesting that there was no preference for specific linkage of the Gal residues (Saffer and Petri, 1991). The purified lectin has the highest affinity for polyvalent N-acetylgalactosaminides which were 140,000 and 500,000 times more potent than the monosaccharides GalNAc and Gal respectively (Adler *et al.*, 1995). Binding to these ligands was enhanced by divalent cations and occurred within a broad pH range of 6 to 9.

The lectin, purified by galactose affinity or immunoaffinity chromatography, is a 260 kDa, membrane-associated heterodimeric glycoprotein, with disulfide-linked subunits of 170 kDa and 31–35 kDa (Petri et al., 1987; Petri et al., 1989). The native protein exists as 440-660 kDa multimeric aggregates (Petri et al., 1989). It is present on the surface membrane of trophozoites and is also secreted into the extracellular medium (Petri et al., 1987). Two distinct gene families, one encoding the 170 kDa heavy subunit and the other the 31-35 kDa light subunit have been identified (Mann et al., 1991; Tannich et al., 1991; Tannich et al., 1992, McCoy et al., 1993; Purdy et al., 1993; Ramakrishnan et al., 1996).

The 170 kDa subunit is encoded by a family of 5 hgl genes located at 5 loci in the genome (Ramakrishnan et al., 1996). Sequence analysis of the 170 kDa subunit revealed multiple domains including an amino-terminal signal sequence, a transmembrane domain, a cytoplasmic tail, and an extracellular domain, which contains cysteine-rich and cysteine-poor regions (Mann et al., 1991). The cysteine-rich region appears to mediate the virulence functions of the lectin (Mann et al., 1993). The extracellular domain contains a number of potential N-linked glycosylation sites. Biochemical studies confirmed that the 170 kDa subunit is glycosylated with ~6% of the molecular mass being carbohydrate (Mann et al., 1991). The promoter driving the 170 kDa subunit gene has been analyzed and found to be located within a 1.35 kb intergenic region upstream of the lectin gene (Buss et al., 1995). Recent studies of lectin gene expression showed that regulation of the hgl5 lectin promoter is mediated by a nuclear protein which recognizes the URE3 DNA sequence motif (Purdy et al., 1996; Gilchrist et al., 1998).

The light subunit is composed of two related isoforms 31 and 35 kDa in molecular weight (McCoy et al., 1993; McCoy et al., 1994b). The 31 kDa isoforms but not the 35 kDa isoform is glycophosphatidylinositol (GPI)-anchored and the latter is more heavily glycosylated than the former (McCoy et al., 1993). The light subunit is encoded by the lgl family of genes, located at 6 loci in the genome (Tannich et al., 1992; McCoy et al., 1993; Ramakrishnan et al., 1996) . While monoclonal antibodies to the heavy subunit inhibit (or enhance) adherence, those to the light subunit do not, suggesting that the former contains the carbohydrate recognition domain (Petri et al., 1990; Petri and Ravdin, 1991; Petri and Mann, 1993; McCoy et al., 1994b).

The primary function of the Gal/GalNAc lectin is to mediate adherence of trophozoites to host cells. This was shown by inhibition of adherence of trophozoites to host cells such as CHO cells, colonic epithelial cells, neutrophils and macrophages, by carbohydrates or antibodies specific for the lectin or the purified lectin itself (Petri et al., 1987; Petri et al., 1989). In addition to mediating attachment of the parasite to cells, the lectin has also been shown to bind specifically to colonic mucins which are rich in Gal and GalNAc residues (Chadee et al., 1987; Chadee et al., 1988; Burchard et al., 1993). This property of the lectin may contribute to colonization of the intestine by the parasite.

Another function of the lectin is to mediate contact dependent cytolysis of host cells, which occurs following adherence of trophozoites. This was shown by the finding that cytolysis of CHO cells by trophozoites could be inhibited by Gal and GalNAc (Ravdin et al., 1989). Furthermore, CHO cell mutants deficient in terminal Gal residues were resistant to cytolysis, whereas mutants with increased Gal residues were more susceptible to amebic killing (Ravdin et al., 1989). Lectin mediated cytolysis is associated with a rise in intracellular calcium (Ravdin et al., 1988) and has been shown to occur via a Bcl-2 independent apoptotic mechanism (Ragland et al., 1994). A study by Leroy et al. suggested that contact-dependent transfer of the lectin from trophozoites to the lateral surface of enterocytes in culture occurred prior to cytolysis (Leroy et al., 1995). In addition to these functions the Gal/GalNAc lectin also confers resistance to lysis of the parasite by complement (Braga et al., 1992). The lectin also appears to be involved in cell signaling as shown by actin polymerization at the site of contact of trophozoites with liposomes containing terminal Gal (Bailey et al., 1990).

The role of this lectin in the pathogenesis of amebiasis has lead to a number of studies investigating its use as a potential vaccine candidate. The 170 kDa subunit is highly immunogenic, being recognized by > 90% of sera from patients with amebic liver abscess (Ravdin et al., 1990). Immunization with the native lectin resulted in protection of 67% of animals in a gerbil model of amebic liver abscess (Petri and Ravdin, 1991). However, animals which were not protected showed exacerbation of disease, a finding that is consistent with an earlier study showing that a subset of monoclonal antibodies to the lectin enhanced adherence (Petri et al., 1990). In two subsequent studies, immunization with recombinant fusion proteins which, included the cysteine-rich domain of the 170 kDa subunit, also resulted in significant protection, (without disease exacerbation in unprotected animals) in the same model (Zhang and Stanley, 1994; Soong et al., 1995). Lotter et al. showed that protective immunity induced by immunization with a recombinant protein containing a portion of the cysteine-rich domain correlated with the development of an antibody response to a 25 amino acid-epitope in this region (Lotter and Tannich, 1997). Conversely, exacerbation of disease was linked to an antibody response to a cysteine-poor amino terminal domain of the lectin.

The mechanisms by which the lectin induces protective immunity is not clearly understood, but appears to involve both cell-mediated as well as humoral responses. Lymphocytes from animals immunized with the lectin or from patients following recovery from amebic infection, proliferate (Salata and Ravdin, 1986; Velazquez et al., 1995) and produce IL-2 and IFN-α in response to stimulation with the lectin in vitro (Schain et al., 1992; Schain et al., 1995). Seguin et al. showed that bone marrow-derived murine macrophages produce TNF-α in response to stimulation with the 170 kDa subunit and identified an amino acid region in the cysteine-rich domain as being responsible for this activity (Seguin et al., 1995). The same investigators later showed that this TNF-α-stimulating region also stimulates IFN-γ-primed macrophages to kill amebic trophozoites via production of nitric oxide (Seguin et al., 1997). The possible role of mucosal antibodies in protection has also been studied. Anti-lectin sIgA has been shown to be present in saliva of patients with amebiasis (Carrero et al., 1994; Kelsall and Ravdin, 1995) and to inhibit trophozoite adherence in vitro (Carrero et al., 1994). Oral immunization of mice with a recombinant cysteine-rich domain of the lectin elicited high titers of sIgA which was also inhibitory for trophozoites adherence in vitro (Beving et al., 1996). Evidence for the role of systemic antibodies to the lectin in protective immunity is provided by the finding that SCID mice passively immunized with serum from rabbits vaccinated with recombinant cysteine-rich domains were protected to the same degree as those actively immunized with the same proteins (Lotter and Tannich, 1997).

2. Chitotriose-binding lectin

The presence of another lectin in *E. histolytica* trophozoites was shown using a hemagglutination assay (Kobiler and Mirelman, 1980). In this system, lectin activity was inhibitable by N-acetylglucosamine (GlcNAc) oligomers such as chitotriose. Adherence of radiolabeled trophozoites to glutaraldehyde-fixed monolayers of intestinal epithelial cells could also be inhibited by GlcNAc-containing glycoconjugates, wheatgerm agglutinin as well as membrane fractions enriched in the lectin, suggesting a possible role for the lectin in mediating attachment (Kobiler and Mirelman, 1981). A lectin with similar carbohydrate specificity was purified by Sepharose-4B gel filtration chromatography followed by elution from SDS polyacrylamide gels and shown to have a molecular weight of 220 kDa (Rosales-Encina *et al.*, 1987). The protein is rich in hydrophobic and acidic amino acids and contains 9% carbohydrate by weight (Rosales-Encina *et al.*, 1987). The purified protein was shown to bind to fixed monolayers of MDCK cells and to inhibit adhesion of trophozoites to the cells (Rosales-Encina *et al.*, 1987). Monoclonal antibodies to the lectin bound to the membrane of live or fixed trophozoites, reduced adherence of the parasite to host cells and inhibited erythrophagocytosis (Meza *et al.*, 1987). Polyclonal murine antibodies to trophozoite membrane preparations and to the purified lectin were used to obtain a recombinant cDNA clone expressing the protein (Talamas-Rohana *et al.*, 1995). Spleen cells from mice immunized with the fusion protein encoded by this clone proliferated in response to the native and recombinant proteins. Cells from mice immunized with chemically obtained peptides from the native protein, which induced a proliferative response, secreted IL-2 and IFN-γ, characteristic of a TH1-type response. However, cells from mice immunized with the intact molecule secreted IL-4 and IL-10, typical of a TH2 response (Talamas-Rohana *et al.*, 1995).

Giardia lamblia

Giardia lamblia is an intestinal protozoan which causes diarrheal disease worldwide (Meyer and Radulescu, 1979). This parasite exists in two developmental forms, trophozoite and cyst. Infection is initiated by ingestion of the infective cyst form, which undergoes excystation in the small intestine to release the trophozoite stage. The trophozoite replicates asexually by binary fission and selectively colonizes the small intestine, where it attaches to intestinal epithelial cells and causes diarrhea by unknown mechanisms. Lectins with specificity for mannosyl and phosphomannosyl residues have been described in *Giardia*.

1. Mannose (Man)/Mannose-6-phosphate (Man-6-P)-binding lectin/s

Surface membrane associated lectin activity, was detected in trophozoites of *Giardia lamblia* using mixed agglutination and hemagglutination assays by several investigators (Lev *et al.*, 1986; Farthing *et al.*, 1986; Ward *et al.*, 1987a; Sreenivas *et al.*, 1995). It is possible that the lectin activity described in these reports represents the same protein, although this remains to be proved. Lev *et al.* found that hemagglutination was specifically activated by limited proteolysis with trypsin, a protease that is present at the site of infection (Lev *et al.*, 1986). The trypsin-activated lectin, named taglin, also agglutinated intestinal epithelial cells which are the cells the parasite adheres to *in vivo*, and in addition, bound to isolated brush border membranes of these cells (Ward *et al.*, 1987b). Sreenivas *et al.* also found that hemagglutination was maximal after trypsinization (Sreenivas *et al.*, 1995). Carbohydrate specificity studies showed hemagglutination inhibition by Man, α-methyl mannoside and Man-6-P (Farthing *et al.*, 1986; Lev *et al.*, 1986; Ward *et al.*, 1987a; Sreenivas *et al.*, 1995). Man-6-phosphate was seven times as potent an inhibitor of taglin as Man (Ward *et al.*, 1987a). Phosphomannose-containing compounds, such as yeast phosphomannans and bovine testicular β-galactosidase inhibited taglin-induced hemagglutination, confirming the specificity of taglin for phosphomannosyl residues (Ward *et al.*, 1987a).

Taglin-induced hemagglutination was optimal at pH 6.5 and was dependent on divalent cations Ca^{++} and Mn^{++} (Ward *et al.*, 1987a). A monoclonal antibody to taglin reacted with the surface membrane of live trophozoites and recognized a protein of 28/30 kDa in lysates of *Giardia* trophozoites, by immunoblotting (Ward *et al.*, 1987a). This finding was confirmed by erythrocyte binding to *G. lamblia* proteins separated by SDS polyacrylamide gel electrophoresis, transferred to nitrocellulose and renatured. There was specific erythrocyte binding to protein bands in the same molecular weight range as those recognized by the monoclonal antibody.

The role of Man/Man-6-P-specific lectin/s in mediating adherence was suggested by studies of trophozoite attachment to intestinal epithelial cells *in vitro*. Attachment of trophozoites to rat intestinal cells was inhibited by Man and Man-containing compounds (Inge *et al.*, 1988). Trophozoite attachment to intestinal IEC-6 cells was also reduced by a monoclonal antibody to taglin (Keusch *et al.*, 1991). In addition, clones of the WB strain of the parasite which attach better to intestinal cells also expressed higher amounts of lectin activity as determined by hemagglutination, as well as by scanning densitometry of immunoblots of lysates of the different clones. (Keusch *et al.*, 1991). Attachment of trophozoites to intestinal epithelial Caco-2 cells was also inhibited by Man and Man-6-P (Katelaris *et al.*, 1995). Taken together, these findings suggest recognition and attachment of trophozoites to host cells may be mediated by Man/Man-6-P-specific lectin/s.

Cryptosporidium parvum

Cryptosporidium parvum is an intestinal protozoan which, is currently recognized as a significant cause of gastrointestinal disease worldwide (Current and Garcia, 1991). Infection with this parasite in immunocompetent individuals is frequently asymptomatic or self-limiting. However in immunocompromised hosts such as patients with the acquired immunodeficiency syndrome (AIDS), *C. parvum* may cause severe and often fatal diarrhea and wasting. The parasite exists in a number of developmental stages and has a complex life cycle. Infection is initiated by the oocyst, which undergoes excystation to release sporozoites. Sporozoites attach to and invade intestinal epithelial cells where the parasite undergoes further intracellular development via sexual as well as asexual cycles. *Cryptosporidium* sporozoites express a Gal/GalNAc-specific lectin.

1. Gal/GalNAc- binding lectin

Lectin activity in *C. parvum* sporozoites was identified using mixed agglutination and hemagglutination assays (Thea *et al.*, 1992; Joe *et al.*, 1994; Keusch *et al.*, 1995; Ward and Cevallos, 1998). Hemagglutination was optimal at a temperature of 4°C, a pH of 7.5, and in the presence of the divalent cations Ca^{++} and Mn^{++}. The sugar specificity of the lectin was determined by hemagglutination inhibition using a wide range of simple and substituted mono-, and disaccharides as well as glycoproteins. Of 12 monosaccharides tested the lectin was most specific for the monosaccharides galactose (Gal) and N-acetyl galactosamine (GalNAc) which inhibited lectin activity at MIC's of 4 and 14 mM, respectively (Joe *et al.*, 1994) . Both α as well as β-linked methyl and paranitrophenyl galactosides inhibited hemagglutination, suggesting that there was no specific preference for linkage of the Gal residues. Gal- and GalNAc-containing disaccharides were potent inhibitors of hemagglutination, the best being Gal (β1-3) GalNAc and Gal (α1-4) Gal (Joe *et al.*, 1994; Ward and Cevallos, 1998). The disaccharide Gal (α1-4) Gal is present in P1 glycoprotein, a mucin-like glycoprotein found in hydatid cyst fluid, which is related to the P1 blood group antigen. This glycoprotein as well as bovine submaxillary mucin and hog gastric mucin inhibited lectin activity at very low concentrations. This was of interest since the parasite exists in a mucin-rich environment. Lectin-mediated interaction of sporozoites with intestinal mucins was studied using a solid-phase binding assay (Bhat and Ward, 1998). Mucins were isolated from intestinal epithelial cells, by gel filtration chromatography and density-gradient centrifugation. Sporozoites bound to purified intestinal mucins and this binding could be inhibited by Gal and GalNAc containing compounds, suggesting that the interaction was mediated by the lectin.

The role of this lectin in mediating attachment to host cells was studied using lectin-specific saccharides and glycoproteins to inhibit adherence of sporozoites to intestinal epithelial cells. The monosaccharides Gal and GalNAc, the disaccharides, lactose and melibiose as well as the mucins bovine submaxillary mucin and hog gastric mucin inhibited attachment, further suggesting the possibility that the lectin is involved (Keusch *et al.*, 1995; Joe *et al.*, 1998). In addition to inhibiting attachment, Gal and GalNAc-containing glycoconjugates inhibited infection of intestinal epithelial cell monolayers by oocysts. The role of Gal residues in mediating attachment was further studied using attachment of sporozoites to CHO cell glycosylation mutants. Attachment of sporozoites to the Lec-2 mutant, which expresses increased terminal Gal residues, was increased compared to the parent strain. In addition, sialidase treatment of intestinal epithelial cells, leading to exposure of terminal Gal residues, resulted in increased attachment of sporozoites to these cells (Joe and Ward, 1998).

Trichomonas vaginalis

Trichomonas vaginalis is a flagellated protozoan, which is responsible for trichomoniasis, one of the most common sexually transmitted diseases which is widely prevalent worldwide (Catterall, 1972). An estimated 180 million cases of trichomoniasis occur annually. The parasite exists in only one developmental stage, which replicates by binary fission. *Trichomonas vaginalis* exerts its pathogenic effects by adhering to the vaginal mucosa and causing a cytopathic effect.

1. GlcNAc /Man-binding lectin

The presence of a carbohydrate-inhibitable cytopathic effect of *T. vaginalis* was described by Roussel *et al.* (Roussel *et al.*, 1991). Using an in vitro model of infection in McCoy cells, these authors showed that the cytopathic effect induced by the parasite could be inhibited by GlcNAc and to a lesser extent by Man, suggesting the presence of a lectin specific for these sugars on the parasite. A subsequent study by Bonilha *et al.* showed the presence of carbohydrate-inhibitable adherence of the parasite to CHO cells (Bonilha *et al.*, 1995). Adherence to CHO cell mutants expressing increased Gal, Man and GlcNAc residues was enhanced compared to the parental cells and this adhesion could be inhibited by the presence of exogenously added Gal, Man and GlcNAc.

Acanthameba castellani

The protozoan parasite *Acanthameba castellani* is a free-living ameba, which is widely distributed in nature and frequently isolated from soil, air, swimming pools, hot tubs etc. (Ma *et al.*, 1990). This parasite causes keratitis, a painful, vision threatening infection of the cornea in contact lens wearers (Auran *et al.*, 1987). A Man-binding protein has been described in this parasite.

1. Man-binding protein

Yang *et al.* showed that *Acanthameba castellani* parasites metabolically labeled with ^{35}S methionine adhered to Man-containing glycoproteins isolated from corneal epithelium and to the neoglycoprotein Man-BSA, using an electrophoresis-blot overlay assay. Adherence also occurred to immobilized Man-BSA in a dose-dependent manner (Yang *et al.*, 1997). Binding of the parasite to Man-containing glycoproteins could be inhibited by methyl-α-mannoside. A 136 kDa Man-binding membrane-associated protein was isolated from extracts of the parasite by affinity chromatography. This protein is postulated to mediate attachment of the parasite to Man-containing glycoproteins of the cornea.

Blood and Tissue Protozoa

Plasmodium

Plasmodium is a mosquito-borne protozoan parasite which, is the causative agent of malaria. This disease which is estimated to affect up to 300 million people worldwide annually, results in the deaths of up to 2 million of these, mainly children, (Greenwood, 1987; Sturchler, 1989). Four species of the parasite infect humans, *Plasmodium falciparum, Plasmodium vivax, Plasmodium ovale* and *Plasmodium malariae*. *Plasmodium falciparum* is the most virulent of these, causing cerebral malaria, which is responsible for much of the mortality. The parasite has a complex life cycle (which varies according to the species) and exists in a number of asexual and sexual developmental stages in the human host as well as in the mosquito vector. Infection is initiated by the bite of an infected mosquito which, releases sporozoites into the blood stream. Within minutes, sporozoites attach to and invade hepatocytes, a primary event in establishing infection. In the hepatocytes, the parasite undergoes an exoerythrocytic cycle of replication resulting in the release of merozoites. Another crucial event in the pathogenesis of malaria is the subsequent invasion of red cells by merozoites, leading to an erythrocytic cycle. Erythrocytes parasitized by *Plasmodium falciparum* adhere to the endothelium of post-capillary venules or to uninfected erythrocytes forming rosettes and both of these events are believed to contribute to the pathogenesis of cerebral malaria (Pasloske and Howard, 1994a). Sialic-acid binding proteins (in *P. falciparum*) as well as those binding to GAGs have been implicated in mediating interactions between merozoites and erythrocytes, sporozoites and hepatocytes as well as between parasitized erythrocytes and endothelium and parasitized and uninfected erythrocytes.

1. Sialic-acid binding proteins

a. EBA-175
EBA-175 is a sialic-acid-binding protein of *P. falciparum* which, is believed to mediate merozoite invasion of erythrocytes (Camus and Hadley, 1985). This protein, (also called *P. falciparum* sialic acid-binding protein) is now considered to be a member of a family of erythrocyte-binding proteins of malaria parasites involved in adhesion (Adams *et al.*, 1992). EBA-175 is released into the culture supernatant of infected erythrocytes and binds to uninfected erythrocytes. EBA-175 also binds to merozoites, leading to the hypothesis that it might function as a "bridge" between merozoite and erythrocytes. The role of sialic acid in binding was initially suggested by the finding that sialidase treatment of erythrocytes reduced invasion and abolished binding of EBA-175. In addition, binding of EBA-175 did not occur to invasion-resistant erythrocytes. Binding could be inhibited by oligosaccharides containing 1Neu5Ac (α2-3) Gal, but not Neu5Ac (α2-6) Gal, indicating the specificity for (α2-3) linkage of Neu5Ac (Orlandi *et al.*, 1992). Selective cleavage of O-linked tetrasaccharides of glycophorin A which contain these Neu5Ac (α2-3) Gal residues resulted in greatly decreased binding of EBA-175 (Orlandi *et al.*, 1992). Binding of EBA-175 to mouse erythrocytes was shown to require N-acetylneuraminic acid but not its 9-O-acetylated form (Klotz *et al.*, 1992). EBA-175 appears to be synthesized initially as a 190 kDa protein in schizonts, which is processed to a 175 kDa form (Orlandi *et al.*, 1990). Initial binding of EBA-175 to sialic acid residues on the erythrocyte surface is followed by proteolytic cleavage of the molecule to a 65 kDa protein which, also binds to erythrocytes, but by a sialic acid-independent mechanism (Kain and Ravdin, 1995).

EBA-175 was purified using human erythrocytes as an affinity matrix (Orlandi *et al.*, 1990; Haynes *et al.*, 1988). The gene encoding EBA-175 was cloned using a mono-specific polyclonal antibody to the affinity-purified protein to screen a genomic DNA library (Sim *et al.*, 1990). Comparison of the deduced amino acid sequence of EBA-175 with that of other members of the erythrocyte-binding family of proteins revealed a common primary structure (Adams *et al.*, 1992). This consists of an N-terminal signal sequence, an extracellular domain containing two cysteine-rich regions, a transmembrane sequence and a C-terminal cytoplasmic domain. Antibodies to synthetic peptides derived from the deduced amino acid sequence inhibited binding of purified EBA-175 to erythrocytes and inhibited merozoite invasion of erythrocytes, confirming the role of this protein in mediating attachment and invasion (Sim *et al.*, 1990). Elegant studies using COS cells expressing truncated portions of the protein have identified the erythrocyte-binding domain of EBA-175 as being on region II of the molecule which contains a cysteine-rich domain (Sim *et al.*, 1994). This erythrocyte-binding domain was found to be highly conserved among a number of different strains and isolates of *P. falciparum* from around the world (Liang and Sim, 1997). The role that, EBA-175 plays in mediating invasion of red cells by merozoites, together with the finding that it is highly conserved have led to the proposal that it may serve as a useful vaccine candidate (Sim *et al.*, 1990; Sim *et al.*, 1994; Liang and Sim, 1997). This is supported by the finding that antibodies in sera from individuals residing in endemic areas recognized fusion proteins of EBA-175 expressed in baculovirus (Daugherty *et al.*, 1997).

b. MSA/-1/MSP-1/pf200
Another protein which has been shown to mediate sialic-acid binding of *P. falciparum* merozoites to erythrocytes is MSA-1 (Perkins and Rocco, 1988). This protein also known as MSP-1 or pf200 is a major surface antigen of merozoites and has been extensively characterized at the protein and molecular level (Holder and Freeman, 1982; Holder *et al.*, 1985). The role of sialic acid in MSA-1-mediated adherence of merozoites to erythrocytes has been shown only in *P. falciparum*. Neuraminidase treatment of erythrocytes abolished binding of the protein. Binding was also abrogated by soluble glycophorin as well as a monoclonal antibody to the glycosylated portion of the molecule (Perkins and Rocco, 1988). Su *et al.*, described a monoclonal antibody, 2B10 to Glycophorin A which apparently binds to the same determinant on erythrocytes as MSA-1 does and showed that the binding of both ligands to erythrocyte receptors was sialic acid-dependent (Su *et al.*, 1993). Using anti-idiotypic antibodies to 2B10 the same authors showed that the C-terminal region of MSA-1 contains the erythrocyte-binding domain (Su *et al.*, 1996).

MSP-1 is present on the surface of merozoites (Holder *et al.*, 1985) and is GPI-anchored (Gerold *et al.*, 1996). The protein is initially synthesized as a high molecular weight (185–200 kDa) precursor which is proteolyticaly cleaved during merozoite maturation, to release fragments which remain associated with the membrane as a multicomponent complex. Subsequently the membrane-bound component is further cleaved, leaving a 19 kDa fragment representing the C-terminal end of the precursor. (Holder *et al.*, 1987; Blackman *et al.*, 1990; Blackman *et al.*, 1991; Blackman and Holder, 1992; Blackman *et al.*, 1993; Stafford *et al.*, 1994). This fragment contains two epidermal growth factor like domains (Blackman *et al.*, 1991) and is highly conserved (Jongwutiwes *et al.*, 1993). Antibodies to this 19 kDa fragment can prevent erythrocyte invasion by merozoites, apparently most effectively by preventing secondary processing of the molecule (Blackman *et al.*, 1994). Other antibodies to MSP-1, including those from humans with naturally acquired immunity, known as blocking antibodies, interfere with the processing-inhibitory antibodies (Blackman *et al.*, 1994; Patino *et al.*, 1997). The 19 kDa fragment appears to be the target of protective immune responses as shown by passive and active immunization studies in murine models of malaria (McKean *et al.*, 1993; Ling *et al.*, 1995; Ling *et al.*, 1997).

In endemic areas, individuals exposed to *P. falciparum* develop humoral and cell-mediated immune responses to MSP-1, notably to the 19 kDa C-terminal fragment (Muller *et al.*, 1989; Riley *et al.*, 1992; Egan *et al.*, 1995; Udhayakumar *et al.*, 1995; Egan *et al.*, 1996; Shi *et al.*, 1996; al-Yaman *et al.*, 1996; Nguer *et al.*, 1997; Egan *et al.*, 1997; Soares *et al.*, 1997; Branch *et al.*, 1998). The role of this protein in mediating erythrocyte invasion together with the findings of naturally acquired immune responses and protective immunity obtained with passive or active immunization in animal models identified MSP-1 as a major target for vaccine development. A number of recent studies have examined the efficacy of this protein as a vaccine candidate in animal models of malaria (Kang and Long, 1995, Daly and Long, 1996, Hui *et al.*, 1996, De Souza *et al.*, 1996, O'Dea *et al.*, 1996, Chang *et al.*, 1996, Burghaus *et al.*, 1996, Tian *et al.*, 1997, Renia *et al.*, 1997, Perera *et al.*, 1998).

A recent study showed that adherence of newly emerged malarial microgametes to neighboring infected or uninfected erythrocytes, during the process of "exflagellation" is also dependent on sialic acid present in glycophorins (Templeton *et al.*, 1998). This study suggested the presence of a putative sialic acid-binding protein in microgametes of the parasite.

2. GAG-binding proteins

a. *Circumsporozoite protein (CSP)*

Binding of plasmodial proteins to cell surface GAGs has been implicated in certain host-parasite interactions, crucial in the pathogenesis of malaria. One of these is the interaction of sporozoites with hepatocytes, which has been shown to be mediated by binding of the circumsporozoite protein (CSP) to heparan sulphate proteoglycans on the basolateral membranes of these cells (Cerami *et al.*, 1992, Frevert *et al.*, 1993). CSP is a major, immunodominant surface protein of malarial sporozoites, which has been extensively characterized (Nussenzweig and Nussenzweig, 1989). CSP contains an immunodominant repeat domain, which is species-specific, a C-terminal region containing two pairs of cysteines and two stretches of conserved amino acid sequences (Dame *et al.*, 1984). One of these conserved sequences, named region II, contains a cell adhesive motif (Rich *et al.*, 1990).

Malarial sporozoites as well as isolated CSP have been shown to bind to sulfated glycoconjugates such as sulfatide, heparin, fucoidan and dextran sulphate (Pancake *et al.*, 1992). These findings were confirmed using CHO cell mutants defective in GAG synthesis. Sporozoites as well as purified CSP bound to the mutant CHO cells to a lesser degree than to the parental cells (Pancake *et al.*, 1992). CSP was shown to bind to the basolateral membrane of hepatocytes via region II plus (Cerami *et al.*, 1992). Binding of this region of the protein occurred to heparan sulphate protcoglycans on the hepatocyte surface membrane (Frevert *et al.*, 1993). This was shown by heparitinase treatment of human liver sections, which resulted in abrogation of binding of the protein to hepatocytes. Specific binding of CSP to HepG2 cells was inhibited by heparin, heparan sulphate, dextran sulphate and fucoidan as well as by heparitinase treatment. Furthermore, recombinant CSP containing region II plus bound specifically to proteoglycans of rat liver and binding was abolished by soluble heparin as well as by heparitinase treatment. Binding of CSP region II plus was shown to be involved in rapid clearance of the protein from hepatocytes (Cerami *et al.*, 1994). Sinnis *et al.* showed that a positively charged stretch of amino acids in region II plus are responsible for binding to heparan sulphate proteoglycans (Sinnis *et al.*, 1994). Recent studies have shown that binding of recombinant CSP occurs to highly sulfated, heparin-like oligosaccharides in heparan sulphate (Ying *et al.,* 1997) and that CSP is released into the host cell cytoplasm during attachment and invasion (Hugel *et al.,* 1996). In addition to binding to heparan sulphate, CSP also interacts with the low-density lipoprotein receptor-related protein (Shakibaei and Frevert, 1996).

The role of CSP in inducing protective immune responses has been well studied (Nardin and Nussenzweig, 1993). Because of its roles in mediating attachment of sporozoites to hepatocytes and in inducing protective immune responses, CSP has been widely studied as a vaccine candidate (Nardin and Nussenzweig, 1993; Pasloske and Howard, 1994b; Good *et al.,* 1998). Recently a successful, preliminary, human trial of CSP vaccine was reported (Stoute *et al.,* 1997).

b. PfEMP-1

Another event, which is believed to be crucial in the pathogenesis of cerebral malaria, is adherence of parasitized erythrocytes to non-infected erythrocytes to form rosettes. Rosette formation is believed to occur via a lectin-type of interaction with blood group specific oligosaccharides and/or GAGs on red cells (Carlson and Wahlgren, 1992; Rogerson *et al.,* 1994; Rowe *et al.,* 1994). Rosetting was inhibited by sulfated glycoconjugates including sulfatide, heparin, fucoidan and dextran sulphate (Rowe *et al.,* 1994). Recently, the parasite ligand mediating rosetting of P. falciparum infected erythrocytes has been shown to be PfEMP-1 (Rowe *et al.,* 1997; Chen *et al.,* 1998). PfEMP-1 is a parasite-derived erythrocyte membrane protein, encoded by the *var* gene family, which is involved in antigenic variation (Baruch *et al.,* 1995; Su *et al.,* 1995; Smith *et al.,* 1995; Baruch *et al.,* 1996; Gardner *et al.,* 1996). PfEMP-1 has been shown to mediate adherence of parasitized erythrocytes to endothelial cells via interaction with multiple ligands (Baruch *et al.,* 1996). PfEMP-1 ranges in size from 200–350 kDa and appears to be linked to the erythrocyte cytoskeleton (Aley *et al.,* 1984; Leech *et al.,* 1984; Howard *et al.,* 1988). In a recent study, PfEMP-1 molecules involved in rosetting were reported to contain clusters of GAG-binding motifs (Chen *et al.,* 1998). PfEMP-1-mediated adhesive interactions could be inhibited by heparan sulphate. In addition, treatment with enzymes, which remove this GAG, also resulted in abrogation of binding, suggesting that heparan sulphate is the host cell receptor for this interaction. The same group of investigators showed that limited proteolysis which differentially cleaved PfEMP-1 from the surface, abrogated rosetting as well as other binding phenotypes (Fernandez *et al.,* 1998).

c. Chondroitin sulphate-A (CSA)-binding protein/s

In addition to a number of other endothelial cell receptors, chondroitin sulphate A (CSA) has been implicated as being involved in adherence of parasitized erythrocytes to host cells (Rogerson *et al.,* 1995; Robert *et al.,* 1995; Cooke *et al.,* 1996; Chaiyaroj *et al.,* 1996; Pouvelle *et al.,* 1997). Binding of parasitized erythrocytes to CHO cell mutants expressing excess chondroitin sulphate A was increased when compared to that of the parent cells, whereas they failed to bind to mutants deficient in this GAG (Rogerson *et al.,* 1995). Low concentrations of CSA but not other GAGs inhibited binding of parasitized erythrocytes to parental CHO cells. Treatment of the CHO cells with chondroitinase markedly decreased binding. Binding of parasitized erythrocytes also occurred to immobilized CSA but not to other GAGs, and this binding could be inhibited by soluble CSA. Subsequent studies showed that adherence of parasitized erythrocytes occurred to CSA chains on the proteoglycan thrombomodulin (Rogerson *et al.*, 1997; Gysin *et al.*, 1997). Parasitized erythrocytes bound to thrombomodulin in both static as well as flow-based systems and this binding could be abrogated by CSA as well as chondroitinase treatment (Rogerson *et al.*, 1997).

In a recent study, parasitized erythrocytes from human placentas (but not those from peripheral blood) were shown to bind to immobilized CSA, but not to other GAGs (Fried and Duffy, 1996). This binding was inhibited in a dose dependent manner by soluble CSA. In addition, binding of parasitized erythrocytes to frozen sections of human placenta could be inhibited by purified CSA or by treatment with chondroitinase AC. These findings were supported by another recent study showing CSA-mediated adherence of parasitized erythrocytes to cultured syncytiotrophoblasts (Maubert *et al.*, 1997). CSA-mediated binding of parasitized erythrocytes is therefore implicated as being responsible for placental sequestration in maternal malaria (Fried and Duffy, 1998). All these studies show that CSA on proteoglycans is a receptor for parasitized erythrocytes on endothelial cells or placental tissue. However, the parasite ligand responsible for CSA binding has not been identified.

3. GlcNAc-binding proteins

GlcNAc-binding proteins have also been implicated in mediating red cell adhesion and invasion. Earlier studies showed inhibition of invasion by GlcNAc, but it was later suggested that this may have been due to toxicity of the carbohydrate for the parasitized erythrocyte (Hadley *et al.*, 1986). Subsequently, three merozoite surface GlcNAc-binding proteins were identified, and implicated in binding of merozoites to red cells (el Moudni *et al.*, 1993). Interaction of *Plasmodium* ookinetes with mosquito midgut glycoproteins has also been implicated as being mediated by GlcNAc residues (Ramasamy *et al.*, 1997).

Trypanosoma cruzi

Trypanosoma cruzi is the causative agent of Chagas' disease which is characterized by multisystem involvement, mainly of the cardiovascular and gastrointestinal systems (Brener, 1973). Millions of people in South America are infected by this parasite annually and Chagas disease is one of the leading causes of death in this region of the world. The parasite exists in three main developmental stages, epimastigote, amastigote and trypomastigote. The epimastigote stage of the parasite replicates extracellularly in the midgut of the insect vector and is transformed to the non-multiplying trypomastigote, which transmits the disease. In the human host trypomastigotes migrate from the skin through the extracellular matrix, via the bloodstream to target organs where they invade host cells and are transformed to the intracellular amastigote form. Two carbohydrate-binding proteins, one a sialic acid-binding enzyme and the other a heparin-binding protein have been implicated in interaction of trypomastigotes with host cells.

1. Sialic acid-binding enzyme (*Trans*-sialidase)

Trans-sialidase is a unique enzyme which catalyses the transfer of α2-3 linked host-derived sialic acid to parasite acceptor molecules and is implicated in mediating a number of host-parasite interactions, including sialic acid-dependent adhesion to host cells. This enzyme was first identified by Pereira as a developmentally regulated neuraminidase activity with low levels in the epimastigote stage, higher levels in the trypomastigote and none in amastigotes (Pereira, 1983). Later, in addition to sialidase activity the enzyme was also shown to catalyze the transfer of sialic acid to terminal β-Gal residues (Schenkman et al., 1991; Schenkman et al., 1992b). This protein has been intensively studied, extensively characterized and shown to be a virulence factor, implicated in a number of pathogenic mechanisms (Colli, 1993; Cross and Takle, 1993; Schenkman et al., 1994). Trans-sialidase is localized on the parasite surface from where it is shed into the external medium (Prioli et al., 1991). All trypomastigotes existing within the cytosol as well as those exiting from host cells express the enzyme (Rosenberg et al., 1991b) . However, only a small proportion (20–30%) of parasites in the extracellular milieu express trans-sialidase, compared to the majority which do not (Cavallesco and Pereira, 1988; Prioli et al., 1990). The protein is polymorphic, ranging from 120–220 kDa in molecular weight (Prioli et al., 1990) and is anchored to the parasite surface by a glycophosphatidylinositol anchor (Rosenberg et al., 1991a; Schenkman et al., 1992b; Agusti et al., 1997). Parasite-derived trans-sialidase as well as antibodies (which inhibited catalytic activity) to it were found to be present in sera of mice infected with Trypanosoma cruzi (Alcantara-Neves and Pontes-de-Carvalho, 1995).

Full length DNA clones encoding trans-sialidase were first isolated by Pereira et al. using polyclonal and monoclonal antibodies to the T. cruzi neuraminidase to screen a trypomastigote genomic DNA library (Pereira et al., 1991) . Analysis of the deduced amino acid sequence revealed a multidomain structure, with regions homologous to bacterial neuraminidases, YWTD repeats of the LDL receptor and Type III modules of fibronectin. Subsequently, sequences of several genes encoding trans-sialidase and related proteins have been reported and found to be encoded by a gene super family present in multiple copies in the genome and on multiple chromosomes (Schenkman et al., 1994). Directed mutagenesis studies identified 2 domains of the protein that appear to be involved in the sialyltransferase activity (Smith and Eichinger, 1997).

Trans-sialidase is thought to mediate trypomastigote attachment and invasion by binding to sialic acid residues on host cells in a lectin-type of interaction (Ming et al., 1993). This was shown using a series of CHO cell glycosylation mutants defective in the synthesis of sialic acid (Ciavaglia et al., 1993; Ming et al., 1993; Schenkman et al., 1993). Attachment to and subsequent infection of the Lec-2 mutant which does not express terminal sialic acid, was decreased compared to the parental K1 cell. Resialylation of the Lec-2 mutant by the trans-sialidase restored attachment and invasion to levels comparable to that of the parent, whereas treatment of the parental cells with exogenous sialidase, decreased attachment and infection (Ciavaglia et al., 1993; Ming et al., 1993). In addition, affinity purified trans-sialidase inhibited attachment and invasion of sialylated parent cells, but not of the sialic acid deficient mutant Lec-2 (Ming et al., 1993). Adhesion and invasion were blocked by α2-3 linked sialyllactose but not its α2-6 linked counterpart, indicating the specificity of the enzyme for α2-3 linkage of the sialic acid (Vandekerckhove et al., 1992; Ferrero-Garcia et al., 1993; Scudder et al., 1993). Trans-sialidase is also believed to mediate attachment and invasion by catalyzing the transfer of α2-3-linked sialic acid to parasite acceptor molecules. These include glycoproteins of tissue culture-derived trypomastigotes which express the Ssp-3 epitope (Schenkman *et al.*, 1991; Schenkman *et al.*, 1992a) and mucin-like glycoproteins of metacyclic trypomastigotes (Yoshida *et al.*, 1989; Schenkman *et al.*, 1993) both of which have been shown to be involved in these processes.

Trans-sialidase is also thought to facilitate invasion and infection by mechanisms other than promoting attachment to host cells. Soluble endogenous or recombinant *trans*-sialidase injected into mice has been shown to enhance parasitemia and mortality in a murine model of Chagas' disease (Chuenkova and Pereira, 1995). A subsequent study by Pereira *et al.* showed that expression of *trans* sialidase confers an invasive phenotype to a sub-population of trypomastigotes (Pereira *et al.*, 1996). Thus, trypomastigotes expressing *trans*-sialidase which were isolated from a mixed population by immuno-magnetic separation were found to be highly invasive, in contrast to parasites which did not express the enzyme. The invasive phenotype could be reconstituted in the population lacking *trans*-sialidase by addition of exogenous *trans*-sialidase. In addition, trypomastigotes expressing *trans*-sialidase were more virulent than those lacking the enzyme in the murine model of Chagas' disease.

2. Heparin-binding protein (Penetrin)

Penetrin is a novel heparin-binding protein of *T. cruzi* trypomastigotes, which is believed to mediate attachment of the parasite to GAGs on the host cell as well as to extracellular matrix components (Ortega-Barria and Pereira, 1991). The presence of a heparin-binding protein in *T. cruzi* was initially shown by the finding that the parasite bound specifically to immobilized heparin, heparan sulphate and collagen, but not to other GAGs or glycoproteins. In addition, trypomastigote adherence to and subsequent infection of host cells could be inhibited by low concentrations of soluble heparin and heparan sulphate. The protein was purified by heparin-affinity chromatography and found to have a molecular weight of 60 kDa. Lectin activity of purified penetrin was shown by hemagglutination, which could be inhibited by very low concentrations of heparin and heparan sulphate. Penetrin is located on the surface of the parasite and promotes adhesion and spreading of cells to the substratum. Binding of purified native as well as recombinant penetrin to host cells could be inhibited by heparin and heparan sulphate. Elegant studies showing that expression of the recombinant protein in *E. coli* conferred these normally non-invasive bacteria with the ability to invade non-phagocytic cells, demonstrated that penetrin is involved in invasion of host cells. The role of penetrin in host cell invasion was further shown by studies using CHO cell mutants deficient in the synthesis and expression of proteoglycans (Herrera *et al.*, 1994). Attachment to and invasion of the proteoglycan-deficient mutant cells was considerably less than that in the parental cells. Inhibition of GAG synthesis in the parent cells by p-nitrophenyl β-D-glycoside also resulted in abrogation of attachment and invasion. In addition, depolymerization of heparin and heparan sulphate on the parent cells by the lyases, heparinase and heparitinase also inhibited interaction of trypomastigotes with these cells.

Leishmania

Protozoan parasites of the genus *Leishmania* cause Leishmaniasis, a diverse group of diseases that are transmitted by sandflies and which, affect millions of people worldwide each year (Peters and Killick-Kendrick, 1987). *Leishmania donovani* causes visceral Leishmaniasis or kala-azar and *L. major, L. tropica*, and *L. braziliensis* cause mucocutaneous Leishmaniasis. *Leishmania* have a relatively simple life cycle. The promastigote, the extracellular flagellated form of the parasite is primarily found in the sandfly vector and is responsible for transmitting the disease to man. While in the sandfly gut, the non-infective procyclic promastigotes undergo metacyclogenesis and are transformed into the metacyclic infective form. These forms are introduced into the skin by the female sandfly during a blood meal. In the human host promastigotes attach to and invade macrophages where they differentiate into amastigotes, the intracellular stage. Heparin-binding proteins, GlcNAc-binding proteins and a lipophosphoglycan (LPG)-binding protein have been described in *Leishmania*.

1. Heparin-binding proteins

Heparin-binding proteins are present in promastigotes as well as amastigotes and are believed to mediate attachment of the parasite to host cell proteoglycans. Heparin-binding activity on the surface of promastigotes was demonstrated using a radioligand assay, electron microscopy with gold-labeled heparin and flow cytometry using FITC-heparin (Mukhopadhyay *et al.*, 1989; Butcher *et al.*, 1990; Butcher *et al.*, 1992). Expression of heparin-binding activity was found to correlate with differentiation of the non-infective promastigote to the infective metacyclic form (Butcher *et al.*, 1992). In addition, adhesion of promastigotes to macrophages was enhanced in the presence of heparin (Butcher *et al.*, 1992).

Heparin-binding proteins in the amastigote form of the parasite was demonstrated using binding of radiolabeled heparin (Love *et al.*, 1993). As in the case of GAG-binding proteins from other protozoa, the role of these compounds in mediating attachment of amastigotes was studied using CHO cell mutants deficient in the production of cell surface proteoglycans. Amastigotes attached to wild type CHO cells, but not to mutants deficient in proteoglycan synthesis or cells treated with heparitinase. Amastigote attachment to cells was inhibited by low doses of soluble heparin but not by other GAGs. These studies demonstrate the presence of heparin-binding proteins in *Leishmania*. However, the parasite ligand responsible for this binding has not been identified yet.

2. GlcNAc binding proteins

The presence of GlcNAc binding proteins in *Leishmania* promastigotes has been suggested by a number of studies. Binding of *Leishmania braziliensis* promastigotes to macrophage-like cells was inhibited by GlcNAc and chitin and specific GlcNAc binding to promastigotes was demonstrated using FITC-labeled GlcNAc-BSA (Hernandez *et al.*, 1986). Neoglycoproteins as well as neoglycoenzymes have also been used to identify GlcNAc-binding proteins on the surface of *Leishmania* promastigotes. In these studies, the neoglycoprotein GlcNAc-BSA bound to promastigotes but not amastigotes as determined by agglutination and fluorescence assays (Schottelius, 1992). Binding was dependent on the presence of calcium, since it could be blocked by EDTA and restored by exogenous calcium. These findings were confirmed by binding of neoglycoenzymes (Schottelius and Gabius, 1992). A recent study also suggested the presence of lectin activity in *Leishmania* amastigotes and promastigotes using a hemagglutination assay (Svobodova *et al.*, 1997a). Amino sugars, LPS, fetuin and heparin inhibited hemagglutination. Attachment of promastigotes and amastigotes to macrophages was inhibited by the same sugars, suggesting a role for the putative carbohydrate-binding protein in attachment (Svobodova *et al.*, 1997b). Another recent study correlated the presence of GlcNAc and heparin-binding activity in *Leishmania* parasites with infectivity (Kock *et al.*, 1997).

3. Lipophosphoglycan (LPG)-binding protein

The major glycoconjugate of promastigotes is LPG, a heterogeneous lipid-containing polysaccharide of *Leishmania* which has been shown to play a major role in mediating attachment by binding to host macrophage receptors (Turco and Descoteaux, 1992, McConville and Ferguson, 1993). A *Leishmania* surface protein, which binds to LPG, has been described (Smith and Rangarajan, 1995). This protein is encoded by Gene B, one of a family of 5 related genes located on a single chromosome. Antibodies to the recombinant protein were used to demonstrate that the protein was localized to the surface of the parasite. A region of the molecule containing a repetitive amino acid motif was found to be homologous to the cell wall peptidoglycan-binding domain of *S. aueus* protein A and led to speculation that the gene B protein was attached to LPG in an analogous fashion (Smith and Rangarajan, 1995). However, the determinant of LPG that is involved in the binding of the gene B protein has not been identified, nor has its role in the host-parasite interaction been shown.

Toxoplasma gondii

Toxoplasma gondii is an obligate intracellular protozoan, which unlike many other parasites exhibits a very broad host and tissue specificity and which infects up to 50% of the population in some areas of the world (Dubey, 1977; Krahenbuhl and Remington, 1982). A number of different vertebrate hosts are infected with the parasite, which is capable of infecting a wide range of cell types. In humans, infection with this parasite is largely asymptomatic. However, in the case of immunocompromised hosts such as patients with AIDS, infection may result in generalized multisystem disease, particularly of the central nervous system. Infection with the parasite in pregnancy may result in congenital toxoplasmosis in the fetus, a devastating disease with multisystem involvement. The parasite exists in a number of developmental stages, including oocysts, tachyzoites, bradyzoites and sexual stages. *Toxoplasma gondii* has a complicated life cycle involving sexual as well as asexual reproduction, which takes place in two hosts, the definitive host being the cat. Consistent with the relative lack of specificity exhibited by this parasite, carbohydrate-binding proteins (including those, which bind to GAGs) do not appear to play a major role in recognition, adhesion or invasion (Mack *et al.*, 1994). There are however, a few reports of carbohydrate-binding proteins in this parasite.

Using neoglycoproteins as probes, deCarvalho *et al.* demonstrated the presence of GlcNAc and Gal-binding sites in rhoptries of tachyzoites, but not on the surface of the parasite (de Carvalho *et al.*, 1991). The authors suggested a possible role for these carbohydrate-binding proteins in sorting of glycoproteins. Another study using neoglycoproteins showed specific binding of BSA-glucosamide to tachyzoites (Robert *et al.*, 1991). The presence of a 15 kDa protein, which is involved in binding and uptake of the glycoprotein fetuin by tachyzoites has also been reported (Gross *et al.*, 1993). The 15 kDa protein was isolated by affinity chromatography using fetuin-Sepharose and elution with sialic acid. The isolated protein inhibited binding of the sialic acid-specific lectin *Sambucus nigra* agglutinin, suggesting that carbohydrate residues mediated binding to fetuin. However, the function of this protein is not known.

SUMMARY AND FUTURE PERSPECTIVES

In conclusion, a number of parasitic protozoa of human importance express surface-associated lectins or carbohydrate-binding proteins which mediate various functional interactions with host cells, crucial in the pathogenesis of disease caused by these organisms. These functions may be mediated by protein-carbohydrate or protein-protein interactions, in accordance with the known bifunctional properties of lectins (Barondes, 1988). Recent studies of these proteins have significantly advanced our understanding of some of the molecular mechanisms, particularly recognition and adhesion, underlying the host-parasite interaction. However, with many host-parasite interactions a number of gaps remain to be filled. In many cases specific carbohydrate binding proteins and/or their cognate ligands have not been identified or fully characterized as yet nor has their functional role been unequivocally proved by gene knockout experiments. These studies should however, be greatly facilitated by the increasingly rapid technological advances in molecular, cellular and glycobiology. Because of their roles as putative or proven virulence factors, many of these proteins are targets for drug and vaccine development. Further investigation of these proteins and elucidation of the molecular mechanisms underlying their role in the host-parasite interaction are therefore of vital importance in developing strategies to combat the diseases caused by these organisms.

REFERENCES

Adams, J.H., Sim, B.K., Dolan, S.A., Fang, X., Kaslow, D.C. and Miller, L.H. (1992) A family of erythrocyte binding proteins of malaria parasites, *Proc. Natl. Acad. Sci. USA*, **89**, 7085–7089.

Adler, P., Wood, S.J., Lee, Y.C., Lee, R.T., Petri, W., Jr. and Schnaar, R.L. (1995) High affinity binding of the Entamoeba histolytica lectin to polyvalent N-acetylgalactosaminides, *J. Biol. Chem.*, **270**, 5164–5171.

Agusti, R., Couto, A.S., Campetella, O.E., Frasch, A.C. and de Lederkremer, R.M. (1997) The trans-sialidase of Trypanosoma cruzi is anchored by two different lipids, *Glycobiology*, **7**, 731–735.

al-Yaman, F., Genton, B., Mokela, D., Narara, A., Raiko, A. and Alpers, M.P. (1996) Resistance of Plasmodium falciparum malaria to amodiaquine, chloroquine and quinine in the Madang Province of Papua New Guinea, 1990–1993, *PNG Med. J.*, **39**, 16–22.

Alcantara-Neves, N.M. and Pontes-de-Carvalho, L.C. (1995) Circulating trans-sialidase activity and trans-sialidase-inhibiting antibodies in Trypanosoma cruzi-infected mice, *Parasitol. Res.*, **81**, 560–564.

Aley, S.B., Barnwell, J.W., Daniel, W. and Howard, R.J. (1984) Identification of parasite proteins in a membrane preparation enriched for the surface membrane of erythrocytes infected with Plasmodium knowlesi, *Mol. Biochem. Parasitol.*, **12**, 69–84.

Auran, J.D., Starr, M.B. and Jakobiec, F.A. (1987) Acanthamoeba keratitis. A review of the literature, *Cornea*, **6**, 2–26.

Bailey, G.B., Nudelman, E.D., Day, D.B., Harper, C.F. and Gilmour, J.R. (1990) Specificity of glycosphingolipid recognition by Entamoeba histolytica trophozoites, *Infect. Immun.*, **58**, 43–47.

Barondes, S.H. (1988) Bifunctional properties of lectins: lectins redefined, *Trends Biochem. Sci.*, **13**, 480–482.

Baruch, D.I., Gormely, J.A., Ma, C., Howard, R.J. and Pasloske, B.L. (1996) Plasmodium falciparum erythrocyte membrane protein 1 is a parasitized erythrocyte receptor for adherence to CD36, thrombospondin, and intercellular adhesion molecule 1, *Proc. Natl. Acad. Sci. USA*, **93**, 3497–3502.

Beving, D.E., Soong, C.J. and Ravdin, J.I. (1996) Oral immunization with a recombinant cysteine-rich section of the Entamoeba histolytica galactose-inhibitable lectin elicits an intestinal secretory immunoglobulin A response that has in vitro adherence inhibition activity, *Infect. Immun.*, **64**, 1473–1476.

Bhat, N. and Ward, H. (1998) unpublished observations

Blackman, M.J., Chappel, J.A., Shai, S. and Holder, A.A. (1993) A conserved parasite serine protease processes the Plasmodium falciparum merozoite surface protein-1, *Mol. Biochem. Parasitol.*, **62**, 103–114.

Blackman, M.J., Heidrich, H.G., Donachie, S., McBride, J.S. and Holder, A.A. (1990) A single fragment of a malaria merozoite surface protein remains on the parasite during red cell invasion and is the target of invasion- inhibiting antibodies, *J. Exp. Med.*, **172**, 379–382.

Blackman, M.J., Ling, I.T., Nicholls, S.C., Holder, A.A., Blackman, M.J., Whittle, H. and Holder, A.A. (1991) Proteolytic processing of the Plasmodium falciparum merozoite surface protein-1 produces a membrane-bound fragment containing two epidermal growth factor-like domains, *Mol. Biochem. Parasitol.*, **49**, 29–33.

Blackman, M.J. and Holder, A.A. (1992) Secondary processing of the Plasmodium falciparum merozoite surface protein-1 (MSP1) by a calcium-dependent membrane-bound serine protease: shedding of MSP133 as a noncovalently associated complex with other fragments of the MSP1, *Mol. Biochem. Parasitol.*, **50**, 307–315.

Blackman, M.J., Scott-Finnigan, T.J., Shai, S. and Holder, A.A. (1994) Antibodies inhibit the protease-mediated processing of a malaria merozoite surface protein, *J. Exp. Med.*, **180**, 389–393.

Bonilha, V.L., Ciavaglia, M.d.C., de Souza, W. and Costa e Silva Filho, F. (1995) The involvement of terminal carbohydrates of the mammalian cell surface in the cytoadhesion of trichomonads, *Parasitol. Res.*, **81**, 121–126.

Braga, L.L., Ninomiya, H., McCoy, J.J., Eacker, S., Wiedmer, T., Pham, C., Wood, S., Sims, P.J. and Petri, W., Jr. (1992) Inhibition of the complement membrane attack complex by the galactose-specific adhesion of Entamoeba histolytica, *J. Clin. Invest.*, **90**, 1131–1137.

Branch, O.H., Udhayakumar, V., Hightower, A.W., Oloo, A.J., Hawley, W.A., Nahlen, B.L., Bloland, P.B., Kaslow, D.C. and Lal, A.A. (1998) A longitudinal investigation of IgG and IgM antibody responses to the merozoite surface protein-1 19–kiloDalton domain of Plasmodium falciparum in pregnant women and infants: associations with febrile illness, parasitemia, and anemia, *Am. J. Trop. Med. Hyg.*, **58**, 211–219.

Brener, Z. (1973) Biology of Trypanosoma cruzi, *Annu. Rev. Microbiol.*, **27**, 349–381.

Burchard, G.D., Prange, G. and Mirelman, D. (1993) Interaction between trophozoites of Entamoeba histolytica and the human intestinal cell line HT-29 in the presence or absence of leukocytes, *Parasitol. Res.*, **79**, 140–145.

Burghaus, P.A., Wellde, B.T., Hall, T., Richards, R.L., Egan, A.F., Riley, E.M., Ballou, W.R. and Holder, A.A. (1996) Immunization of Aotus nancymai with recombinant C terminus of Plasmodium falciparum merozoite surface protein 1 in liposomes and alum adjuvant does not induce protection against a challenge infection, *Infect. Immun.*, **64**, 3614–3619.

Buss, H., Lioutas, C., Dobinsky, S., Nickel, R. and Tannich, E. (1995) Analysis of the 170–kDa lectin gene promoter of Entamoeba histolytica, *Mol. Biochem. Parasitol.*, **72**, 1–10.

Butcher, B.A., Shome, K., Estes, L.W., Choay, J., Petitou, M., Sie, P. and Glew, R.H. (1990) Leishmania donovani cell surface heparin receptors of promastigotes are recruited from an internal pool after trypsinization, *Exp. Parasitol.*, **71**, 49–.

Butcher, B.A., Sklar, L.A., Seamer, L.C. and Glew, R.H. (1992) Heparin enhances the interaction of infective Leishmania donovani promastigotes with mouse peritoneal macrophages. A fluorescence flow cytometric analysis, *J. Immunol.*, **148**, 2879–2886.

Camus, D. and Hadley, T.J. (1985) A Plasmodium falciparum antigen that binds to host erythrocytes and merozoites., *Science*, **230**, 553–556.

Carlson, J. and Wahlgren, M. (1992) Plasmodium falciparum erythrocyte rosetting is mediated by promiscuous lectin-like interactions, *J. Exp. Med.*, **176**, 1311–1317.

Carrero, J.C., Diaz, M.Y., Viveros, M., Espinoza, B., Acosta, E. and Ortiz-Ortiz, L. (1994) Human secretory immunoglobulin A anti-Entamoeba histolytica antibodies inhibit adherence of amebae to MDCK cells, *Infect. Immun.*, **62**, 764–767.

Catterall, R.D. (1972) Trichomonal infections of the genital tract, *Med. Clin. North Am.*, **56**, 1203–1209.

Cavallesco, R. and Pereira, M.E. (1988) Antibody to Trypanosoma cruzi neuraminidase enhances infection in vitro and identifies a subpopulation of trypomastigotes, *J. Immunol.*, **140**, 617–625.

Cerami, C., Frevert, U., Sinnis, P., Takacs, B., Clavijo, P., Santos, M.J. and Nussenzweig, V. (1992) The basolateral domain of the hepatocyte plasma membrane bears receptors for the circumsporozoite protein of Plasmodium falciparum sporozoites., *Cell*, **70**, 14021–14033.

Cerami, C., Frevert, U., Sinnis, P., Takacs, B. and Nussenzweig, V. (1994) Rapid clearance of malaria circumsporozoite protein (CS) by hepatocytes, *J. Exp. Med.*, **179**, 695–701.

Chadee, K., Johnson, M.L., Orozco, E., Petri, W., Jr. and Ravdin, J.I. (1988) Binding and internalization of rat colonic mucins by the galactose/N-acetyl-D-galactosamine adherence lectin of Entamoeba histolytica, *J. Infect. Dis.*, **158**, 398–406.

Chadee, K., Petri, W., Jr., Innes, D.J. and Ravdin, J.I. (1987) Rat and human colonic mucins bind to and inhibit adherence lectin of Entamoeba histolytica., *J. Clin. Invest.*, **80**, 1245–1254.

Chaiyaroj, S.C., Angkasekwinai, P., Buranakiti, A., Looareesuwan, S., Rogerson, S.J. and Brown, G.V. (1996) Cytoadherence characteristics of Plasmodium falciparum isolates from Thailand: evidence for chondroitin sulfate a as a cytoadherence receptor, *Am. J. Trop. Med. Hyg.*, **55**, 76–80.

Chang, S.P., Case, S.E., Gosnell, W.L., Hashimoto, A., Kramer, K.J., Tam, L.Q., Hashiro, C.Q., Nikaido, C.M., Gibson, H.L., Lee-Ng, C.T., Barr, P.J., Yokota, B.T. and Hut, G.S. (1996) A recombinant baculovirus 42–kilodalton C-terminal fragment of Plasmodium falciparum merozoite surface protein 1 protects Aotus monkeys against malaria, *Infect. Immun.*, **64**, 253–261.

Chen, Q., Barragan, A., Fernandez, V., Sundstrom, A., Schlichtherle, M., Sahlen, A., Carlson, J., Datta, S. and Wahlgren, M. (1998) Identification of Plasmodium falciparum erythrocyte membrane protein 1 (PfEMP1) as the rosetting ligand of the malaria parasite P. falciparum, *J. Exp. Med.*, **187**, 15–23.

Chuenkova, M. and Pereira, M.E. (1995) Trypanosoma cruzi trans-sialidase: enhancement of virulence in a murine model of Chagas' disease., *J. Exp. Med.*, **181**, 1693–1703.

Ciavaglia, M., de Carvalho, T.U. and de Souza, W. (1993) Interaction of Trypanosoma cruzi with cells with altered glycosylation patterns., *Biochem. Biophys. Res. Commun.*, **193**, 718–721.

Colli, W. (1993) Trans-sialidase: a unique enzyme activity discovered in the protozoan Trypanosoma cruzi, *Faseb J.*, **7**, 1257–1264.

Cooke, B.M., Rogerson, S.J., Brown, G.V. and Coppel, R.L. (1996) Adhesion of malaria-infected red blood cells to chondroitin sulfate A under flow conditions, *Blood*, **88**, 4040–4044.

Cross, G.A. and Takle, G.B. (1993) The surface trans-sialidase family of Trypanosoma cruzi, *Annu. Rev. Microbiol.*, **47**, 385–411.

Current, W.L. and Garcia, L.S. (1991) Cryptosporidiosis, *Clin. Lab. Med.*, **11**, 873–897.

Daly, T.M. and Long, C.A. (1996) Influence of adjuvants on protection induced by a recombinant fusion protein against malarial infection, *Infect. Immun.*, **64**, 2602–2608.

Dame, J.B., Williams, J.L., McCutchan, T.F., Weber, J.L., Wirtz, R.A., Hockmeyer, W.T., Maloy, W.L., Haynes, J.D., Schneider, I., Roberts, D. and *et al.* (1984) Structure of the gene encoding the immunodominant surface antigen on the sporozoite of the human malaria parasite Plasmodium falciparum, *Science*, **225**, 593–599.

Daugherty, J.R., Murphy, C.I., Doros-Richert, L.A., Barbosa, A., Kashala, L.O., Ballou, W.R., Snellings, N.J., Ockenhouse, C.F. and Lanar, D.E. (1997) Baculovirus-mediated expression of Plasmodium falciparum erythrocyte binding antigen 175 polypeptides and their recognition by human antibodies, *Infect. Immun.*, **65**, 3631–3637.

de Carvalho, L., Souto-Padron, T. and de Souza, W. (1991) Localization of lectin-binding sites and sugar-binding proteins in tachyzoites of Toxoplasma gondii, *J. Parasitol.*, **77**, 156–161.

De Souza, J.B., Ling, I.T., Ogun, S.A., Holder, A.A. and Playfair, J.H. (1996) Cytokines and antibody subclass associated with protective immunity against blood-stage malaria in mice vaccinated with the C terminus of merozoite surface protein 1 plus a novel adjuvant, *Infect. Immun.*, **64**, 3532–3536.

Dubey, J.P. (Ed.) (1977) *Toxoplasma, Hammondia, Besnotia, Sarcocystis and other tissue cyst-forming coccidia of man and animals*, Academic Press, New York.

Egan, A., Waterfall, M., Pinder, M., Holder, A. and Riley, E. (1997) Characterization of human T- and B-cell epitopes in the C terminus of Plasmodium falciparum merozoite surface protein 1: evidence for poor T- cell recognition of polypeptides with numerous disulfide bonds, *Infect. Immun.*, **65**, 3024–3031.

Egan, A.F., Chappel, J.A., Burghaus, P.A., Morris, J.S., McBride, J.S., Holder, A.A., Kaslow, D.C. and Riley, E.M. (1995) Serum antibodies from malaria-exposed people recognize conserved epitopes formed by the two epidermal growth factor motifs of MSP1(19), the carboxy-terminal fragment of the major merozoite surface protein of Plasmodium falciparum, *Infect. Immun.*, **63**, 456–466.

Egan, A.F., Morris, J., Barnish, G., Allen, S., Greenwood, B.M., Kaslow, D.C., Holder, A.A. and Riley, E.M. (1996) Clinical immunity to Plasmodium falciparum malaria is associated with serum antibodies to the 19–kDa C-terminal fragment of the merozoite surface antigen, PfMSP-1, *J. Infect. Dis.*, **173**, 765–769.

el Moudni, B., Philippe, M., Monsigny, M. and Schrevel, J. (1993) N-acetylglucosamine-binding proteins on Plasmodium falciparum merozoite surface, *Glycobioogy*, **3**, 305–312.

Farthing, M.J., Pereira, M.E. and Keusch, G.T. (1986) Description and characterization of a surface lectin from Giardia lamblia, *Infect. Immun.*, **51**, 661–667.

Fernandez, V., Treutiger, C.J., Nash, G.B. and Wahlgren, M. (1998) Multiple adhesive phenotypes linked to rosetting binding of erythrocytes in plasmodium falciparum malaria [In Process Citation], *Infect. Immun.*, **66**, 2969–2975.

Ferrero-Garcia, M.A., Trombetta, S.E., Sanchez, D.O., Reglero, A., Frasch, A.C. and Parodi, A.J. (1993) The action of Trypanosoma cruzi trans-sialidase on glycolipids and glycoproteins, *Eur. J. Biochem.*, **213**, 765–771.

Frevert, U., Sinnis, P., Cerami, C., Shreffler, W., Takacs, B. and Nussenzweig, V. (1993) Malaria circumsporozoite protein binds to heparan sulphate proteoglycans associated with the surface membrane of hepatocytes., *J. Exp. Med.*, **177**, 1287–1298.

Fried, M. and Duffy, P.E. (1996) Adherence of Plasmodium falciparum to chondroitin sulfate A in the human placenta [see comments], *Science*, **272**, 1502–1504.

Fried, M. and Duffy, P.E. (1998) Maternal malaria and parasite adhesion, *J. Mol. Med.*, **76**, 162–171.

Gardner, J.P., Pinches, R.A., Roberts, D.J. and Newbold, C.I. (1996) Variant antigens and endothelial receptor adhesion in Plasmodium falciparum, *Proc. Natl. Acad. Sci. USA*, **93**, 3503–3508.

Gerold, P., Schofield, L., Blackman, M.J., Holder, A.A. and Schwarz, R.T. (1996) Structural analysis of the glycosyl-phosphatidylinositol membrane anchor of the merozoite surface proteins-1 and -2 of Plasmodium falciparum, *Mol. Biochem. Parasitol.*, **75**, 131–143.

Gilchrist, C.A., Mann, B.J. and Petri, W.A., Jr. (1998) Control of ferredoxin and Gal/GalNAc lectin gene expression in Entamoeba histolytica by a cis-acting DNA sequence, *Infect. Immun.*, **66**, 2383–2386.

Good, M.F., Kaslow, D.C. and Miller, L.H. (1998) Pathways and strategies for developing a malaria blood-stage vaccine [In Process Citation], *Annu. Rev. Immunol.*, **16**, 57–87.

Greenwood, B.M., Bradley, A.K., Greenwood, A.M. *et al.* (1987) Mortality and morbidity from malaria among children in a rural area of The Gambia., *Trans. R. Soc. Trop. Med. Hyg.*, **81**, 478–486.

Gross, U., Hambach, C., Windeck, T. and Heesemann, J. (1993) Toxoplasma gondii: uptake of fetuin and identification of a 15–kDa fetuin-binding protein, *Parasitol. Res.*, **79**, 191–194.

Gysin, J., Pouvelle, B., Le Tonqueze, M., Edelman, L. and Boffa, M.C. (1997) Chondroitin sulfate of thrombomodulin is an adhesion receptor for Plasmodium falciparum-infected erythrocytes, *Mol. Biochem. Parasitol.*, **88**, 267–271.

Hadley, T.J., Klotz, F.W. and Miller, L.H. (1986) Invasion of erythrocytes by malaria parasites: a cellular and molecular overview. [Review], *Annu. Rev. Microbiol.*, **40**, 451–477.

Haynes, J.D., Dalton, J.P., Klotz, F.W., McGinniss, M.H., Hadley, T.J., Hudson, D.E. and Miller, L.H. (1988) Receptor-like specificity of a Plasmodium knowlesi malarial protein that binds to Duffy antigen ligands on erythrocytes, *J. Exp. Med.*, **167**, 1873–1881.

Hernandez, A.G., Rodriguez, N., Stojanovic, D. and Candelle, D. (1986) The localization of a lectin-like component on the Leishmania cell surface, *Mol. Biol. Rep.*, **11**, 149–153.

Herrera, E.M., Ming, M., Ortega-Barria, E. and Pereira, M.E. (1994) Mediation of Trypanosoma cruzi invasion by heparan sulfate receptors on host cells and penetrin counter-receptors on the trypanosomes, *Mol. Biochem. Parasitol.*, **65**, 73–83.

Holder, A.A. and Freeman, R.R. (1982) Biosynthesis and processing of a Plasmodium falciparum schizont antigen recognized by immune serum and a monoclonal antibody, *J. Exp. Med.*, **156**, 1528–1538.

Holder, A.A., Freeman, R.R., Uni, S. and Aikawa, M. (1985) Isolation of a Plasmodium falciparum rhoptry protein, *Mol. Biochem. Parasitol.*, **14**, 293–303.

Holder, A.A., Sandhu, J.S., Hillman, Y., Davey, L.S., Nicholls, S.C., Cooper, H. and Lockyer, M.J. (1987) Processing of the precursor to the major merozoite surface antigens of Plasmodium falciparum, *Parasitology*, **94**, 199–208.

Howard, R.J., Barnwell, J.W., Rock, E.P., Neequaye, J., Ofori-Adjei, D., Maloy, W.L., Lyon, J.A. and Saul, A. (1988) Two approximately 300 kilodalton Plasmodium falciparum proteins at the surface membrane of infected erythrocytes, *Mol. Biochem. Parasitol.*, **27**, 207–223.

Hugel, F.U., Pradel, G. and Frevert, U. (1996) Release of malaria circumsporozoite protein into the host cell cytoplasm and interaction with ribosomes, *Mol. Biochem. Parasitol.*, **81**, 151–170.

Hui, G.S., Nikaido, C., Hashiro, C., Kaslow, D.C. and Collins, W.E. (1996) Dominance of conserved B-cell epitopes of the Plasmodium falciparum merozoite surface protein, MSP1, in blood-stage infections of naive Aotus monkeys, *Infect. Immun.*, **64**, 1502–1509.

Inge, P.M., Edson, C.M. and Farthing, M.J. (1988) Attachment of Giardia lamblia to rat intestinal epithelial cells, *Gut*, **29**, 795–801.

Jackson, R.L., Busch, S.J. and Cardin, A.D. (1991) Glycosaminoglycans: molecular properties, protein interactions, and role in physiological processes, *Physiol. Rev.*, **71**, 481–539.

Joe, A., Hamer, D.H., Kelley, M.A., Pereira, M.E.A., Keusch, G.T., Tzipori, S. and Ward, H.D. (1994) Role of a Gal/GalNAc specific sporozoite surface lectin in C. parvum-host cell interaction., *J. Euk. Microbiol.*, , 44S.

Joe, A., Verdon, R., Tzipori, S., Keusch, G.T. and Ward, H.D. (1998) Attachment of Cryptosporidium parvum sporozoites to human intestinal epithelial cells., *Infect. Immun.*, **66**, 3429–3432.

Joe, A. and Ward, H. (1998) unpublished observations,.

Jongwutiwes, S., Tanabe, K. and Kanbara, H. (1993) Sequence conservation in the C-terminal part of the precursor to the major merozoite surface proteins (MSP1) of Plasmodium falciparum from field isolates, *Mol. Biochem. Parasitol.*, **59**, 95–100.

Kain, K.C. and Ravdin, J.I. (1995) Galactose-specific adhesion mechanisms of Entamoeba histolytica: model for study of enteric pathogens, *Methods Enzymol.*, **253**, 424–439.

Kang, Y. and Long, C.A. (1995) Sequence heterogeneity of the C-terminal, Cys-rich region of the merozoite surface protein-1 (MSP-1) in field samples of Plasmodium falciparum, *Mol. Biochem. Parasitol.*, **73**, 103–110.

Katelaris, P.H., Naeem, A. and Farthing, M.J. (1995) Attachment of Giardia lamblia trophozoites to a cultured human intestinal cell line, *Gut*, **37**, 512–518.

Kelsall, B.L. and Ravdin, J.I. (1995) Immunization of rats with the 260–kilodalton *Entamoeba histolytica* galactose-inhibitable lectin elicits an intestinal secretory immunoglobulin A response that has in vitro adherence-inhibitory activity, *Infect. Immun.*, **63**, 686–689.

Keusch, G.T., Hamer, D., Joe, A., Kelley, M., Griffiths, J. and Ward, H. (1995) Cryptosporidia—who is at risk?, *J. Suisse de Med.*, **125**, 899–908.

Keusch, G.T., Ward, H.D., Ortega-Barria, E., Galindo, N. and Pereira, M.E.A. (1991), Molecular pathogenesis of Giardia lamblia: adherence and encystation In *Molecular pathogenesis of gastrointestinal infections*(Eds, Wadstrom, T., Makela, P. H., Svennerholm, A. M. and Wolf-Watz, H.) Plenum Press, New York, pp. 237–246.

Klotz, F.W., Orlandi, P.A., Reuter, G., Cohen, S.J., Haynes, J.D., Schauer, R., Howard, R.J., Palese, P. and Miller, L.H. (1992) Binding of Plasmodium falciparum 175–kilodalton erythrocyte binding antigen and invasion of murine erythrocytes requires N-acetylneuraminic acid but not its O-acetylated form, *Mol. Biochem. Parasitol.*, **51**, 49–54.

Kobiler, D. and Mirelman, D. (1980) Lectin activity in Entamoeba histolytica trophozoites, *Infect. Immun.*, **29**, 221–225.

Kobiler, D. and Mirelman, D. (1981) Adhesion of Entamoeba histolytica trophozoites to monolayers of human cells, *Journal of Infectious Diseases*, **144**, 539–546.

Kock, N.P., Gabius, H.J., Schmitz, J. and Schottelius, J. (1997) Receptors for carbohydrate ligands including heparin on the cell surface of Leishmania and other trypanosomatids, *Trop. Med. Int. Health*, **2**, 863–874.

Krahenbuhl, J.L. and Remington, J.S. (1982) *The immunology of Toxoplasma and toxoplasmosis*, Blackwell Science, Oxford.

Leech, J.H., Barnwell, J.W., Miller, L.H. and Howard, R.J. (1984) Identification of a strain-specific malarial antigen exposed on the surface of Plasmodium falciparum-infected erythrocytes, *J. Exp. Med.*, **159**, 1567–1575.

Leroy, A., De Bruyne, G., Mareel, M., Nokkaew, C., Bailey, G. and Nelis, H. (1995) Contact-dependent transfer of the galactose-specific lectin of Entamoeba histolytica to the lateral surface of enterocytes in culture, *Infect. Immun.*, **63**, 4253–4260.

Lev, B., Ward, H., Keusch, G.T. and Pereira, M.E. (1986) Lectin activation in Giardia lamblia by host protease: a novel host-parasite interaction, *Science*, **232**, 71–73.

Li, E., Becker, A. and Stanley, S., Jr. (1988) Use of Chinese hamster ovary cells with altered glycosylation patterns to define the carbohydrate specificity of Entamoeba histolytica adhesion, *J. Exp. Med.*, **167**, 1725–1730.

Li, E., Becker, A. and Stanley, S., Jr. (1989) Chinese hamster ovary cells deficient in N-acetylglucosaminyltransferase I activity are resistant to Entamoeba histolytica-mediated cytotoxicity, *Infect. Immun.*, **57**, 8–12.

Liang, H. and Sim, B.K. (1997) Conservation of structure and function of the erythrocyte-binding domain of Plasmodium falciparum EBA-175, *Mol. Biochem. Parasitol.*, **84**, 241–245.

Ling, I.T., Ogun, S.A. and Holder, A.A. (1995) The combined epidermal growth factor-like modules of Plasmodium yoelii Merozoite Surface Protein-1 are required for a protective immune response to the parasite, *Parasite Immunol.*, **17**, 425–433.

Ling, I.T., Ogun, S.A., Momin, P., Richards, R.L., Garcon, N., Cohen, J., Ballou, W.R. and Holder, A.A. (1997) Immunization against the murine malaria parasite Plasmodium yoelii using a recombinant protein with adjuvants developed for clinical use, *Vaccine*, **15**, 1562–1567.

Lotter, H. and Tannich, E. (1997) The galactose-inhibitable surface lectin of Entamoeba histolytica, a possible candidate for a subunit vaccine to prevent amoebiasis, *Behring Inst Mitt*, , 112–116.

Love, D.C., Esko, J.D. and Mosser, D.M. (1993) A heparin-binding activity on Leishmania amastigotes which mediates adhesion to cellular proteoglycans, *J. Cell Biol.*, **123**, 759–766.

Ma, P., Visvesvara, G.S., Martinez, A.J., Theodore, F.H., Daggett, P.M. and Sawyer, T.K. (1990) Naegleria and Acanthamoeba infections: review, *Rev. Infect. Dis.*, **12**, 490–513.

Mack, D., Kasper, L. and McLeod, R. (1994) Alterations in cell surface glycosylation, heparin, and chondroitin sulfate do not modify invasion of CHO cells by the PTg B strain of Toxoplasma gondii, *J. Eukaryot. Microbiol.*, **41**, 14S.

Mann, B.J., Chung, C.Y., Dodson, J.M., Ashley, L.S., Braga, L.L. and Snodgrass, T.L. (1993) Neutralizing monoclonal antibody epitopes of the Entamoeba histolytica galactose adhesin map to the cysteine-rich extracellular domain of the 170–kilodalton subunit, *Infect. Immun.*, **61**, 1772–1778.

Mann, B.J., Torian, B.E., Vedvick, T.S. and Petri, W., Jr. (1991) Sequence of a cysteine-rich galactose-specific lectin of Entamoeba histolytica, *Proc. Natl. Acad. Sci. U.*, **88**, 3248–3252.

Maubert, B., Guilbert, L.J. and Deloron, P. (1997) Cytoadherence of Plasmodium falciparum to intercellular adhesion molecule 1 and chondroitin-4–sulfate expressed by the syncytiotrophoblast in the human placenta, *Infect. Immun.*, **65**, 1251–1257.

McConville, M.J. and Ferguson, A.J. (1993) The structure, biosynthesis and function of glycosylated phosphatidylinositols in the parasitic protozoa and higher eukaryotes., *Biochem. J.*, **294**, 305–324.

McCoy, J.J., Mann, B.J. and Petri, W.A., Jr. (1994a) Adherence and cytotoxicity of Entamoeba histolytica or how lectins let parasites stick around [published erratum appears in Infect. Immun. 1994 Dec;62(12):5707], *Infect. Immun.*, **62**, 3045–3050.

McCoy, J.J., Mann, B.J., Vedvick, T.S., Pak, Y., Heimark, D.B. and Petri, W.A., Jr. (1993) Structural analysis of the light subunit of the Entamoeba histolytica galactose-specific adherence lectin, *J. Biol. Chem.*, **268**, 24223–24231.

McCoy, J.J., Weaver, A.M. and Petri, W., Jr. (1994b) Use of monoclonal anti-light subunit antibodies to study the structure and function of the Entamoeba histolytica Gal/GalNAc adherence lectin, *Glycoconjugate J.*, **11**, 432–436.

McKean, P.G., K, O.D. and Brown, K.N. (1993) Nucleotide sequence analysis and epitope mapping of the merozoite surface protein 1 from Plasmodium chabaudi chabaudi AS, *Mol. Biochem. Parasitol.*, **62**, 199–209.

Meyer, E.A. and Radulescu, S. (1979), Giardia and giardiasis In *Advances in Parasitology*, Vol. 17 (Eds, Lumsden, W. H. R., Muller, R. and Baber, J. R.) Academic Press, New York, pp. 1–47.

Meza, I., Cazares, F., Rosales-Encina, J.L., Talamas-Rohana, P. and Rojkind, M. (1987) Use of antibodies to characterize a 220–kilodalton surface protein from Entamoeba histolytica, *J. Infect. Dis.*, **156**, 798–805.

Ming, M., Chuenkova, M., Ortega-Barria, E. and Pereira, M.E. (1993) Mediation of Trypanosoma cruzi invasion by sialic acid on the host cell and trans-sialidase on the trypanosome, *Mol. Biochem. Parasitol.*, **59**, 243–252.

Mukhopadhyay, N.K., Shome, K., Saha, A.K., Hassell, J.R. and Glew, R.H. (1989) Heparin binds to Leishmania donovani promastigotes and inhibits protein phosphorylation., *Biochem. J.*, **264**, 517–.

Muller, H.M., Fruh, K., von Brunn, A., Esposito, F., Lombardi, S., Crisanti, A. and Bujard, H. (1989) Development of the human immune response against the major surface protein (gp190) of Plasmodium falciparum, *Infect. Immun.*, **57**, 3765–3769.

Nardin, E.H. and Nussenzweig, R.S. (1993) T cell responses to pre-erythrocytic stages of malaria: role in protection and vaccine development against pre-erythrocytic stages, *Annu. Rev. Immunol.*, **11**, 687–727.

Nguer, C.M., Diallo, T.O., Diouf, A., Tall, A., Dieye, A., Perraut, R. and Garraud, O. (1997) Plasmodium falciparum- and merozoite surface protein 1–specific antibody isotype balance in immune Senegalese adults, *Infect. Immun.*, **65**, 4873–4876.

Nussenzweig, R.S. and Nussenzweig, V. (1989) Antisporozoite vaccine for malaria: experimental basis and current status, *Rev. Infect. Dis.*, **11 Suppl 3**, S579–585.

O'Dea, K., McKean, P.G., Jarra, W. and Brown, K.N. (1996) A single gene copy merozoite surface antigen and immune evasion?, *Parasite Immunol.*, **18**, 165–172.

Orlandi, P.A., Klotz, F.W. and Haynes, J.D. (1992) A malaria invasion receptor, the 175–kilodalton erythrocyte binding antigen of Plasmodium falciparum recognizes the terminal Neu5Ac(alpha 2–3)Gal-sequences of glycophorin A, *J. Cell Biol.*, **116**, 901–909.

Orlandi, P.A., Sim, B.K., Chulay, J.D. and Haynes, J.D. (1990) Characterization of the 175–kilodalton erythrocyte binding antigen of Plasmodium falciparum, *Mol. Biochem. Parasitol.*, **40**, 285–294.

Ortega-Barria, E. and Pereira, M.E. (1991) A novel T. cruzi heparin-binding protein promotes fibroblast adhesion and penetration of engineered bacteria and trypanosomes into mammalian cells, *Cell*, **67**, 411–421.

Pancake, S.J., Holt, G.D., Mellouk, S. and Hoffman, S.L. (1992) Malaria sporozoites bind specifically to sulphated glycoconjugates., *J. Cell. Biol.*, **117**.

Pasloske, B.L. and Howard, R.J. (1994a) Malaria, the red cell, and the endothelium. [Review], *Annu. Rev. Med.*, **45**, 283–295.

Pasloske, B.L. and Howard, R.J. (1994b) The promise of asexual malaria vaccine development, *Am. J. Trop. Med. Hyg.*, **50**, 3–10.

Patino, J.A., Holder, A.A., McBride, J.S. and Blackman, M.J. (1997) Antibodies that inhibit malaria merozoite surface protein-1 processing and erythrocyte invasion are blocked by naturally acquired human antibodies, *J. Exp. Med.*, **186**, 1689–1699.

Pereira, M.E. (1983) A developmentally regulated neuraminidase activity in *Trypanosoma cruzi*, *Science*, **219**, 1444–1446.

Pereira, M.E., Mejia, J.S., Ortega-Barria, E., Matzilevich, D. and Prioli, R.P. (1991) The Trypanosoma cruzi neuraminidase contains sequences similar to bacterial neuraminidases, YWTD repeats of the low density lipoprotein receptor, and type III modules of fibronectin, *J. Exp. Med.*, **174**, 179–191.

Pereira, M.E., Zhang, K., Gong, Y., Herrera, E.M. and Ming, M. (1996) Invasive phenotype of Trypanosoma cruzi restricted to a population expressing trans-sialidase, *Infect. Immun.*, **64**, 3884–3892.

Pereira, M.E.A. (1986), Lectins and agglutinins in protozoa In *Microbial lectins and agglutinins:properties and biological activity*(Ed, Mirelman, D.) John Wiley and sons, New York, pp. 297–300.

Perera, K.L., Handunnetti, S.M., Holm, I., Longacre, S. and Mendis, K. (1998) Baculovirus merozoite surface protein 1 C-terminal recombinant antigens are highly protective in a natural primate model for human Plasmodium vivax malaria, *Infect. Immun.*, **66**, 1500–1506.

Perkins, M.E. and Rocco, L.J. (1988) Sialic acid-dependent binding of Plasmodium falciparum merozoite surface antigen, Pf200, to human erythrocytes, *J. Immunol.*, **141**, 3190–3196.

Peters, W. and Killick-Kendrick, R. (1987) *The leishmaniases in biology and medicine*, Academic Press, London.

Petri, W., Jr. (1996) Amebiasis and the Entamoeba histolytica Gal/GalNAc lectin: from lab bench to bedside, *J Investig Med*, **44**, 24–36.

Petri, W., Jr., Chapman, M.D., Snodgrass, T., Mann, B.J., Broman, J. and Ravdin, J.I. (1989) Subunit structure of the galactose and N-acetyl-D-galactosamine-inhibitable adherence lectin of Entamoeba histolytica, *J. Biol. Chem.*, **264**, 3007–3012.

Petri, W., Jr. and Mann, B.J. (1993) Molecular mechanisms of invasion by Entamoeba histolytica. [Review], *Seminars in Cell Biology*, **4**, 305–313.

Petri, W., Jr. and Ravdin, J.I. (1991) Protection of gerbils from amebic liver abscess by immunization with the galactose-specific adherence lectin of Entamoeba histolytica, *Infect. Immun.*, **59**, 97–101.

Petri, W., Jr., Smith, R.D., Schlesinger, P.H., Murphy, C.F. and Ravdin, J.I. (1987) Isolation of the galactose-binding lectin that mediates the in vitro adherence of Entamoeba histolytica, *J. Clin. Invest.*, **80**, 1238–1244.

Petri, W., Jr., Snodgrass, T.L., Jackson, T.F., Gathiram, V., Simjee, A.E., Chadee, K. and Chapman, M.D. (1990) Monoclonal antibodies directed against the galactose-binding lectin of Entamoeba histolytica enhance adherence, *J. Immunol.*, **144**, 4803–4809.

Pouvelle, B., Meyer, P., Robert, C., Bardel, L. and Gysin, J. (1997) Chondroitin-4–sulfat · impairs in vitro and in vivo cytoadherence of Plasmodium falciparum infected erythrocytes, *Mol. Med.*, **3**, 508–518.

Prioli, R.P., Mejia, J.S., Aji, T., Aikawa, M. and Pereira, M.E. (1991) Trypanosoma cruzi: localization of neuraminidase on the surface of trypomastigotes, *Trop. Med. Parasitol.*, **42**, 146–150.

Prioli, R.P., Mejia, J.S. and Pereira, M.E. (1990) Monoclonal antibodies against Trypanosoma cruzi neuraminidase reveal enzyme polymorphism, recognize a subset of trypomastigotes, and enhance infection in vitro, *J. Immunol.*, **144**, 4384–4391.

Purdy, J.E., Mann, B.J., Shugart, E.C. and Petri Jr., W.A. (1993) Analysis of the gene family encoding the Entamoeba histolytica galactose-specific adhesin 170–kDa subunit, *Mol. Biochem. Parasitol.*, **62**, 53–60.

Purdy, J.E., Pho, L.T., Mann, B.J. and Petri, W.A., Jr. (1996) Upstream regulatory elements controlling expression of the Entamoeba histolytica lectin, *Mol. Biochem. Parasitol.*, **78**, 91–103.

Ragland, B.D., Ashley, L.S., Vaux, D.L. and Petri, W., Jr. (1994) Entamoeba histolytica: target cells killed by trophozoites undergo DNA fragmentation which is not blocked by Bcl-2, *Exp. Parasitol.*, **79**, 460–467.

Ramakrishnan, G., Ragland, B.D., Purdy, J.E. and Mann, B.J. (1996) Physical mapping and expression of gene families encoding the N-acetyl D-galactosamine adherence lectin of Entamoeba histolytica, *Mol. Microbiol.*, **19**, 91–100.

Ramasamy, R., Wanniarachchi, I.C., Srikrishnaraj, K.A. and Ramasamy, M.S. (1997) Mosquito midgut glycoproteins and recognition sites for malaria parasites, *Biochim. Biophys. Acta*, **1361**, 114–122.

Ravdin, J.I., Guerrant, R.L. (1981) Role of adherence in cytopathic mechanisms of Entamoeba histolytica. Study with mammalian tissue culture cells and human erythrocytes., *J. Clin. Invest.*, **68**, 1305–1313.

Ravdin, J.I., Jackson, T.F., Petri, W.A., Jr., Murphy, C.F., Ungar, B.L., Gathiram, V., Skilogiannis, J. and Simjee, A.E. (1990) Association of serum antibodies to adherence lectin with invasive amebiasis and asymptomatic infection with pathogenic Entamoeba histolytica, *J. Infect. Dis.*, **162**, 768–772.

Ravdin, J.I., Moreau, F., Sullivan, J.A., Petri, W., Jr. and Mandell, G.L. (1988) Relationship of free intracellular calcium to the cytolytic activity of Entamoeba histolytica, *Infect. Immun.*, **56**, 1505–1512.

Ravdin, J.I. and Murphy, C.F. (1992) Characterization of the galactose-specific binding activity of a purified soluble *Entamoeba histolytica* adherence lectin, *J. Protozool.*, **39**, 319–323.

Ravdin, J.I., Stanley, P., Murphy, C.F. and Petri, W., Jr. (1989) Characterization of cell surface carbohydrate receptors for Entamoeba histolytica adherence lectin, *Infect. Immun.*, **57**, 2179–2186.

Renia, L., Ling, I.T., Marussig, M., Miltgen, F., Holder, A.A. and Mazier, D. (1997) Immunization with a recombinant C-terminal fragment of Plasmodium yoelii merozoite surface protein 1 protects mice against homologous but not heterologous P. yoelii sporozoite challenge, *Infect. Immun.*, **65**, 4419–4423.

Rich, K.A., George, F.W.T., Law, J.L. and Martin, W.J. (1990) Cell-adhesive motif in region II of malarial circumsporozoite protein, *Science*, **249**, 1574–1577.

Riley, E. M., Olerup, O., Bennett, S., Rowe, P., Allen, S.J., Blackman, M.J., Troye-Blomberg, M., Holder, A.A. and Greenwood, B.M. (1992) MHC and malaria: the relationship between HLA class II alleles and immune responses to Plasmodium falciparum, *Int Immunol*, **4**, 1055–1063.

Robert, C., Pouvelle, B., Meyer, P., Muanza, K., Fujioka, H., Aikawa, M., Scherf, A. and Gysin, J. (1995) Chondroitin-4–sulphate (proteoglycan), a receptor for Plasmodium falciparum-infected erythrocyte adherence on brain microvascular endothelial cells, *Res Immunol*, **146**, 383–393.

Robert, R., de la Jarrige, P.L., Mahaza, C., Cottin, J., Marot-Leblond, A. and Senet, J.M. (1991) Specific binding of neoglycoproteins to Toxoplasma gondii tachyzoites, *Infect. Immun.*, **59**, 4670–4673.

Rogerson, S.J., Chaiyaroj, S.C., Ng, K., Reeder, J.C. and Brown, G.V. (1995) Chondroitin sulfate A is a cell surface receptor for Plasmodium falciparum-infected erythrocytes, *J. Exp. Med.*, **182**, 15–20.

Rogerson, S.J., Novakovic, S., Cooke, B.M. and Brown, G.V. (1997) Plasmodium falciparum-infected erythrocytes adhere to the proteoglycan thrombomodulin in static and flow-based systems, *Exp. Parasitol.*, **86**, 8–18.

Rogerson, S.J., Reeder, J.C., al-Yaman, F. and Brown, G.V. (1994) Sulfated glycoconjugates as disrupters of Plasmodium falciparum erythrocyte rosettes, *Am. J. Trop. Med. Hyg.*, **51**, 198–203.

Rosales-Encina, J.L., Meza, I., Lopez-De-Leon, A., Talamas-Rohana, P. and Rojkind, M. (1987) Isolation of a 220–kilodalton protein with lectin properties from a virulent strain of Entamoeba histolytica, *J. Inf. Dis.*, **156**, 790–797.

Rosenberg, I., Prioli, R.P., Ortega-Barria, E. and Pereira, M.E. (1991a) Stage-specific phospholipase C-mediated release of Trypanosoma cruzi neuraminidase, *Mol. Biochem. Parasitol.*, **46**, 303–305.

Rosenberg, I.A., Prioli, R.P., Mejia, J.S. and Pereira, M.E. (1991b) Differential expression of Trypanosoma cruzi neuraminidase in intra- and extracellular trypomastigotes, *Infect. Immun.*, **59**, 464–466.

Roussel, F., De Carli, G. and Brasseur, P. (1991) A cytopathic effect of Trichomonas vaginalis probably mediated by a mannose/N-acetyl-glucosamine binding lectin, *Int J. Parasitol.*, **21**, 941–944.

Rowe, A., Berendt, A.R., Marsh, K. and Newbold, C.I. (1994) Plasmodium falciparum: a family of sulphated glycoconjugates disrupts erythrocyte rosettes, *Exp. Parasitol.*, **79**, 506–516.

Rowe, J.A., Moulds, J.M., Newbold, C.I. and Miller, L.H. (1997) P. falciparum rosetting mediated by a parasite-variant erythrocyte membrane protein and complement-receptor 1, *Nature*, **388**, 292–295.

Saffer, L.D. and Petri, W., Jr. (1991) Entamoeba histolytica: recognition of alpha- and beta-galactose by the 260–kDa adherence lectin, *Exp. Parasitol.*, **72**, 106–108.

Salata, R.A. and Ravdin, J.I. (1986) The interaction of human neutrophils and Entamoeba histolytica increases cytopathogenicity for liver cell monolayers, *J. Infect. Dis.*, **154**, 19–26.

Schain, D.C., Salata, R.A. and Ravdin, J.I. (1992) Human T-lymphocyte proliferation, lymphokine production, and amebicidal activity elicited by the galactose-inhibitable adherence protein of Entamoeba histolytica, *Infect. Immun.*, **60**, 2143–2146.

Schain, D.C., Salata, R.A. and Ravdin, J.I. (1995) Development of amebicidal cell-mediated immunity in gerbils (Meriones unguiculatus) immunized with the galactose-inhibitable adherence lectin of Entamoeba histolytica, *J. Parasitol.*, **81**, 563–568.

Schenkman, R.P., Vandekerckhove, F. and Schenkman, S. (1993) Mammalian cell sialic acid enhances invasion by Trypanosoma cruzi, *Infect. Immun.*, **61**, 898–902.

Schenkman, S., Eichenger, D., Pereira, M.E.A. and Nussenzweig, V. (1994) Structural and functional properties of Trypanosoma trans-sialidase, *Annu. Rev. Microbiol.*, **48**, 499–523.

Schenkman, S., Jiang, M.S., Hart, G.W. and Nussenzweig, V. (1991) A novel cell surface trans-sialidase of Trypanosoma cruzi generates a stage-specific epitope required for invasion of mammalian cells, *Cell*, **65**, 1117–1125.

Schenkman, S., Kurosaki, T., Ravetch, J.V. and Nussenzweig, V. (1992a) Evidence for the participation of the Ssp-3 antigen in the invasion of non-phagocytic mammalian cells by Trypanosoma cruzi, *J. Exp. Med.*, **175**, 1635–1641.

Schenkman, S., Pontes-de-Carvalho, L. and Nussenzweig, V. (1992b) *Trypanosoma cruzi* trans-sialidase and neuraminidase activities can be mediated by the same enzymes, *J. Exp. Med.*, **175**, 567–575.

Schottelius, J. (1992) Neoglycoproteins as tools for the detection of carbohydrate-specific receptors on the cell surface of Leishmania, *Parasitol. Res.*, **78**, 309–315.

Schottelius, J. and Gabius, H.J. (1992) Detection and quantitation of cell-surface sugar receptor(s) of Leishmania donovani by application of neoglycoenzymes, *Parasitol. Res.*, **78**, 529–533.

Scudder, P., Doom, J.P., Chuenkova, M., Manger, I.D. and Pereira, M.E. (1993) Enzymatic characterization of beta-D-galactoside alpha 2,3–trans-sialidase from *Trypanosoma cruzi*, *J. Biol. Chem.*, **268**, 9886–9891.

Seguin, R., Mann, B.J., Keller, K. and Chadee, K. (1995) Identification of the galactose-adherence lectin epitopes of Entamoeba histolytica that stimulate tumor necrosis factor-alpha production by macrophages, *Proc. Natl. Acad. Sci. USA*, **92**, 12175–12179.

Seguin, R., Mann, B.J., Keller, K. and Chadee, K. (1997) The galactose adherence lectin of Entamoeba histolytica activates primed macrophages for amebicidal activity mediated by nitric oxide, *Arch Med Res*, **28 Spec No**, 228–229.

LECTINS IN PARASITES 199

Shakibaei, M. and Frevert, U. (1996) Dual interaction of the malaria circumsporozoite protein with the low density lipoprotein receptor-related protein (LRP) and heparan sulfate proteoglycans, *J. Exp. Med.*, **184**, 1699–1711.

Sharon, N. and Lis, H. (1989) Lectins as cell recognition molecules, *Science*, **246**, 227–234.

Shi, Y.P., Sayed, U., Qari, S.H., Roberts, J.M., Udhayakumar, V., Oloo, A.J., Hawley, W.A., Kaslow, D.C., Nahlen, B.L. and Lal, A.A. (1996) Natural immune response to the C-terminal 19–kilodalton domain of Plasmodium falciparum merozoite surface protein 1, *Infect. Immun.*, **64**, 2716–2723.

Sim, B.K., Chitnis, C.E., Wasniowska, K., Hadley, T.J. and Miller, L.H. (1994) Receptor and ligand domains for invasion of erythrocytes by Plasmodium falciparum, *Science*, **264**, 1941–1944.

Sim, B.K., Orlandi, P.A., Haynes, J.D., Klotz, F.W., Carter, J.M., Camus, D., Zegans, M.E. and Chulay, J.D. (1990) Primary structure of the 175K Plasmodium falciparum erythrocyte binding antigen and identification of a peptide which elicits antibodies that inhibit malaria merozoite invasion, *J. Cell Biol.*, **111**, 1877–1884.

Sinnis, P., Clavijo, P., Fenyo, D., Chait, B.T., Cerami, C. and Nussenzweig, V. (1994) Structural and functional properties of region II-plus of the malaria circumsporozoite protein, *J. Exp. Med.*, **180**, 297–306.

Smith, D.F. and Rangarajan, D. (1995) Cell surface components of Leishmania: identification of a novel parasite lectin?, *Glycobiology*, **5**, 161–166.

Smith, L.E. and Eichinger, D. (1997) Directed mutagenesis of the Trypanosoma cruzi trans-sialidase enzyme identifies two domains involved in its sialyltransferase activity, *Glycobiology*, **7**, 445–451.

Soares, I.S., Levitus, G., Souza, J.M., Del Portillo, H.A. and Rodrigues, M.M. (1997) Acquired immune responses to the N- and C-terminal regions of Plasmodium vivax merozoite surface protein 1 in individuals exposed to malaria, *Infect. Immun.*, **65**, 1606–1614.

Soong, C.J., Kain, K.C., Abd-Alla, M., Jackson, T.F. and Ravdin, J.I. (1995) A recombinant cysteine-rich section of the Entamoeba histolytica galactose-inhibitable lectin is efficacious as a subunit vaccine in the gerbil model of amebic liver abscess, *J. Inf. Dis.*, **171**, 645–651.

Sreenivas, K., Ganguly, N.K., Ghosh, S., Sehgal, R. and Mahajan, R.C. (1995) Identification of a 148–kDa surface lectin from Giardia lamblia with specificity for alpha-methyl-D-mannoside, *FEMS Microbiol. Lett.*, **134**, 33–37.

Stafford, W.H., Blackman, M.J., Harris, A., Shai, S., Grainger, M. and Holder, A.A. (1994) N-terminal amino acid sequence of the Plasmodium falciparum merozoite surface protein-1 polypeptides, *Mol. Biochem. Parasitol.*, **66**, 157–160.

Stoute, J.A., Slaoui, M., Heppner, D.G., Momin, P., Kester, K.E., Desmons, P., Wellde, B.T., Garcon, N., Krzych, U. and Marchand, M. (1997) A preliminary evaluation of a recombinant circumsporozoite protein vaccine against Plasmodium falciparum malaria. RTS,S Malaria Vaccine Evaluation Group [see comments], *N. Engl. J. Med.*, **336**, 86–91.

Sturchler, D. (1989) How much malaria is there worldwide?, *Parasitol. Today*, **5**, 39–40.

Su, S., Sanadi, A.R., Ifon, E. and Davidson, E.A. (1993) A monoclonal antibody capable of blocking the binding of Pf200 (MSA-1) to human erythrocytes and inhibiting the invasion of Plasmodium falciparum merozoites into human erythrocytes, *J. Immunol.*, **151**, 2309–2317.

Su, S., Yang, S., Ding, R. and Davidson, E.A. (1996) Primary structure of the variable region of monoclonal antibody 2B10, capable of inducing anti-idiotypic antibodies that recognize the C-terminal region of MSA-1 of Plasmodium falciparum, *Infect. Immun.*, **64**, 326–331.

Su, X.Z., Heatwole, V.M., Wertheimer, S.P., Guinet, F., Herrfeldt, J.A., Peterson, D.S., Ravetch, J.A. and Wellems, T.E. (1995) The large diverse gene family var encodes proteins involved in cytoadherence and antigenic variation of Plasmodium falciparum-infected erythrocytes [see comments], *Cell*, **82**, 89–100.

Svobodova, M., Bates, P.A. and Volf, P. (1997a) Detection of lectin activity in Leishmania promastigotes and amastigotes, *Acta Trop.*, **68**, 23–35.

Svobodova, M., Capo, C. and Mege, J.M. (1997b) A biological role for haemagglutination activity of Leishmania promastigotes and amastigotes *Parasite*, **4**, 245–251.

Talamas-Rohana, P., Schlie-Guzman, M.A., Hernandez-Ramirez, V.I. and Rosales-Encina, J.L. (1995) T-cell suppression and selective in vivo activation of TH2 subpopulation by the Entamoeba histolytica 220–kilodalton lectin, *Infect. Immun.*, **63**, 3953–3958.

Tannich, E., Ebert, F. and Horstmann, R.D. (1991) Primary structure of the 170–kDa surface lectin of pathogenic Entamoeba histolytica, *Proc. Natl. Acad. Sci. USA*, **88**, 1849–1853.

Tannich, E., Ebert, F. and Horstmann, R.D. (1992) Molecular cloning of cDNA and genomic sequences coding for the 35–kilodalton subunit of the galactose-inhibitable lectin of pathogenic Entamoeba histolytica, *Mol. Biochem. Parasitol.*, **55**, 225–227.

Templeton, T.J., Keister, D.B., Muratova, O., Procter, J.L. and Kaslow, D.C. (1998) Adherence of erythrocytes during exflagellation of plasmodium falciparum microgametes is dependent on erythrocyte surface sialic acid and glycophorins [In Process Citation], *J. Exp. Med.*, **187**, 1599–1609.

Thea, D.M., Pereira, M.E., Kotler, D., Sterling, C.R. and Keusch, G.T. (1992) Identification and partial purification of a lectin on the surface of the sporozoite of Cryptosporidium parvum, *J. Parasitol.*, **78**, 886–893.

Tian, J.H., Kumar, S., Kaslow, D.C. and Miller, L.H. (1997) Comparison of protection induced by immunization with recombinant proteins from different regions of merozoite surface protein 1 of Plasmodium yoelii, *Infect. Immun.*, **65**, 3032–3036.

Turco, S.J. and Descoteaux, A. (1992) The lipophosphoglycan of *Leishmania* parasites. [Review], *Annu. Rev. Microbiol.*, **46**, 65–94.

Udhayakumar, V., Anyona, D., Kariuki, S., Shi, Y.P., Bloland, P.B., Branch, O.H., Weiss, W., Nahlen, B.L., Kaslow, D.C. and Lal, A.A. (1995) Identification of T and B cell epitopes recognized by humans in the C- terminal 42–kDa domain of the Plasmodium falciparum merozoite surface protein (MSP)-1, *J. Immunol.*, **154**, 6022–6030.

Vandekerckhove, F., Schenkman, S., Pontes de Carvalho, L., Tomlinson, S., Kiso, M., Yoshida, M., Hasegawa, A. and Nussenzweig, V. (1992) Substrate specificity of the Trypanosoma cruzi trans-sialidase, *Glycobiology*, **2**, 541–548.

Velazquez, C., Valette, I., Cruz, M., Labra, M.L., Montes, J., Stanley, S.L., Jr. and Calderon, J. (1995) Identification of immunogenic epitopes of the 170–kDa subunit adhesin of Entamoeba histolytica in patients with invasive amebiasis, *J. Eukaryot. Microbiol.*, **42**, 636–641.

Walsh, J.A. (1986) Problems in recognition and diagnosis of amebiasis:estimation of the global magnitude of morbidity and mortality., *Rev. Infect. Dis.*, **8**, 228–238.

Ward, H. and Cevallos, A.M. (1998) Cryptosporidium: molecular basis of host-parasite interaction, *Adv. Parasitol.*, **40**, 151–185.

Ward, H.D. (1996), Glycobiology of parasites: Role of carbohydrate-binding proteins in the host-parasite interaction In *Glycosciences. Status and Perspectives* (Eds, Gabius, H.-J. and Gabius, S.) Chapman and Hall GmbH, Weienheim, pp. 399–409.

Ward, H.D., Lev, B.I., Kane, A.V., Keusch, G.T. and Pereira, M.E. (1987a) Identification and characterization of taglin, a mannose 6–phosphate binding, trypsin-activated lectin from Giardia lamblia, *Biochem.*, **26**, 8669–8675.

Ward, H.D., Lev, B.I., Keusch G.T. and Pereira, M.E.A. (1987b), Induction of lectin activity in Giardia In *Molecular strategies of parasite invasion*Alan R. Liss, pp. 521–530.

Yang, Z., Cao, Z. and Panjwani, N. (1997) Pathogenesis of Acanthamoeba keratitis: carbohydrate-mediated host- parasite interactions, *Infect. Immun.*, **65**, 439–445.

Ying, P., Shakibaei, M., Patankar, M.S., Clavijo, P., Beavis, R.C., Clark, G.F. and Frevert, U. (1997) The malaria circumsporozoite protein: interaction of the conserved regions I and II-plus with heparin-like oligosaccharides in heparan sulfate, *Exp. Parasitol.*, **85**, 168–182.

Yoshida, N., Mortara, R.A., Araguth, M.F., Gonzalez, J.C. and Russo, M. (1989) Metacyclic neutralizing effect of monoclonal antibody 10D8 directed to the 35– and 50–kilodalton surface glycoconjugates of Trypanosoma cruzi, *Infect. Immun.*, **57**, 1663–1667.

Zhang, T. and Stanley, S.L., Jr. (1994) Protection of gerbils from amebic liver abscess by immunization with a recombinant protein derived from the 170–kilodalton surface adhesin of Entamoeba histolytica, *Infect. Immun.*, **62**, 2605–2608.

INDEX

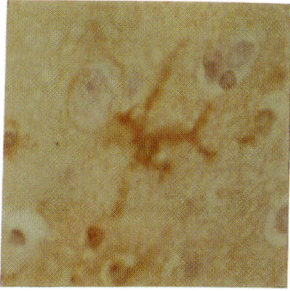

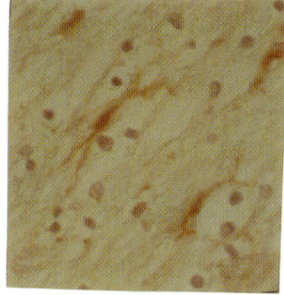

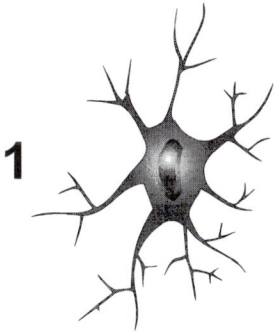

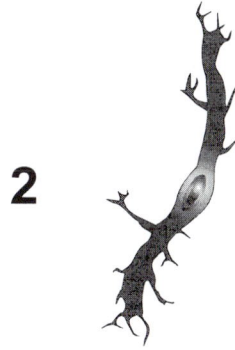

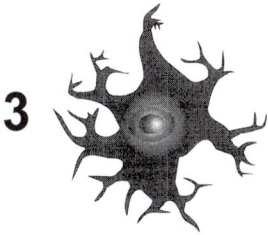

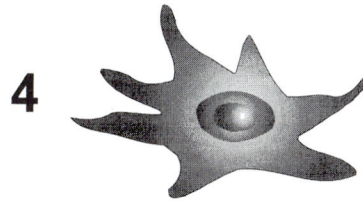

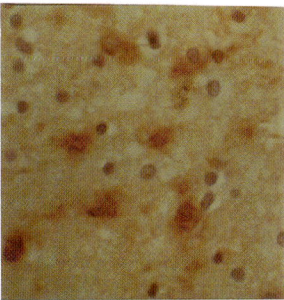

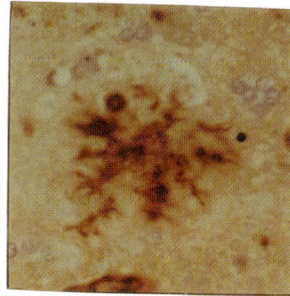

COLOUR PLATE I.   *See* P. Zatta *et al.*, Figure 2, page 40.

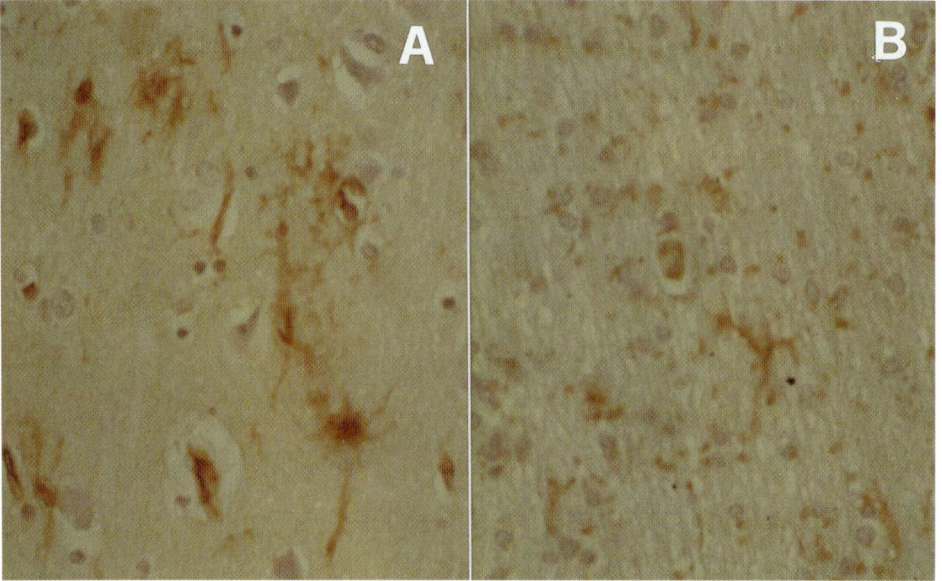

COLOUR PLATE II. *See* P. Zatta *et al.*, Figure 3, page 44.

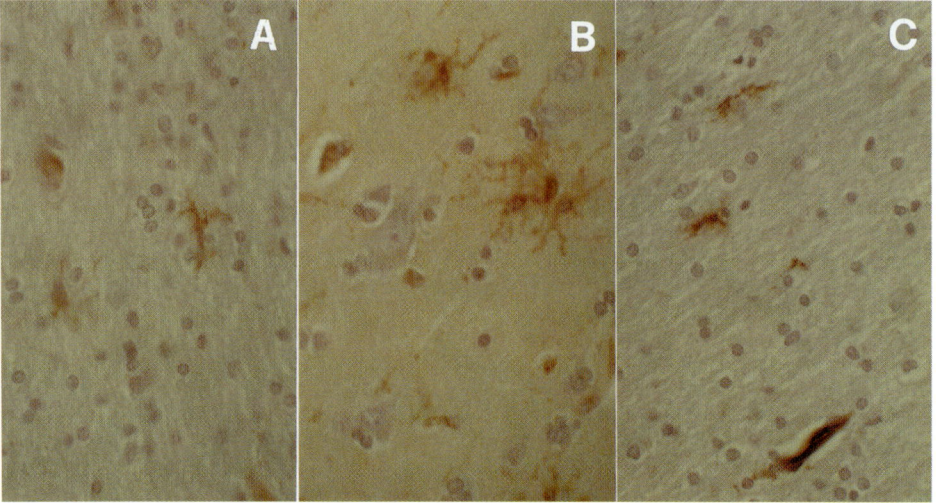

COLOUR PLATE III. *See* P. Zatta *et al.*, Figure 4, page 44.

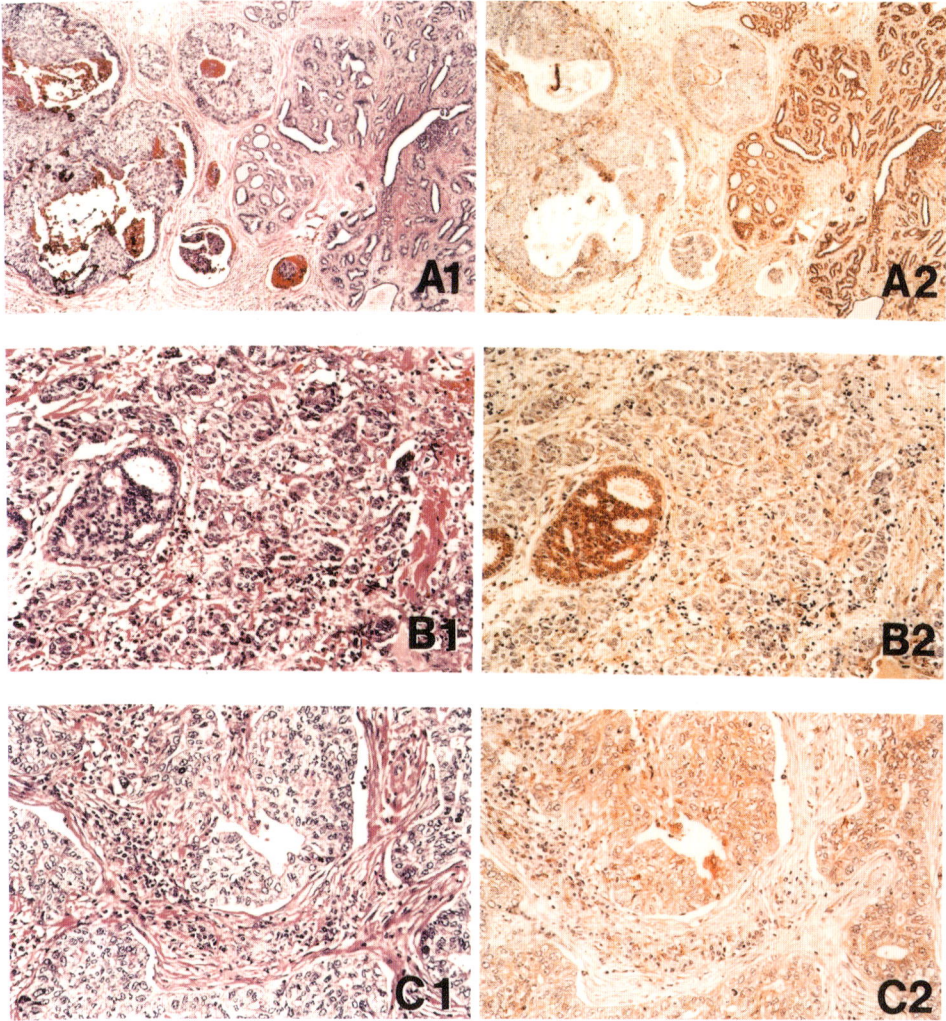

COLOUR PLATE IV. *See* F.A. Van Den Brûle and V. Castronovo, Figure 9, page 106.